Second Edition

# QUALITATIVE
# INQUIRY&
# RESEARCH DESIGN

*I dedicate this book to Uncle Jim (James W. Marshall, M.D., 1915–1977) who provided love, support, and inspiration.*

Second Edition

# QUALITATIVE INQUIRY & RESEARCH DESIGN
## Choosing Among Five Approaches

# John W. Creswell

*University of Nebraska, Lincoln*

**SAGE** Publications
Thousand Oaks ▪ London ▪ New Delhi

*For information:*

Sage Publications, Inc.
2455 Teller Road
Thousand Oaks, California 91320
E-mail: order@sagepub.com

Sage Publications Ltd.
1 Oliver's Yard
55 City Road
London EC1Y 1SP
United Kingdom

Sage Publications India Pvt. Ltd.
B-42, Panchsheel Enclave
Post Box 4109
New Delhi 110 017  India

Printed in the United States of America

**Library of Congress Cataloging-in-Publication Data**

Creswell, John W.
Qualitative inquiry and research design: Choosing among five approaches /
John W. Creswell.—2nd ed.
    p. cm.
Includes bibliographical references and index.
ISBN  978-1-4129-1606-6 (cloth)
ISBN  978-1-4129-1607-3 (pbk.)
    1. Social sciences—Methodology.  I. Title.

H61.C73 2007
300.72—dc22

                                                2006031956

This book is printed on acid-free paper.

                09  10  11  10  9  8  7  6  5

| Acquisitions Editor: | Lisa Cuevas Shaw |
| Editorial Assistant: | Karen Greene |
| Production Editor: | Denise Santoyo |
| Copy Editor: | Jamie Robinson |
| Typesetter: | C&M Digitals (P) Ltd. |
| Indexer: | Kathy Paparchontis |

# Contents

# Analytic Table of Contents by Approach

## Grounded Theory

## Ethnography

## Case Study

# List of Tables and Figures

## Tables

## Figures

# Acknowledgements

I am most thankful to the many students in my "Seminar in Qualitative Research" at the University of Nebraska-Lincoln who helped to shape this book over the years. They offered suggestions, provided examples, and discussed the material in this book. Also, I benefited from capable scholars who helped to shape and form this book in the first edition: Paul Turner, Ken Robson, Dana Miller, Diane Gillespie, Gregory Schraw, Sharon Hudson, Karen Eifler, Neilida Aguilar, and Harry Wolcott. Ben Crabtree and Rich Hofmann helped form the first edition text significantly and encouraged me to proceed, and they diligently and timely responded to Sage's request to be first edition external reviewers. In addition, Keith Pezzoli, Kathy O'Byrne, Joanne Cooper, and Phyllis Langton served as first edition reviewers for Sage and added insight into content and structure that I could not see because of my closeness to the material. As always, I am indebted to C. Deborah Laughton, who served as my acquisition editor for the first edition, and to Lisa Cuevas Shaw who served in this role for the second edition. Also, for the second edition, members of my Office of Qualitative and Mixed Methods Research all provided valuable input. I especially single out Dr. Vicki Plano Clark and Dr. Ron Shope, who have been instrumental in refining and shaping my ideas about qualitative research. Also, I am grateful to the Department of Educational Psychology, which provided me with a full sabbatical year during 2005–2006 to work on this book. Finally, to members of my family—Karen, David, and Johanna—thanks for providing me with time to spend long hours writing and revising this book. Thank you all.

# 1

# Introduction

Work on the first edition of this book initially began during a 1994 summer qualitative seminar in Vail, Colorado, sponsored by the University of Denver under the able guidance of Edith King of the College of Education. One morning, I facilitated the discussion about qualitative data analysis. I began on a personal note, introducing one of my recent qualitative studies—a case study of a campus response to a student gun incident (Asmussen & Creswell, 1995). I knew this case might provoke some discussion and present some complex analysis issues. It involved a Midwestern university's reaction to a gunman who attempted to fire on students in his undergraduate class. Standing before the group, I chronicled the events of the case, the themes, and the lessons we learned about a university reaction to a near tragic event. Then, unplanned, Harry Wolcott of the University of Oregon, another resource person for our seminar, raised his hand and asked for the podium. He explained how *he* would approach the study as a cultural anthropologist. To my surprise, he "turned" our case study into ethnography, framing the study in an entirely new way. After Harry had concluded, Les Goodchild, then of Denver University, spoke, and he turned the gunman case into a historical study. I delighted in these surprise turns of my initial case study. This unforeseen set of events kindled an idea I had long harbored—that one designed a study differently depending on the type of qualitative research. I began to write the first edition of this book, guided by a single, compelling question: How does the type or approach of qualitative inquiry shape the design or procedures of a study?

# Purpose

Both editions of this book are my attempt to answer this question. My primary intent is to examine five different approaches to qualitative inquiry—narrative, phenomenology, grounded theory, ethnography, and case studies—and to discuss their procedures for conducting a qualitative study. The conduct of a study includes the introduction to a study, including the formation of the purpose and research questions; data collection; data analysis; report writing; and standards of validation and evaluation. In the process of providing procedures for conducting a study, I introduce a comparative analysis of the five approaches so that researchers can make an informed choice as to which approach best suits their research problems.

Because the procedures for conducting research evolve from a researcher's philosophical and theoretical stances, I begin with these stances. Then, to set the stage for discussing each of the five approaches, I summarize the major characteristics and provide an example of each from a published journal article. With this understanding, I next go through the steps in the process of conducting a study and illustrate how this might proceed for each of the five types of qualitative research. Throughout the book, I provide tables that summarize major differences among the approaches. I end the book by taking the qualitative case study presented at the beginning of the book in Chapter 5 and "turn" the type of study from the original case study to a narrative study, a phenomenology, a grounded theory study, and an ethnography. By reading this book, I hope that you will gain a better understanding of the steps in the process of research, five qualitative approaches to inquiry, and the differences and similarities among the five *approaches to inquiry* (see the glossary in Appendix A for definitions of terms in bold italics).

# What Is New in This Edition

Since I wrote the first edition of this book, many changes have occurred on the landscape of qualitative research, and these changes and my thinking about them are reflected in this second edition. Qualitative research has become more accepted as a legitimate mode of inquiry in the social behavioral and health sciences than it was 10 years ago. Courses on qualitative research, funding invitations for qualitative projects, and the emergence of qualitative journals all speak to an increased acceptance of qualitative research within the social and human sciences. Thus, I provide references to new books that have captured the attention of the qualitative community since I wrote the first edition during the mid-1990s.

Since then, the *interpretive qualitative research* approach, focusing on the self-reflective nature of how qualitative research is conducted, read, and advanced, has become much more dominant in the qualitative discourse, and has, in many ways, been integrated into the core of qualitative inquiry. The role of the researcher, the person reading a textual passage, and the individuals from whom qualitative data are collected play a more central role in researchers' design decisions (Denzin & Lincoln, 2005). Some researchers have called for a methodological dialogue to address questions of disciplinary power, theoretical future of the field, alternative theoretical approaches, discontinuance of conceptual traditions, new methods of training and preparation, and alternative writing and publication possibilities (Koro-Ljungberg & Greckhamer, 2005). I see this trend coming largely from ethnography, but writers in grounded theory (Charmaz, 2006), narrative research (Clandinin & Connelly, 2000), and phenomenology (van Manen, 1990) have certainly embraced this interpretive "turn." To incorporate different theoretical approaches and to speak to the power of discourse in qualitative research is certainly necessary. Not all writers, however, have embraced the self-referential component of the interpretive approach. For example, Atkinson, Coffey, and Delamont (2003) have recently written about the dangers of forgetting the disciplinary traditions of ethnography: "We believe that too much contemporary work advocates and celebrates self-referential work, with little relevance to our understanding of actual social worlds." (p. xi). I agree. The focus of all qualitative research needs to be on understanding the phenomenon being explored rather than solely on the reader, the researcher, or the participants being studied. But the interpretive stance has much to offer. Thus, for each of the approaches discussed in this book, I now reflect on interpretive elements of procedures. These interpretive aspects also inform how I view the basic design of qualitative research found in Chapter 3. In addition, I brought up to the front of the book the philosophical and theoretical discussion (Chapter 2) so that it can help frame all other discussions about qualitative research.

Some have argued that the purpose of qualitative research should be to advance a social justice agenda (Denzin & Lincoln, 2005). While one needs to acknowledge that our society has become more diverse, cognizant of underrepresented groups, and educated about racial and ethnic tensions, not all qualitative projects *must* have this agenda as a central feature. All studies should acknowledge and recognize these issues as part of all inquiry and actively write about them. The passages on data collection in this book focus on the sensitivity required to collect data among diverse samples and the strategies that inform these procedures.

I have become much more cognizant of the variations within each of the five approaches (Creswell & Maietta, 2002). Partly this has developed because readers have called it to my attention (for example, by saying that "there are several ways to approach grounded theory"), and partly it is due to the increasing fragmentation and diversity that now exists in qualitative research. Book writers on the various approaches have contributed to this development as well. For example, I now see biography (Denzin, 1989a), described in detail in my first edition, as one of only many approaches to *narrative research* (Clandinin & Conoley, 2000), a broader more inclusive term. So narrative research is now one of the five approaches highlighted in this book. Narrative research incorporates many forms, such as autobiography, life stories, and personal stories, as well as biographies. Phenomenology, as I view it now, has several approaches, such as hermeneutical phenomenology (van Manen, 1990) and transcendental or psychological phenomenology (Moustakas, 1994). Grounded theory, for years dominated by Strauss and Corbin (1990, 1998), now has the strong constructivist, less structured approach advanced by Charmaz (2005), and ethnography has taken a turn from description and the objective, realist orientation to an openly ideological production of cultures (Koro-Ljungberg & Greckhamer, 2005). Case study research has the voice of Yin (2003), a more structured approach to research than the earlier Stake (1995) approach. I have inserted new passages addressing alternative types of procedures within each of the five approaches, and I now discuss specific steps in conducting a study within each of the five approaches.

The qualitative enterprise is much more fragmented than it was, and it is being challenged by writings that advocate for a return to the experimental model of inquiry, such as those found in the No Child Left Behind Act (Maxwell, 2005) and the National Research Council's monograph on scientific research in education (National Academy of Sciences, 2000). The "camps" in qualitative research seem to be the methodologists, who embrace rigorous methods; the philosophical advocates, who seek to identify and expand the number of paradigmatic and theoretical lenses used in qualitative research; the social justice researchers, largely drawn from ethnography, who advocate the social ends for qualitative research; and those in the health science group, who look to qualitative research to augment their experimental, intervention trials and their correlational designs. Today, individuals teaching, writing, and discussing qualitative research need to be clear about their stance and share it with their audiences. My attempt has been to honor all of these diverse perspectives in qualitative research, but my strong background in applied methods has led to an overall methods orientation to this text.

The data analysis has become more sophisticated as many qualitative software programs vie for a privileged status in qualitative research and incorporate more sophisticated subprograms that enable researchers to output qualitative codes to spreadsheets, statistical programs, or to concept maps. In this edition, I introduce several computer programs being used to analyze qualitative data, and thus extend the discussion of options available.

The term that I used in the first edition, "traditions," has now been replaced by "approaches," signaling that I not only want to respect past approaches, but I also want to encourage current practices in qualitative research. Other writers have referred to the approaches as "strategies of inquiry" (Denzin & Lincoln, 2005), "varieties" (Tesch, 1990), or "methods" (Morse & Richards, 2002). By *research design*, I refer to the entire process of research from conceptualizing a problem to writing research questions, and on to data collection, analysis, interpretation, and report writing (Bogdan & Taylor, 1975). Yin (2003) commented, "The design is the logical sequence that connects the empirical data to a study's initial research questions and, ultimately, to its conclusions" (p. 20). Hence, I include in the specific design features from the broad philosophical and theoretical perspectives to the quality and validation of a study.

## Rationale for This Book

Since the 1994 Vail seminar, I have been asking individuals who approached me with their qualitative studies, "What type are you doing?" Since the publication of the first edition of this book in 1998, I have increasingly heard from individuals that they are doing ethnography, or grounded theory, case study research. The types of qualitative approaches that I wrote about in the first edition have now become part of the lexicon of qualitative research, and researchers are much more aware of the type of designs they are using than they were in the 1990s.

My intent is that this book will include several features:

• It highlights the procedures of actually doing qualitative research. For years, the actual "doing" of qualitative research has been relegated to secondary status, behind the philosophical ideas and the important research questions. Of course, philosophy and the guiding questions are important, but the methods and procedures are important, too, and cannot be overlooked in conducting scholarly qualitative research.

• It provides qualitative researchers with options for conducting qualitative inquiry and helps them with decisions about what approach is best to use in studying their research problems. With so many books on qualitative research in general and on the various approaches of inquiry, qualitative research students are often at a loss for understanding what options (i.e., approaches) exist and how one makes an informed choice of an option for research. To clarify this, I limit the options to five major types.

- The book also provides a comparison of approaches so that readers can weigh the options and decide which option is right for their research problem. The way a researcher writes a qualitative study, for example, will differ depending on the approach the researcher chooses. Research questions framed from grounded theory *look* different than questions framed from a phenomenological study. When the approaches are displayed side by side throughout the phases of the design of a study, the reader not only gains perspective on their differences but also develops an applied perspective as to how the approaches inform each phase of the inquiry. In talking about seeing different perspectives, I am reminded of the delightful little book by Redfield (1963), who explored diverse anthropological and social views of a "village" by devoting a chapter to each view. In the discussion here of the different approaches to qualitative research, the reader will find different views of conducting qualitative research.

- Regardless of approach, all qualitative research tends to follow the basic process of research (e.g., introduction, questions, methods of data collection and analysis, etc.). Thus, this book introduces the reader to or reinforces the basic procedures of inquiry and serves as a reminder of the importance of these steps.

- Finally, for individuals trained or socialized in a specific approach, this comparative analysis can enlarge their scope of inquiry methods and encourage them to seek out alternative procedures. Students in my class who have used the book may start with a grounded theory project, then change their approach, and end with phenomenology. I hope that this book will encourage inquirers to experiment with different forms of inquiry.

## Selection of the Five Approaches

Those undertaking qualitative studies have a baffling number of choices of approaches. One can gain a sense of this diversity by examining several classifications or typologies. One of the more popular classifications is provided by Tesch (1990), who organized 28 approaches into four branches of a flowchart, sorting out these approaches based on the central interest of the investigator. Wolcott (1992) classified approaches in a "tree" diagram with branches of the tree designating strategies for data collection. Miller and Crabtree (1992) organized 18 types according to the "domain" of human life of primary concern to the researcher, such as a focus on the individual, the social world, or the culture. In the field of education, Jacob (1987) categorized all qualitative research into "traditions" such as ecological psychology, symbolic interactionism, and holistic ethnography. Lancy (1993) organized qualitative inquiry into discipline perspectives such as anthropology, sociology, biology, cognitive psychology, and history. Denzin and Lincoln (2005) organized their types of qualitative strategies of inquiry into ethnography (performance and ethnographic representation), case studies, grounded theory, life and narrative approaches, participatory action research, and clinical research. In short, there is no lack for classification systems for types of qualitative research, and Table 1.1 lists these systems and notes their interdisciplinary roots.

| Table 1.1 | Qualitative Approaches Mentioned by Authors | |
|---|---|---|
| *Authors* | *Qualitative Approaches* | *Discipline/Field* |
| Jacob (1987) | Ecological Psychology | Education |
| | Holistic Ethnography | |
| | Cognitive Anthropology | |
| | Ethnography of Communication | |
| | Symbolic Interactionism | |
| Munhall & Oiler (1986) | Phenomenology | Nursing |
| | Grounded Theory | |
| | Ethnography | |
| | Historical Research | |
| Lancy (1993) | Anthropological Perspectives | Education |
| | Sociological Perspectives | |
| | Biological Perspectives | |
| | Case Studies | |
| | Personal Accounts | |
| | Cognitive Studies | |
| | Historical Inquiries | |
| Strauss & Corbin (1990) | Grounded Theory | Sociology, Nursing |
| | Ethnography | |
| | Phenomenology | |
| | Life Histories | |
| | Conversational Analysis | |
| Morse (1994) | Phenomenology | Nursing |
| | Ethnography | |
| | Ethnoscience | |
| | Grounded Theory | |

*(Continued)*

**Table 1.1** (Continued)

| Authors | Qualitative Approaches | Discipline/Field |
|---|---|---|
| Moustakas (1994) | Ethnography<br>Grounded Theory<br>Hermeneutics<br>Empirical Phenomenological Research<br>Heuristic Research<br>Transcendental Phenomenology | Psychology |
| Denzin & Lincoln (1994) | Case Studies<br>Ethnography<br>Phenomenology<br>Ethnomethodology<br>Interpretative Practices<br>Grounded Theory<br>Biographical<br>Historical<br>Clinical Research | Social Sciences |
| Miles & Huberman (1994) | Approaches to Qualitative Data Analysis:<br>Interpretivism<br>Social Anthropology<br>Collaborative Social Research | Social Sciences |
| Slife & Williams (1995) | Categories of Qualitative Methods:<br>Ethnography<br>Phenomenology<br>Studies of Artifacts | Psychology |
| Denzin & Lincoln (2005) | Performance, Critical, and Public Ethnography<br>Interpretive Practices<br>Case Studies<br>Grounded Theory<br>Life History<br>Narrative Authority<br>Participatory Action Research<br>Clinical Research | Social Sciences |

With so many possibilities, how did I decide on the five approaches presented in this book? My choice of the five approaches resulted from following personal interests, selecting different foci, and electing to choose representative discipline orientations. I have personal experience with each of the five, as an advisor in counseling students and as a researcher in conducting qualitative studies. Beyond this personal experience, I have been reading the qualitative literature since my initial teaching assignment in the area in 1985. The five approaches discussed in this book reflect the types of qualitative research that I most frequently see in the social, behavioral, and health science literature. It is not unusual, too, for authors to state that certain approaches are most important in their fields (e.g., Morse & Field, 1995). Also, I prefer approaches with systematic procedures for inquiry. That I could find books that espouse rigorous data collection and analysis methods also contributed to the selection of the five. These books were also useful in that they represented different discipline perspectives in the social, behavioral, and health sciences. For example, narrative originates from the humanities and social sciences, phenomenology from psychology and philosophy, grounded theory from sociology, ethnography from anthropology and sociology, and case studies from the human and social sciences and applied areas such as evaluation research.

The primary ideas for this book came from several books that I have synthesized to reflect scholarly, rigorous approaches to qualitative research. In contrast to relying on one book per approach as in the first edition, in this edition I rely on several books for each approach. On narrative research, I refer to the educational perspective of Clandinin and Connelly (2000) but also consider the organizational approach of Czarniawska (2004) and the biographical approach of Denzin (1989a). In my discussion of phenomenology, I largely advance a psychological perspective based on Moustakas (1994) and also include the interpretive approach of van Manen (1990). In describing grounded theory, my approach relies on the systematic approach of the sociologists Strauss and Corbin (1990) but also incorporates ideas from the more recent sociological constructivist approach of Charmaz (2006). In discussing ethnography, I rely on the educational anthropology perspective of Wolcott (1999) and incorporate other perspectives from LeCompte and Schensul (1999) and the interpretive stances of Atkinson, Coffey, and Delamont (2003). In my description of case study research, I rely on an evaluation perspective from Stake (1995) but also include the applied social science and cognitive science orientation of Yin (2003).

To enumerate further, the reader may appreciate a listing of the core books I used in developing this discussion of approaches of inquiry and research design.

## Narrative Research

- Clandinin, D. J., & Connelly, F. M. (2000). *Narrative inquiry: Experience and story in qualitative research.* San Francisco: Jossey-Bass.
- Czarniawska, B. (2004). *Narratives in social science research.* London: Sage.
- Denzin, N. K. (1989a). *Interpretive biography.* Newbury Park, CA: Sage.

## Phenomenology

- Moustakas, C. (1994). *Phenomenological research methods.* Thousand Oaks, CA: Sage.
- van Manen, M. (1990). *Researching lived experience: Human science for an action sensitive pedagogy.* Albany: State University of New York Press.

## Grounded Theory

- Charmaz, K. (2006). *Constructing grounded theory.* London: Sage.
- Strauss, A., & Corbin, J. (1990). *Basics of qualitative research: Grounded theory procedures and techniques.* Newbury Park, CA: Sage.

## Ethnography

- Atkinson, P., Coffey, A., & Delamont, S. (2003). *Key themes in qualitative research: Continuities and changes.* Walnut Creek, CA: AltaMira.
- LeCompte, M. D., & Schensul, J. J. (1999). *Designing and conducting ethnographic research.* Walnut Creek, CA: AltaMira.
- Wolcott, H. F. (1994b). *Transforming qualitative data: Description, analysis, and interpretation.* Thousand Oaks, CA: Sage.
- Wolcott, H. F. (1999). *Ethnography: A way of seeing.* Walnut Creek, CA: AltaMira.

## Case Study

- Stake, R. (1995). *The art of case study research.* Thousand Oaks, CA: Sage.
- Yin, R. K. (2003). *Case study research: Design and methods* (3rd ed.). Thousand Oaks, CA: Sage.

# Positioning Myself

My approach is to present the five approaches as "pure" approaches to research design, when, in fact, authors may integrate them within a single study. But before blending them, I find it useful as a heuristic to separate them out, to see them as distinct approaches and visit each one, individually, as a procedural guide for research. For beginning researchers, I would not recommend choosing more than one approach. Separating them out as I do in this book should help beginning researchers focus on a single approach for their studies. For more advanced qualitative researchers, this book can serve as a reminder of the many options available and the current writers about the different approaches to qualitative research.

I limit the design discussion to specific components of the research design process. I leave for others additional aspects of research design, such as defining terms, addressing the significance of the study, posing limitations and delimitations, and advancing the role of the researcher (Marshall & Rossman, 2006).

In a book of this scope, I cannot undertake an examination and comparison of all types of qualitative inquiry. For example, I have not addressed two approaches in this book. At the macro-community level, participatory action research, aimed at social change and examining the political structures that deprive and oppress groups of people, is a major approach to qualitative inquiry (Kemmis & Wilkinson, 1998). At the micro-level, discourse and conversational analysis involves analyzing the content of text for syntax, semantics, and social and historical situatedness (Cheek, 2004). The basic premise is that language is not transparent or value free. In order to limit the scope of this discussion, both approaches will not be addressed in detail. However, some of the underlying principles of both approaches (e.g., the collaborative nature of inquiry and the historical context for reading, writing, and understanding research) are features emphasized here within the five types.

I situate this book within my thinking about qualitative research, and I hope to model *reflexivity*, or self-awareness. Although I have been referred to as associated with the postpositivist writers in qualitative research (Denzin & Lincoln, 2005), my perspective tends to vary from social justice (e.g., see my study of the homeless, Miller & Creswell, 1998) to realist accounts (e.g., see my study of department chairpersons, Creswell & Brown, 1992). I do tend to hold a more objective, scientific approach to qualitative research, as is documented in my realist projects, use of analytic tools such as computer programs, and emphasis on rigorous and thorough qualitative data collection and analysis. This is not to suggest that I am advocating the acceptance of qualitative research in a "quantitative" world (Ely, Anzul, Friedman, Garner, & Steinmetz, 1991). Qualitative inquiry represents a legitimate mode of social and human science exploration, without apology or comparisons to quantitative research.

Unquestionably, too, as an applied research methodologist my focus is on research designs or procedures, not on philosophical assumptions. Granted, these assumptions cannot be separated from procedures, but I position these assumptions in the background rather than the foreground, admitting openly that I am not a philosopher of education but rather a research methodologist, and my orientation in this book reflects this orientation. Throughout the book, the writing of research is featured, and I continually bring to the front the architecture or compositional approaches authors used in their qualitative studies. I place emphasis on the terms used by authors in each of the approaches and on the way in which the authors use *encoding* of significant passages with these terms to make the text a distinct illustration of an approach. I highlight *foreshadowing* information early in a study to hint at topics of ideas to come later. Along this line, I concur with Agger (1991), who says that readers and writers can understand methodology in less technical ways, thereby affording greater access to scholars and democraticizing science. Finally, my approach presents not a "lock-step" procedural guide but rather a general direction, offering alternatives for the researcher and advancing my preferred stance. In many ways, I see this book as a "quest" (Edel, 1984) for materials and ideas to best display and convey design within the five approaches.

## Audience

Although multiple audiences exist for any text (Fetterman, 1998), I direct this book toward academics and scholars affiliated with the social and human sciences. Examples throughout the book illustrate the diversity of disciplines and fields of study including sociology, psychology, education, the health sciences, urban studies, marketing, communication and journalism, educational psychology, family science and therapy, and other social and human science areas.

My aim is to provide a useful text for those who produce scholarly qualitative research in the form of journal articles, theses, or dissertations. The focus on a single type of qualitative research is ideal for shorter forms of scholarly communication; longer works, such as books or monographs, may employ multiple types. The level of discussion here is suitable for upper division students and graduate students. For graduate students writing master's theses or doctoral dissertations, I compare and contrast the five approaches in the hope that such analysis helps in establishing a rationale for the choice of a type to use. For beginning qualitative researchers, I provide Chapter 2 on the philosophical and theoretical lens that shapes qualitative research and Chapter 3 on the basic elements in designing a qualitative study. While discussing the basic elements, I suggest several books aimed at the beginning qualitative researcher that can provide a more extensive review of the basics of qualitative research. Such basics are necessary before delving into the five approaches. For both inexperienced and experienced researchers, I supply recommendations for further reading that can extend the material in this book. A focus on comparing the five approaches throughout this book provides an introduction for experienced researchers to approaches that build on their training and research experiences.

## Organization

The basic premise of this book is that different forms of qualitative approaches exist and that the design of research within each has distinctive features. In Chapter 2, I provide an introduction to the philosophical assumptions, worldviews or paradigms, and theoretical lenses used in qualitative research. These broad perspectives guide all aspects of qualitative research designs. Then, in Chapter 3, I review the basic elements of designing a qualitative study. These elements begin with a definition of qualitative research, the reasons for using this approach, and the phases in the process of research. In Chapter 4, I provide an introduction to each of the five approaches of inquiry: narrative research, phenomenology, grounded theory, ethnography, and case study research. Chapter 5 continues this discussion by presenting five published journal articles (with the complete articles in the appendices), which provide good illustrations of each of the approaches. By reading my overview and then reading for yourself the complete article, you can gain a deeper understanding of each of the five approaches.

These five chapters form an introduction to the five types and an overview of the process of research design. They set the stage for the remaining chapters, which relate research design to each approach: writing introductions to studies (Chapter 6), collecting data (Chapter 7), analyzing and representing data (Chapter 8), writing qualitative studies (Chapter 9), and the validation of results and the use of evaluation standards (Chapter 10). In all of these design chapters, I continually compare the five types of qualitative inquiry.

As a final experience to sharpen distinctions made among the five types, I present Chapter 11, in which I return to the gunman case study (Asmussen & Creswell, 1995), first introduced in Chapter 5, and "turn" the story from a case study into a narrative biography, a phenomenology, a grounded theory study, and an ethnography. This culminating chapter brings the reader full circle to examining the gunman case in several ways, an extension of my earlier Vail seminar experience.

Throughout the book, I provide several aids to help the reader. At the beginning of each chapter, I offer several conceptual questions to guide the reading. At the end of each chapter, I provide further readings and sample exercises. At least one of the exercises encourages the reader to design and conduct an entire qualitative study, with phases in this study identified progressively throughout the book. Also, in most of the chapters, I present comparison tables that show the differences among the five approaches to inquiry as well as figures to visualize distinctions and major design processes. Finally, each approach comes with distinct terms that may be unfamiliar to the reader. I provide a glossary of terms in Appendix A to facilitate the reading and understanding of the material in this book.

# 2

# Philosophical, Paradigm, and Interpretive Frameworks

The research design process in qualitative research begins with philosophical assumptions that the inquirers make in deciding to undertake a qualitative study. In addition, researchers bring their own worldviews, paradigms, or sets of beliefs to the research project, and these inform the conduct and writing of the qualitative study. Further, in many approaches to qualitative research, the researchers use interpretive and theoretical frameworks to further shape the study. Good research requires making these assumptions, paradigms, and frameworks explicit in the writing of a study, and, at a minimum, to be aware that they influence the conduct of inquiry. The purpose of this chapter is to make explicit the assumptions made when one chooses to conduct qualitative research, the worldviews or paradigms available in qualitative research, and the diverse interpretive and theoretical frameworks that shape the content of a qualitative project.

Five philosophical assumptions lead to an individual's choice of qualitative research: ontology, epistemology, axiology, rhetorical, and methodological assumptions. The qualitative researcher chooses a stance on each of these assumptions, and the choice has practical implications for designing and conducting research. Although the paradigms of research continually evolve, four will be mentioned that represent the beliefs of researchers that they bring to qualitative research: postpositivism, constructivism, advocacy/participatory, and pragmatism. Each represents a different paradigm for making claims about knowledge, and the characteristics of each differ considerably. Again, the practice of research is informed. Finally, the chapter will address theoretical frameworks, those interpretive communities that have developed within qualitative research that informs specific procedures of research. Several of these frameworks will be discussed: postmodern theories, feminist research, critical theory and critical race theory, queer theory, and disability inquiry. The three elements discussed above—assumptions, paradigms, and interpretive frameworks—often overlap and reinforce each other. For the purposes of our discussion, they will be discussed separately.

## Questions for Discussion

- When qualitative researchers chose a qualitative study, what philosophical assumptions are being implicitly acknowledged?
- When qualitative researchers bring their beliefs to qualitative research, what alternative paradigm stances are they likely to use?
- When qualitative researchers select a framework as a lens for their study, what interpretive or theoretical frameworks are they likely to use?
- In the practice of designing or conducting qualitative research, how are assumptions, paradigms, and interpretive and/or theoretical frameworks used?

## Philosophical Assumptions

In the choice of qualitative research, inquirers make certain assumptions. These philosophical assumptions consist of a stance toward the nature of reality (ontology), how the researcher knows what she or he knows (epistemology), the role of values in the research (axiology), the language of research (rhetoric), and the methods used in the process (methodology) (Creswell, 2003). These assumptions, shown in Table 2.1, are adapted from the "axiomatic" issues advanced by Guba and Lincoln (1988). However, my discussion departs from their analysis in three ways. I do not contrast qualitative or naturalistic assumptions with conventional or positive assumptions as they do, acknowledging that today qualitative research is legitimate in its own right and does not need to be compared to achieve respectability. I add to their issues one of my own concerns, the rhetorical assumption, recognizing that one needs to attend to the language and terms of qualitative inquiry. Finally, I discuss the practical implications of each assumption in an attempt to bridge philosophy and practice.

The *ontological* issue relates to the nature of reality and its characteristics. When researchers conduct qualitative research, they are embracing the idea of multiple realities. Different researchers embrace different realities, as

**Table 2.1**     Philosophical Assumptions With Implications for Practice

| Assumption | Question | Characteristics | Implications for Practice (Examples) |
|---|---|---|---|
| Ontological | What is the nature of reality? | Reality is subjective and multiple, as seen by participants in the study | Researcher uses quotes and themes in words of participants and provides evidence of different perspectives |
| Epistemological | What is the relationship between the researcher and that being researched? | Researcher attempts to lessen distance between himself or herself and that being researched | Researcher collaborates, spends time in field with participants, and becomes an "insider" |
| Axiological | What is the role of values? | Researcher acknowledges that research is value-laden and that biases are present | Researcher openly discusses values that shape the narrative and includes his or her own interpretation in conjunction with the interpretations of participants |
| Rhetorical | What is the language of research? | Researcher writes in a literary, informal style using the personal voice and uses qualitative terms and limited definitions | Researcher uses an engaging style of narrative, may use first-person pronoun, and employs the language of qualitative research |
| Methodological | What is the process of research? | Researcher uses inductive logic, studies the topic within its context, and uses an emerging design | Researcher works with particulars (details) before generalizations, describes in detail the context of the study, and continually revises questions from experiences in the field |

do also the individuals being studied and the readers of a qualitative study. When studying individuals, qualitative researchers conduct a study with the intent of reporting these multiple realities. Evidence of multiple realities includes the use of multiple quotes based on the actual words of different individuals and presenting different perspectives from individuals. When writers compile a phenomenology, they report how individuals participating in the study view their experiences differently (Moustakas, 1994).

With the *epistemological* assumption, conducting a qualitative study means that researchers try to get as close as possible to the participants being studied. In practice, qualitative researchers conduct their studies in the "field," where the participants live and work—these are important contexts for understanding what the participants are saying. The longer researchers stay in the "field" or get to know the participants, the more they "know what they know" from firsthand information. A good ethnography requires prolonged stay at the research site (Wolcott, 1999). In short, the researcher tries to minimize the "distance" or "objective separateness" (Guba & Lincoln, 1988, p. 94) between himself or herself and those being researched.

All researchers bring values to a study, but qualitative researchers like to make explicit those values. This is the *axiological* assumption that characterizes qualitative research. How does the researcher implement this assumption in practice? In a qualitative study, the inquirers admit the value-laden nature of the study and actively report their values and biases as well as the value-laden nature of information gathered from the field. We say that they "position themselves" in a study. In an interpretive biography, for example, the researcher's presence is apparent in the text, and the author admits that the stories voiced represent an interpretation and presentation of the author as much as the subject of the study (Denzin, 1989a).

Researchers are notorious for providing labels and names for aspects of qualitative methods (Koro-Ljungberg & Greckhamer, 2005). There is a rhetoric for the discourse of qualitative research that has evolved over time. Qualitative researchers tend to embrace the *rhetorical* assumption that the writing needs to be personal and literary in form. For example, they use metaphors, they refer to themselves using the first-person pronoun, "I," and they tell stories with a beginning, middle, and end, sometimes crafted chronologically, as in narrative research (Clandinin & Connelly, 2000). Instead of using quantitative terms such as "internal validity," "external validity," "generalizability," and "objectivity," the qualitative researcher writing a case study may employ terms such as "credibility," "transferability," "dependability," and "confirmability" (Lincoln & Guba, 1985) or "validation" (Angen, 2000), as well as naturalistic generalizations (Stake, 1995). Words such as "understanding," "discover," and "meaning" form the glossary of emerging qualitative terms (see Schwandt, 2001) and are important rhetorical markers in writing purpose statements and research questions (as discussed later). The language of the qualitative researcher becomes personal, literary, and based on definitions that evolve during a study rather than being defined by the researcher. Seldom does one see an extensive "Definition of Terms" section in a qualitative study, because the terms as defined by participants are of primary importance.

The procedures of qualitative research, or its methodology, are characterized as inductive, emerging, and shaped by the researcher's experience in collecting and analyzing the data. The logic that the qualitative researcher follows is inductive, from the ground up, rather than handed down entirely from a theory or from the perspectives of the inquirer. Sometimes the research questions change in the middle of the study to reflect better the types of questions needed to understand the research problem. In response, the data collection strategy, planned before the study, needs to be modified to accompany the new questions. During the data analysis, the researcher follows a path of analyzing the data to develop an increasingly detailed knowledge of the topic being studied.

## Paradigms or Worldviews

The assumptions reflect a particular stance that researchers make when they choose qualitative research. After researchers make this choice, they then further shape their research by bringing to the inquiry paradigms or worldviews. A *paradigm or worldview* is "a basic set of beliefs that guide action" (Guba, 1990, p. 17). These beliefs have been called paradigms (Lincoln & Guba, 2000; Mertens, 1998); philosophical assumptions, epistemologies, and ontologies (Crotty, 1998); broadly conceived research methodologies (Neuman, 2000); and alternative knowledge claims (Creswell, 2003). Paradigms used by qualitative researchers vary with the set of beliefs they bring to research, and the types have continually evolved over time (contrast the paradigms of Denzin and Lincoln, 1994, with the paradigms of Denzin and Lincoln, 2005). Individuals may also use multiple paradigms in their qualitative research that are compatible, such as constructionist and participatory worldviews (see Denzin & Lincoln, 2005).

In this discussion, I focus on four worldviews that inform qualitative research and identify how these worldviews shape the practice of research. The four are postpositivism, constructivism, advocacy/participatory, and pragmatism (Creswell, 2003). It is helpful to see the major elements of each paradigm, and how they inform the practice of research differently.

## Postpositivism

Those who engage in qualitative research using a belief system grounded in postpositivism will take a scientific approach to research. The approach has the elements of being reductionistic, logical, an emphasis on empirical data collection, cause-and-effect oriented, and deterministic based on a priori theories. We can see this approach at work among individuals with prior quantitative research training, and in fields such as the health sciences in which qualitative research is a new approach to research and must be couched in terms acceptable to quantitative researchers and funding agents (e.g., the a priori use of theory; see Barbour, 2000). A good overview of post-postivist approaches is available in Phillips and Burbules (2000).

In terms of practice, postpositivist researchers will likely view inquiry as a series of logically related steps, believe in multiple perspectives from participants rather than a single reality, and espouse rigorous methods of qualitative data collection and analysis. They will use multiple levels of data analysis for rigor, employ computer programs to assist in their analysis, encourage the use of validity approaches, and write their qualitative studies in the form of scientific reports, with a structure resembling quantitative approaches (e.g., problem, questions, data collection, results, conclusions). My approach to qualitative research has been identified as belonging to post-positivism (Denzin & Lincoln, 2005), as have the approaches of others (e.g., Taylor & Bogdan, 1998). I do tend to use this belief system, although I would not characterize all of my research as framed within a postpositivist qualitative orientation (e.g., see the constructivist approach in McVea, Harter, McEntarffer, and Creswell, 1999, and the social justice perspective in Miller and Creswell, 1998). In their discussion here of the five approaches, for example, I emphasize the systematic procedures of grounded theory found in Strauss and Corbin (1990), the analytic steps in phenomenology (Moustakas, 1994), and the alternative analysis strategies of Yin (2003).

## Social Constructivism

Social constructivism (which is often combined with interpretivism; see Mertens, 1998) is another worldview. In this worldview, individuals seek understanding of the world in which they live and work. They develop subjective meanings of their experiences—meanings directed toward certain objects or things. These meanings are varied and multiple, leading the researcher to look for the complexity of views rather than narrow the meanings into a few categories or ideas. The goal of research, then, is to rely as much as possible on the participants' views of the situation. Often these subjective meanings are negotiated socially and historically. In other words, they are not simply imprinted on individuals but are formed through inter-action with others (hence social constructivism) and through historical and cultural norms that operate in individuals' lives. Rather than starting with a theory (as in postpositivism), inquirers generate or inductively develop a theory or pattern of meaning. Examples of recent writers who have summa-rized this position are Crotty (1998), Lincoln and Guba (2000), Schwandt (2001), and Neuman (2000).

In terms of practice, the questions become broad and general so that the participants can construct the meaning of a situation, a meaning typically forged in discussions or interactions with other persons. The more open-ended the questioning, the better, as the researcher listens carefully to what people say or do in their life setting. Thus, constructivist researchers often address the "processes" of interaction among individuals. They also focus on the specific contexts in which people live and work in order to understand the historical and cultural settings of the participants. Researchers recognize that their own background shapes their interpretation, and they "position themselves" in the research to acknowledge how their interpretation flows from their own personal, cultural, and historical experiences. Thus the researchers make an interpretation of what they find, an interpretation shaped by their own expe-riences and background. The researcher's intent, then, is to make sense (or interpret) the meanings others have about the world. This is why qualitative research is often called "interpretive" research.

In the discussion here of the five approaches, we will see the constructivist worldview manifest in phenomenological studies, in which individuals describe their experiences (Moustakas, 1994), and in the grounded theory perspective of Charmaz (2006), in which she grounds her theoretical orien-tation in the views or perspectives of individuals.

## Advocacy/Participatory

Researchers might use an alternative worldview, advocacy/participatory, because the postpositivist imposes structural laws and theories that do not fit marginalized individuals or groups and the constructivists do not go far enough in advocating for action to help individuals. The basic tenet of this worldview is that research should contain an action agenda for reform that may change the lives of participants, the institutions in which they live and work, or even the researchers' lives. The issues facing these marginalized groups are of paramount importance to study, issues such as oppression, domination, suppression, alienation, and hegemony. As these issues are studied and exposed, the researchers provide a voice for these participants, raising their consciousness and improving their lives. Kemmis and Wilkinson (1998) summarize the key features of advocacy/participatory practice:

- Participatory action is recursive or dialectical and is focused on bringing about change in practices. Thus, at the end of advocacy/participatory studies, researchers advance an action agenda for change.
- It is focused on helping individuals free themselves from constraints found in the media, in language, in work procedures, and in the relationships of power in educational settings. Advocacy/participatory studies often begin with an important issue or stance about the problems in society, such as the need for empowerment.
- It is emancipatory in that it helps unshackle people from the constraints of irrational and unjust structures that limit self-development and self-determination. The aim of advocacy/participatory studies is to create a political debate and discussion so that change will occur.
- It is practical and collaborative because it is inquiry completed "with" others rather than "on" or "to" others. In this spirit, advocacy/participatory authors engage the participants as active collaborators in their inquiries.

Other researchers that embrace this worldview are Fay (1987) and Heron and Reason (1997).

In practice, this worldview has shaped several approaches to inquiry. Specific social issues (e.g., domination, oppression, inequity) help frame the research questions. Not wanting to further marginalize the individuals participating in the research, advocacy/participatory inquirers collaborate with research participants. They may ask participants to help with designing the questions, collecting the data, analyzing it, and shaping the final report of the research. In this way, the "voice" of the participants becomes heard throughout the research process. The research also contains an action agenda for reform, a specific plan for addressing the injustices of the marginalized group. These practices will be seen in the ethnographic approaches to research found in Denzin and Lincoln (2005) and in the advocacy tone of some forms of narrative research (Angrosino, 1994).

## Pragmatism

There are many forms of pragmatism. Individuals holding this worldview focus on the outcomes of the research—the actions, situations, and consequences of inquiry—rather than antecedent conditions (as in postpositivism). There is a concern with applications—"what works"—and solutions to problems (Patton, 1990). Thus, instead of a focus on methods, the important aspect of research is the problem being studied and the questions asked about this problem (see Rossman & Wilson, 1985). Cherryholmes (1992) and Murphy (1990) provide direction for the basic ideas:

- Pragmatism is not committed to any one system of philosophy and reality.
- Individual researchers have a freedom of choice. They are "free" to choose the methods, techniques, and procedures of research that best meet their needs and purposes.
- Pragmatists do not see the world as an absolute unity. In a similar way, mixed methods researchers look to many approaches to collecting and analyzing data rather than subscribing to only one way (e.g., quantitative or qualitative).
- Truth is what works at the time; it is not based in a dualism between reality independent of the mind or within the mind.
- Pragmatist researchers look to the "what" and "how" to research based on its intended consequences—where they want to go with it.
- Pragmatists agree that research always occurs in social, historical, political, and other contexts.
- Pragmatists have believed in an external world independent of the mind as well as those lodged in the mind. But they believe (Cherryholmes, 1992) that we need to stop asking questions about reality and the laws of nature. "They would simply like to change the subject" (Rorty, 1983, p. xiv.)
- Recent writers embracing this worldview include Rorty (1990), Murphy (1990), Patton (1990), Cherryholmes (1992), and Tashakkori and Teddlie (2003).

In practice, the individual using this worldview will use multiple methods of data collection to best answer the research question, will employ both quantitative and qualitative sources of data collection, will focus on the practical implications of the research, and will emphasize the importance of conducting research that best addresses the research problem. In the discussion here of the five approaches to research, you will see this worldview at work when ethnographers employ both quantitative (e.g., surveys) and qualitative data collection (LeCompte & Schensul, 1999) and when case study researchers use both quantitative and qualitative data (Luck, Jackson, & Usher, 2006; Yin, 2003).

## Interpretive Communities

Operating at a less philosophical level are various interpretive communities for qualitative researchers (Denzin & Lincoln, 2005). Each community mentioned below is a community with a distinct body of literature and unique issues of discussion. Space does not permit doing justice here to the scope and issues raised by interpretive communities. However, at the end of this chapter, I advance several readings that can extend and probe in more detail the interpretive communities' stances. Also, throughout the approaches to qualitative research discussed in this book, I will interweave research procedures and specific journal articles that use interpretive approaches. Our focus in this discussion will be on how interpretive lenses impact the process of research across the different interpretive communities. Although qualitative researchers use *social sciences theories* to frame their theoretical lens in studies, such as the use of these theories in ethnography (see Chapter 4), our discussion will be limited to the interpretive lens related to societal issues and issues influencing marginalized or underrepresented groups.

Interpretive positions provide a pervasive lens or perspective on all aspects of a qualitative research project. The participants in these interpretive projects represent underrepresented or marginalized groups, whether those differences take the form of gender, race, class, religion, sexuality, and geography (Ladson-Billings & Donnor, 2005) or some intersection of these differences. The problems and the research questions explored aim to understanding specific issues or topics—the conditions that serve to disadvantage and exclude individuals or cultures, such as hierarchy, hegemony, racism, sexism, unequal power relations, identity, or inequities in our society.

In addition, the procedures of research, such as data collection, data analysis, representing the material to audiences, and standards of evaluation and ethics, emphasize an interpretive stance. During data collection, the researcher does not further marginalize the participants, but respects the participants and the sites for research. Further, researchers provide reciprocity by giving or paying back those who participate in research, and they focus on the multiple-perspective stories of individuals and who tells the stories. Researchers are also sensitive to power imbalances during all facets of the research process. They respect individual differences rather than employing the traditional aggregation of categories such as men and women, or Hispanics or African Americans. Ethical practices of the researchers recognize the importance of the subjectivity of their own lens, acknowledge the powerful position they have in the research, and admit that the participants or the co-construction of the account between the researchers and the participants are the true owners of the information collected.

How the research is presented and used also is important. The research may be presented in traditional ways, such as journal articles, or in experimental approaches, such as theater or poetry. Using an interpretive lens may also lead to the call for action and transformation—the aims of social justice—in which the qualitative project ends with distinct steps of reform and an incitement to action.

Based on these core ideas, several theoretical perspectives will be reviewed: the postmodern perspective, feminist theories, critical theory and critical race theory (CRT), queer theory, and disability theories.

## Postmodern Perspectives

Thomas (1993) calls postmodernists "armchair radicals" (p. 23) who focus their critiques on changing ways of thinking rather than on calling for action based on these changes. Rather than viewing *postmodernism* as a theory, it might be considered a family of theories and perspectives that have something in common (Slife & Williams, 1995). The basic concept is that knowledge claims must be set within the conditions of the world today and in the multiple perspectives of class, race, gender, and other group affiliations. These conditions are well articulated by individuals such as Foucault, Derrida, Lyotard, Giroux, and Freire (Bloland, 1995). These are negative conditions, and they show themselves in the presence of hierarchies, power and control by individuals in these hierarchies, and the multiple meanings of language. The conditions include the importance of different discourses, the importance of marginalized people and groups (the "other"), and the presence of "meta-narratives" or universals that hold true regardless of the social conditions. Also included are the need to "deconstruct" texts in terms of language, their reading and their writing, and the examining and bringing to the surface concealed hierarchies as well as dominations, oppositions, inconsistencies, and contradictions (Bloland, 1995; Clarke, 2005; Stringer, 1993). Denzin's (1989a) approach to "interpretive" biography, Clandinin and Connelly's (2000) approach to narrative research, and Clarke's (2005) perspective on grounded theory draw on postmodernism in that researchers study turning points, or problematic situations in which people find themselves during transition periods (Borgatta & Borgatta, 1992). Regarding a "postmodern-influenced ethnography," Thomas (1993) writes that such a study might "confront the centrality of media-created realities and the influence of information technologies" (p. 25). Thomas also comments that narrative texts need to be challenged (and written), according to the postmodernists, for their "subtexts" of dominant meanings.

## Feminist Theories

Feminism draws on different theoretical and pragmatic orientations, different national contexts, and dynamic developments (Olesen, 2005). *Feminist research approaches* center and make problematic women's diverse situations and the institutions that frame those situations. Research topics may include policy issues related to realizing social justice for women in specific contexts and knowledge about oppressive situations for women (Olesen, 2005). The theme of domination prevails in the feminist literature as well, but the subject matter is gender domination within a patriarchal society. Feminist research also embraces many of the tenets of postmodern critiques as a challenge to current society. In feminist research approaches, the goals are to establish collaborative and nonexploitative relationships, to place the researcher within the study so as to avoid objectification, and to conduct research that is transformative. It is a complex area of inquiry, with numerous frameworks (e.g., male oriented, white feminist oriented, able-bodied female oriented) and difficult issues (e.g., the absence and invisibility of women, who can be "knowers") (Olesen, 2005).

One of the leading scholars of this approach, Lather (1991), comments on the essential perspectives of this framework. Feminist researchers see gender as a basic organizing principle that shapes the conditions of their lives. It is "a lens that brings into focus particular questions" (Fox-Keller, 1985, p. 6). The questions feminists pose relate to the centrality of gender in the shaping of our consciousness. The aim of this ideological research is to "correct both the invisibility and distortion of female experience in ways relevant to ending women's unequal social position" (Lather, 1991, p. 71). Another writer, Stewart (1994), translates feminist critiques and methodology into procedural guides. She suggests that researchers need to look for what has been left out in social science writing, and to study women's lives and issues such as identities, sex roles, domestic violence, abortion activism, comparable worth, affirmative action, and the way in which women struggle with their social devaluation and powerlessness within their families. Also, researchers need to consciously and systematically include their own roles or positions and assess how they impact their understandings of a woman's life. In addition, Stewart views women as having agency, the ability to make choices and resist oppression, and she suggests that researchers need to inquire into how a woman understands her gender, acknowledging that gender is a social contract that differs for each individual. Stewart highlights the importance of studying power relationships and individuals' social position and how they impact women. Finally, she sees each woman as different and recommends that scholars avoid the search for a unified or coherent self or voice.

Recent discussions indicate that the approach of finding appropriate methods for feminist research has given way to the thought that any method can be made feminist (Deem, 2002; Moss, 2006). The focus on feminist-oriented methods is a fruitless one; rather, the focus, as noted by Olesen (2005), needs to be on topics such as what feminist knowledge might look like, with questions including whose knowledge it is and where and how is it obtained, by whom, and for what purposes. Olesen further explains some of the issues feminist researchers are addressing today, such as the feminist researcher as objective with insider knowledge; the need to uncover the hidden or unrecognized elements in a researcher's background; the credibility, trustworthiness, and validity of researchers' accounts; the reporting of women's voices without exploiting or distorting them; the use of experimentation in presentation, such as in performance pieces, dramatic readings, and plays; and ethical issues of care, establishing positive relationships with participants, and recognizing power and ownership of materials. In short, rather than a focus on methods, the discussions have now turned to how to use the methods in a self-disclosing and respectful way.

## Critical Theory and Critical Race Theory (CRT)

*Critical theory* perspectives are concerned with empowering human beings to transcend the constraints placed on them by race, class, and gender (Fay, 1987). Researchers need to acknowledge their own power, engage in dialogues, and use theory to interpret or illuminate social action (Madison, 2005). Central themes that a critical researcher might explore include the scientific study of social institutions and their transformations through interpreting the meanings of social life; the historical problems of domination, alienation, and social struggles; and a critique of society and the envisioning of new possibilities (Fay, 1987; Morrow & Brown, 1994).

In research, critical theory can be "defined by the particular configuration of methodological postures it embraces" (p. 241). The critical researcher might design, for example, an ethnographic study to include changes in how people think; encourage people to interact, form networks, become activists, and action-oriented groups; and help individuals examine the conditions of their existence (Madison, 2005; Thomas, 1993). The end goal of the study might be social theorizing, which Morrow and Brown (1994) define as "the desire to comprehend and, in some cases, transform (through praxis) the underlying orders of social life—those social and systemic relations that constitute society" (p. 211). The investigator accomplishes this, for example, through an intensive case study or across a small number of historically comparable cases of specific actors (biographies), mediations, or systems and through "ethnographic accounts (interpretive social psychology), componential taxonomies (cognitive anthropology), and formal models (mathematical sociology)" (p. 212). In critical action research in teacher education, for example, Kincheloe (1991) recommends that the "critical teacher" exposes the assumptions of existing research orientations, critiques of the knowledge base, and through these critiques reveals ideological effects on teachers, schools, and the culture's view of education. The design of research within a critical theory approach, according to sociologist Agger (1991), falls into two broad categories: *methodological,* in that it affects the ways in which people write and read, and *substantive,* in the theories and topics of the investigator (e.g., theorizing about the role of the state and culture in advanced capitalism). An often-cited classic of critical theory is the ethnography from Willis (1977) of the "lads" who participated in behavior as opposition to authority, as informal groups "having a laff" (p. 29) as a form of resistance to their school. As a study of the manifestations of resistance and state regulation, it highlights ways in which actors come to terms with and struggle against cultural forms that dominate them (Morrow & Brown, 1994). Resistance is also the theme addressed in the ethnography of a subcultural group of youths highlighted as an example of ethnography in this book (see Haenfler, 2004).

*Critical race theory* (CRT) focuses theoretical attention on race and how racism is deeply embedded within the framework of American society (Parker & Lynn, 2002). Racism has directly shaped the U.S. legal system and the ways people think about the law, racial categories, and privilege (Harris, 1993). According to Parker and Lynn (2002), CRT has three main goals. Its first goal is to present stories about discrimination from the perspective of people of color. These may be qualitative case studies of descriptions and interviews. These cases may then be drawn together to build cases against racially biased officials or discrminatory practices. Since many stories advance White privilege through "majoritiarian" master narratives, counterstories by people of color can help to shatter the complacency that may accompany such privilege and challenge the dominant discourses that serve to suppress people on the margins of society (Solorzano & Yosso, 2002). As a second goal, CRT argues for the eradication of racial subjugation while simultaneously recognizing that race is a social construct (Parker & Lynn, 2002). In this view, race is not a fixed term, but one that is fluid and continually shaped by political pressures and informed by individual lived experiences. Finally, the third goal of CRT addresses other areas of difference, such as gender, class, and any inequities experienced by individuals. As Parker and Lynn (2002) comment: "In the case of Black women, race does not exist outside of gender and gender does not exist outside of race" (p. 12). In research, the use of CRT methodology means that the researcher foregrounds race and racism in all aspects of the research process; challenges the traditional research paradigms, texts, and theories used to explain the experiences of people of color; and offers transformative solutions to racial, gender, and class subordination in our societal and institutional structures.

## Queer Theory

*Queer theory* is characterized by a variety of methods and strategies relating to individual identity (Watson, 2005). As a body of literature continuing to evolve, it explores the myriad complexities of the construct, identity, and how identities reproduce and "perform" in social forums. Writers also use a postmodern or poststructural orientation to critique and deconstruct dominant theories (a "radical deconstruction," Plummer, 2005, p. 359) related to identity (Watson, 2005). They focus on how it is culturally and historically constituted, linked to discourse, and overlaps gender and sexuality. The term itself—"queer theory," rather than gay, lesbian, or homosexual theory—allows for keeping open to question the elements of race, class, age, and anything else (Turner, 2000). Most queer theorists work to challenge and undercut identity as singular, fixed, or normal (Watson, 2005). They also seek to challenge categorization processes and their deconstructions, rather than focus on specific populations. The historical binary distinctions are inadequate to describe sexual identity. Plummer (2005) provides a concise overview of the queer theory stance:

- Both the heterosexual/homosexual binary and the sex/gender split are challenged.
- There is a decentering of identity.
- All sexual categories (lesbian, gay, bisexual, transgender, heterosexual) are open, fluid, and nonfixed.
- Mainstream homosexuality is critiqued.
- Power is embodied discursively.
- All normalizing strategies are shunned.
- Academic work may become ironic, and often comic and paradoxical.
- Versions of homosexual subject positions are inscribed everywhere.
- Deviance is abandoned, and interest lies in insider and outsider perspectives and transgressions.
- Common objects of study are films, videos, novels, poetry, and visual images.
- The most frequent interests include the social worlds of the so-called radical sexual fringe (e.g, drag kings and queens, sexual playfulness).

Although queer theory is less a methodology and more a focus of inquiry, queer methods often find expression in a rereading of cultural texts (e.g., films, literature); ethnographies and case studies of sexual worlds that challenge assumptions; data sources that contain multiple texts; documentaries that include performances; and projects that focus on individuals (Plummer, 2005). Queer theorists have engaged in research and/or political activities such as ACT-UP and QUEER NATION around HIV/AIDS awareness, as well as artistic and cultural representations of art and theater aimed at disrupting or rendering unnatural and strange practices that are taken for granted. These representations convey the voices and experiences of individuals who have been suppressed (Gamson, 2000). Useful readings about queer theory are found in the journal article overview provided by Watson

(2005) and the chapter by Plummer (2005), and in key books, such the book by Tierney (1997).

## Disability Theories

Disability inquiry addresses the meaning of inclusion in schools and encompasses administrators, teachers, and parents who have children with disabilities (Mertens, 1998). Mertens recounts how disability research has moved through stages of development, from the medical model of disability (sickness and the role of the medical community in threatening it) to an environmental response to individuals with a disability. Now, researchers focus more on disability as a dimension of human difference and not as a defect. As a human difference, its meaning is derived from social construction (i.e., society's response to individuals) and it is simply one dimension of human difference (Mertens, 2003). Viewing individuals with disabilities as different is reflected in the research process, such as in the types of questions asked, the labels applied to these individuals, considerations of how the data collection will benefit the community, the appropriateness of communication methods, and how the data are reported in a way that is respectful of power relationships.

## Summary

In this chapter, I situated qualitative research within the larger discussion about philosophical, paradigmatic, and interpretive frameworks that investigators bring to their studies. It is a complex area, and one that I can only begin to sketch with some clarity. I see, however, that the basic philosophical assumptions relate to ontology, epistemology, axiology, rhetoric, and methodology as central features of all qualitative studies. Researchers take a philosophical stance on each of these assumptions when they decide to undertake a qualitative study. They also bring to the research their paradigms or worldviews, and those frequently used by qualitative researchers consist of postpositivist, constructivist, advocacy/participatory, and pragmatist. These worldviews, in turn, narrow to interpretive or theoretical stances taken by the researcher. These interpretive stances shape the individuals studied; the types of questions and problems examined; the approaches to data collection, data analysis, writing, and evaluation; and the use of the information to change society or add to social justice. Some of the interpretive stances used in qualitative research include postmodernism, feminist research, critical theory and critical race theory, queer theory, and disability theory. Thinking related to the philosophical assumptions, paradigms or worldview, and interpretive stances will be threaded throughout our exploration of the five approaches.

## Additional Readings

Several writers, in addition to Guba and Lincoln (1988, 2005), discuss the paradigm assumptions of qualitative research. In counseling psychology, Hoshmand (1989) reviews these assumptions. In education, see Sparkes (1992) or Cunningham and Fitzgerald (1996). In management, see Burrell and Morgan (1979) or Gioia and Pitre (1990).

Burrell, G., & Morgan, G. (1979). *Sociological paradigms and organizational analysis*. London: Heinemann.

Cunningham, J. W., & Fitzgerald, J. (1996). Epistemology and reading. *Reading Research Quarterly, 31*(1), 36–60.

Gioia, D. A., & Pitre, E. (1990). Multiparadigm perspectives on theory building. *Management Review, 15,* 584–602.

Guba, E., & Lincoln, Y. S. (1988). Do inquiry paradigms imply inquiry methodologies? In D. M. Fetterman (Ed.), *Qualitative approaches to evaluation in education* (pp. 89–115). New York: Praeger.

Guba, E., & Lincoln, Y. S. (2005). Paradigmatic controversies, contradictions, and emerging confluences. In N. K. Denzin & Y. S. Lincoln, *The Sage handbook of qualitative research* (3rd ed., pp. 191–215). Thousand Oaks, CA: Sage.

Hoshmand, L. L. S. T. (1989). Alternative research paradigms: A review and teaching proposal. *The Counseling Psychologist, 17*(1), 3–79.

Sparkes, A. C. (1992). The paradigms debate: An extended review and celebration of differences. In A. C. Sparkes (Ed.), *Research in physical education and sport: Exploring alternative visions* (pp. 9–60). London: Falmer Press.

For an introduction to postmodern thinking in the social sciences, see Rosenau (1992), Slife and Williams (1995), Clarke (2005), and the journal article by Bloland (1995).

Bloland, H. G. (1995). Postmodernism and higher education. *Journal of Higher Education, 66,* 521–559.

Clarke, A. E. (2005). *Situational analysis: Grounded theory after the postmodern turn*. Thousand Oaks, CA: Sage.

Rosenau, P. M. (1992). *Post-modernism and the social sciences: Insights, inroads, and intrusions*. Princeton, NJ: Princeton University Press.

Slife, B. D., & Williams, R. N. (1995). *What's behind the research? Discovering hidden assumptions in the behavioral sciences*. Thousand Oaks, CA: Sage.

For critical theory and critical race theory, see the following articles, which provide an introduction to the subject: Bloland (1995), Agger (1991), and Carspecken and Apple (1992). For book-length works, see Morrow and Brown (1994), a useful book for drawing the connection between critical theory and methodology. Other book-length works that take the critical theory discussion into ethnography are Thomas (1993) and Madison (2005). For critical race theory, examine Parker and Lynn (2002) and Solorzano and Yosso (2002).

Agger, B. (1991). Critical theory, poststructuralism, postmodernism: Their sociological relevance. In W. R. Scott & J. Blake (Eds.), *Annual review of sociology* (Vol. 17, pp. 105–131). Palo Alto, CA: Annual Reviews.

Bloland, H. G. (1995). Postmodernism and higher education. *Journal of Higher Education, 66,* 521–559.

Carspecken, P. F., & Apple, M. (1992). Critical qualitative research: Theory, methodology, and practice. In M. L. LeCompte, W. L. Millroy, & J. Preissle (Eds.), *The handbook of qualitative research in education* (pp. 507–553). San Diego, CA: Academic Press.

Madison, D. S. (2005). *Critical ethnography: Method, ethics, and performance.* Thousand Oaks, CA: Sage.

Morrow, R. A., & Brown, D. D. (1994). *Critical theory and methodology.* Thousand Oaks, CA: Sage.

Parker, L., & Lynn, M. (2002). What race got to do with it? Critical race theory's conflicts with and connections to qualitative research methodology and epistemology. *Qualitative Inquiry, 8*(1), 7–22.

Solorzano, D. G., & Yosso, T. J. (2002). Critical race methodology: Counter-storytelling as an analytical framework for education research. *Qualitative Inquiry, 8*(1), 23–44.

Thomas, J. (1993). *Doing critical ethnography.* Newbury Park, CA: Sage.

For an introduction to feminist research and social science methods, see the articles or chapters by Roman (1992), Olesen (1994, 2005), Stewart (1994), and Moss (2006). For book-length works, examine Harding (1987), Nielsen (1990), Lather (1991), Reinharz (1992), and Ferguson and Wicke (1994).

Ferguson, M., & Wicke, J. (1994). *Feminism and postmodernism.* Durham, NC: Duke University Press.

Harding, S. (1987). *Feminism and methodology.* Bloomington: Indiana University Press.

Lather, P. (1991). *Getting smart: Feminist research and pedagogy with/in the postmodern.* New York: Routledge.

Moss, P. (2006). Emergent methods in feminist research. In S. N. Hesse-Biber (Ed.), *Handbook of feminist research methods.* Thousand Oaks, CA: Sage.

Nielsen, J. M. (Ed.). (1990). *Feminist research methods: Exemplary readings in the social sciences.* Boulder, CO: Westview Press.

Olesen, V. (1994). Feminisms and models of qualitative research. In N. K. Denzin & Y. S. Lincoln (Eds.), *The handbook of qualitative research* (pp. 158–174). Thousand Oaks, CA: Sage.

Olesen, V. (2005). Early millennial feminist qualitative research: Challenges and contours. In N. K. Denzin & Y. S. Lincoln (Eds.), *The Sage handbook of qualitative research* (3rd ed., pp. 235–278). Thousand Oaks, CA: Sage.

Reinharz, S. (1992). *Feminist methods in social research.* New York: Oxford University Press.

Roman, L. G. (1992). The political significance of other ways of narrating ethnography: A feminist materialist approach. In M. L. LeCompte, W. L. Millroy, & J. Preissle (Eds.), *The handbook of qualitative research in education* (pp. 555–594). San Diego, CA: Academic Press.

Stewart, A. J. (1994). Toward a feminist strategy for studying women's lives. In C. E. Franz & A. J. Stewart (Eds.), *Women creating lives: Identities, resilience and resistance* (pp. 11–35). Boulder, CO: Westview Press.

For a recent introduction to queer theory and its applications in the social sciences and sociology, see:

Tierney, W. G. (1997). *Academic outlaws: Queer theory and cultural studies in the academy.* London: Sage.

Valocchi, S. (2005). Not yet queer enough: The lessons of queer theory for the sociology of gender and sexuality. *Gender & Society, 19*(6), 750–770.

Watson, K. (2005). Queer theory. *Group Analysis, 38*(1), 67–81.

For an overview of disability theories, see:

Mertens, D. M. (2003). Mixed methods and the politics of human research: The transformative-emancipatory perspective. In A. Tashakkori & C. Teddlie (Eds.), *Handbook of mixed methods in social and behavioral research* (pp. 135–164). Thousand Oaks, CA: Sage.

## Exercises

1. In the study you are planning to conduct, you may or may not use an interpretive perspective. It is good practice to consider how you might design this component into your proposed study. Take the study that you would like to design, and select a postmodern, feminist, critical race theory, queer theory, or disability perspective. Discuss how this interpretive stance will shape the

participants selected, the issues explored, the modes of data collection, and the use of the study.

2. Take the five philosophical assumptions and design a matrix like Table 2.1 that includes a column for how you plan to address each assumption in your proposed study.

3. Select a postpositivist, constructivist, advocacy/participatory, or pragmatic worldview for your study. Discuss the ways that this worldview will inform the design of your study.

# 3

# Designing a Qualitative Study

I think metaphorically of qualitative research as an intricate fabric composed of minute threads, many colors, different textures, and various blends of material. This fabric is not explained easily or simply. Like the loom on which fabric is woven, general worldviews and perspectives hold qualitative research together. To describe these frameworks, qualitative researchers use terms—constructivist, interpretivist, feminist, methodology, postmodernist, and naturalistic research. Within these worldviews and through these lenses are approaches to qualitative inquiry, such as narrative research, phenomenology, grounded theory, ethnography, and case studies. This field has many different individuals with different perspectives who are on their own looms creating the fabric of qualitative research. Aside from these differences, the creative artists are all at work making a fabric. In other words, there are characteristics common to all forms of qualitative research, and the different characteristics will receive different emphases depending on the qualitative project.

The basic intent of this chapter is to provide an overview of and introduction to qualitative research so that we can see the common characteristics of qualitative research before we explore the different threads of it. I begin with a general definition of qualitative research and highlight the essential characteristics of conducting this form of inquiry. I then discuss the types of research problems and issues best suited for a qualitative study and emphasize the requirements needed to conduct this rigorous, time-consuming research. Given that you have the essentials (the problem, the time) to engage in this inquiry, I then sketch out the overall process involved in designing and planning a study. I end by suggesting several outlines that you might consider as the overall structure for planning or proposing a qualitative research study. The chapters to follow will then address the different types of inquiry approaches. The general design features, outlined here, will be refined for the five approaches emphasized in this book.

# Questions for Discussion

- What are the key characteristics of qualitative research?
- Why do researchers conduct a qualitative study?
- What is required to undertake this type of research?
- How do researchers design a qualitative study?
- What topics should be addressed in a plan or proposal for a qualitative study?

## The Characteristics of Qualitative Research

I typically begin talking about qualitative research by posing a definition for it. This seemingly uncomplicated approach has become more difficult in recent years. I note that some extremely useful introductory books to qualitative research these days do not contain a definition that can be easily located (Morse & Richards, 2002, 2007; Weis & Fine, 2000). Perhaps this has less to do with the authors' decision to convey the nature of this inquiry and more to do with a concern about advancing a "fixed" definition. It is interesting, however, to look at the evolving definition by Denzin and Lincoln (1994, 2000, 2005) as their *Handbook of Qualitative Research* has moved through time. Their definition conveys the ever-changing nature of qualitative inquiry from social construction, to interpretivist, and on to social justice. I include their latest definition here:

> Qualitative research is a situated activity that locates the observer in the world. It consists of a set of interpretive, material practices that make the world visible. These practices transform the world. They turn the world into a series of representations, including fieldnotes, interviews, conversations, photographs, recordings, and memos to the self. At this level, qualitative research involves an interpretive, naturalistic approach to the world. This means that qualitative researchers study things in their natural settings, attempting to make sense of, or interpret, phenomena in terms of the meanings people bring to them. (Denzin & Lincoln, 2005, p. 3)

Although some of the traditional approaches to qualitative research, such as the "interpretive, naturalistic approach" and "meanings," are evident in this definition, the definition also has a strong orientation toward the impact of qualitative research and in transforming the world.

As an applied research methodologist, my working definition of qualitative research emphasizes the design of research and the use of distinct approaches to inquiry (e.g., ethnography, narrative). At this time, I provide this definition:

*Qualitative research* begins with assumptions, a worldview, the possible use of a theoretical lens, and the study of research problems inquiring into the meaning individuals or groups ascribe to a social or human problem. To study this problem, qualitative researchers use an emerging qualitative approach to inquiry, the collection of data in a natural setting sensitive to the people and places under study, and data analysis that is inductive and establishes patterns or themes. The final written report or presentation includes the voices of participants, the reflexivity of the researcher, and a complex description and interpretation of the problem, and it extends the literature or signals a call for action.

Notice in this definition that I place emphasis on the *process* of research as flowing from philosophical assumptions, to worldviews and through a theoretical lens, and on to the procedures involved in studying social or human problems. Then, a framework exists for the procedures—the approach to inquiry, such as grounded theory, or case study research. At a more micro level are the procedures that are common to all forms of qualitative research.

Examine Table 3.1 for three recent introductory qualitative research books and the characteristics they espouse for doing a qualitative study. As compared to a similar table I designed almost 10 years ago in the first edition of this book (drawing on other authors), qualitative research today involves closer attention to the interpretive nature of inquiry and situating the study within the political, social, and cultural context of the researchers, the participants, and the readers of a study. By examining Table 3.1, one can arrive at several common characteristics of qualitative research. These are presented in no specific order of importance:

• *Natural setting*—Qualitative researchers tend to collect data in the field at the site where participants' experience the issue or problem under study. They do not bring individuals into a lab (a contrived situation), nor do they typically send out instruments for individuals to complete. This up-close information gathered by actually talking directly to people and seeing them behave and act within their context is a major characteristic of qualitative research. In the natural setting, the researchers have face-to-face interaction over time.

**Table 3.1**    Characteristics of Qualitative Research

| Characteristics | LeCompte & Schensul (1999) | Marshall & Rossman (2006) | Hatch (2002) |
|---|---|---|---|
| Natural setting (field focused), a source of data for close interaction | Yes | Yes | Yes |
| Researcher as key instrument of data collection | | | Yes |
| Multiple data sources in words or images | Yes | Yes | |
| Analysis of data inductively, recursively, interactively | Yes | Yes | Yes |
| Focus on participants' perspectives, their meanings, their subjective views | Yes | | Yes |
| Framing of human behavior and belief within a social-political/historical context or through a cultural lens | Yes | | |
| Emergent rather than tightly prefigured design | | Yes | Yes |
| Fundamentally interpretive inquiry— researcher reflects on her or his role, the role of the reader, and the role of the participants in shaping the study | | Yes | |
| Holistic view of social phenomena | | Yes | Yes |

- *Researcher as key instrument.* The qualitative researchers collect data themselves through examining documents, observing behavior, and interviewing participants. They may use a protocol—an instrument for collecting data— but the researchers are the ones who actually gather the information. They do not tend to use or rely on questionnaires or instruments developed by other researchers.

- *Multiple sources of data.* Qualitative researchers typically gather multiple forms of data, such as interviews, observations, and documents, rather than rely on a single data source. Then the researchers review all of the data and make sense of them, organizing them into categories or themes that cut across all of the data sources.

- *Inductive data analysis.* Qualitative researchers build their patterns, categories, and themes from the "bottom-up," by organizing the data into increasingly more abstract units of information. This inductive process involves researchers working back and forth between the themes and the database until they establish a comprehensive set of themes. It may also involve collaborating with the participants interactively, so that they have a chance to shape the themes or abstractions that emerge from the process.

- *Participants' meanings.* In the entire qualitative research process, the researchers keep a focus on learning the meaning that the participants hold about the problem or issue, not the meaning that the researchers bring to the research or writers from the literature.

- *Emergent design.* The research process for qualitative researchers is emergent. This means that the initial plan for research cannot be tightly prescribed, and that all phases of the process may change or shift after the researchers enter the field and begin to collect data. For example, the questions may change, the forms of data collection may shift, and the individuals studied and the sites visited may be modified. The key idea behind qualitative research is to learn about the problem or issue from participants and to address the research to obtain that information.

- *Theoretical lens.* Qualitative researchers often use a lens to view their studies, such as the concept of culture, central to ethnography, or gendered, racial, or class differences from the theoretical orientations discussed in Chapter 2. Sometimes, the study may be organized around identifying the social, political, or historical context of the problem under study.

- *Interpretive inquiry.* Qualitative research is a form of inquiry in which researchers make an interpretation of what they see, hear, and understand. The researchers' interpretations cannot be separated from their own background, history, context, and prior understandings. After a research report is issued, the readers make an interpretation as well as the participants, offering yet other interpretations of the study. With the readers, the participants, and the researchers all making an interpretation, we can see how multiple views of the problem can emerge.

- *Holistic account.* Qualitative researchers try to develop a complex picture of the problem or issue under study. This involves reporting multiple perspectives, identifying the many factors involved in a situation, and generally sketching the larger picture that emerges. Researchers are bound not by tight cause-and-effect relationships among factors, but rather by identifying the complex interactions of factors in any situation.

## When to Use Qualitative Research

When is it appropriate to use qualitative research? We conduct qualitative research because a problem or issue needs to be explored. This exploration is needed, in turn, because of a need to study a group or population, identify variables that can then be measured, or hear silenced voices. These are all good reasons to explore a problem rather than to use predetermined information from the literature or rely on results from other research studies. We also conduct qualitative research because we need a *complex,* detailed understanding of the issue. This detail can only be established by talking directly with people, going to their homes or places of work, and allowing them to tell the stories unencumbered by what we expect to find or what we have read in the literature. We conduct qualitative research when we want to empower individuals to share their stories, hear their voices, and minimize the power relationships that often exist between a researcher and the participants in a study. To further de-emphasize a power relationship, we may collaborate directly with participants by having them review our research questions, or by having them collaborate with us during the data analysis and interpretation phases of research. We conduct qualitative research when we want to write in a literary, flexible style that conveys stories, or theater, or poems, without the restrictions of formal academic structures of writing. We conduct qualitative research because we want to understand the contexts or settings in which participants in a study address a problem or issue. We cannot separate what people say from the context in which they say it—whether this context is their home, family, or work. We use qualitative research to follow up quantitative research and help explain the mechanisms or linkages in causal theories or models. These theories provide a general picture of trends, associations, and relationships, but they do not tell us about why people responded as they did, the context in which they responded, and their deeper thoughts and behaviors that governed their responses. We use qualitative research to develop theories when partial or inadequate theories exist for certain populations and samples or existing theories do not adequately capture the complexity of the problem we are examining. We also use qualitative research because quantitative measures and the statistical analyses simply do not *fit* the problem. Interactions among people, for example, are difficult to capture with existing measures, and these measures may not be sensitive to issues such as gender differences, race, economic status, and individual differences. To level all individuals to a statistical mean overlooks the uniqueness of individuals in our studies. Qualitative approaches are simply a better fit for our research problem.

What does it take to engage in this form of research? To undertake qualitative research requires a strong commitment to study a problem and demands time and resources. Qualitative research keeps good company with the most rigorous quantitative research, and it should not be viewed as an easy substitute for a "statistical" or quantitative study. Qualitative inquiry is for the researcher who is willing to do the following:

- Commit to extensive time in the field. The investigator spends many hours in the field, collects extensive data, and labors over field issues of trying to gain access, rapport, and an "insider" perspective.
- Engage in the complex, time-consuming process of data analysis through the ambitious task of sorting through large amounts of data and reducing them to a few themes or categories. For a multidisciplinary team of qualitative researchers, this task can be shared; for most researchers, it is a lonely, isolated time of struggling with the data. The task is challenging, especially because the database consists of complex texts and images.
- Write long passages, because the evidence must substantiate claims and the writer needs to show multiple perspectives. The incorporation of quotes to provide participants' perspectives also lengthens the study.
- Participate in a form of social and human science research that does not have firm guidelines or specific procedures and is evolving and constantly changing. This guideline complicates telling others how one plans to conduct a study and how others might judge it when the study is completed.

## The Process of Designing a Qualitative Study

At the outset, I need to say that there is no agreed upon structure for how to design a qualitative study. Books on qualitative research vary. Some authors believe that by reading about a study, discussing the procedures, and pointing out issues that emerged, the aspiring qualitative researcher will have a sense of how to conduct this form of inquiry (see Weis & Fine, 2000). That may be true for some individuals. For others, understanding the broader issues may suffice (see Morse & Richards, 2002, 2007), or guidance from a "how to" book may be better (see Hatch, 2002). I am not sure whether I write from exactly a "how to" perspective; my approach is more in line with creating options for qualitative researchers (hence, the five approaches), weighing the options given my experiences, and then letting readers choose for themselves.

There are certain design principles that I work from when I design my own qualitative research studies. First, I do find that qualitative research generally falls within the process of scientific research, with common phases whether one is writing qualitatively or quantitatively. All researchers seem to start with an issue or problem, examine the literature in some way related to the problem, pose questions, gather data and then analyze them, and write up their reports. Qualitative research fits within this structure, and I have accordingly organized the chapters in this book to reflect this process. Second, several aspects of a qualitative project vary from study to study as to the amount of detail developed by researchers. For example, stances on the use of the literature vary widely, as do the stances on using an a priori theory. The literature may be fully reviewed and used to inform the questions actually asked, it may be reviewed late in the process of research, or it may be used solely to help document the importance of the research problem. Other options may also exist, but these possibilities point to the varied uses of literature in qualitative research. Similarly, the use of theory varies extensively. For example, cultural theories form the basic building blocks of a good qualitative ethnography (LeCompte & Schensul, 1999), whereas in grounded theory, the theories are developed or generated during the process of research (Strauss & Corbin, 1990). In health science research, I find the use of a priori theories common practice, and a key element that must be included in a rigorous qualitative investigation (Barbour, 2000). Another consideration in qualitative research is the writing format for the qualitative project. It varies considerably from scientific-oriented approaches, to storytelling, and on to performances, such as theater, plays, or poems. There is no one standard or accepted structure as one typically finds in quantitative research.

Given these differences, we still are left with the graduate student who needs to organize a qualitative thesis or dissertation, researchers who need to submit a proposal for state or federal funding, and the research team that seeks to investigate a timely issue in the social, behavioral, or health sciences. All of these individuals will probably profit from having some structure to their qualitative writing. Thus, I would like to discuss a general approach to designing a qualitative study and then begin to shape this design as we visit the five approaches to qualitative research in this book. I like the concept of "methodological congruence" advanced by Morse and Richards (2002, 2007)—that the purposes, questions, and methods of research are all interconnected and interrelated so that the study appears as a cohesive whole rather than as fragmented, isolated parts.

The process of designing a qualitative study begins not with the methods—which is actually the easiest part of research, I believe—but instead with the broad assumptions central to qualitative inquiry, a worldview consistent with it, and in many cases, a theoretical lens that shapes the study. In addition, the researcher arrives at the doorstep of qualitative research with a topic or substantive area of investigation, and perhaps has reviewed the literature about the topic and knows that a problem or issue exists that needs to be studied. This problem may be one in the "real world" or it may be a deficiency in the literature or past investigations on a topic. Problems in qualitative research span the topics in the social and human sciences, and a hallmark of qualitative research today is the deep involvement in issues of gender, culture, and marginalized groups. The topics about which we write are emotion laden, close to people, and practical.

To study these topics, we ask open-ended research questions, wanting to listen to the participants we are studying and shaping the questions after we "explore," and we refrain from assuming the role of the expert researcher with the "best" questions. Our questions change during the process of research to reflect an increased understanding of the problem. Furthermore, we take these questions out to the field to collect either "words" or "images." I like to think in terms of four basic types of information: interviews, observations, documents, and audiovisual materials. Certainly, new forms emerge that challenge this traditional categorization. Where do we place sounds, e-mail messages, and computer software? Unquestionably, the backbone of qualitative research is extensive collection of data, typically from multiple sources of information. After organizing and storing our data, we analyze them by carefully masking the names of respondents, and we engage in the perplexing (and "lonely" if we are the sole researcher) exercise of trying to make sense of the data. We examine the qualitative data working inductively from particulars to more general perspectives, whether these perspectives are called themes, dimensions, codes, or categories. One helpful way to see this process is to recognize it as working through multiple levels of abstraction, starting with the raw data and forming larger and larger categories. Recognizing the highly interrelated set of activities of data collection, analysis, and report writing, we do not always know clearly which stage we are in. I remember working on a case study (Asmussen & Creswell, 1995) as interviewing, analyzing, and writing the case study—all intermingled processes, not distinct phases in the process. Also, we experiment with many forms of analysis—making metaphors, developing matrices and tables, and using visuals—to convey simultaneously breaking down the data and reconfiguring them into new forms. We (re)present our data, partly based on participants' perspectives and partly based on our own interpretation, never clearly escaping our own personal stamp on a study.

Throughout the slow process of collecting data and analyzing them, we shape our narrative—a narrative with many forms in qualitative research. We tell a story that unfolds over time. We present the study following the traditional approach to scientific research (i.e., problem, question, method, findings). We talk about our experiences in conducting the study, and how they shape our interpretations of the results. We let the voices of our participants speak and carry the story through dialogue, perhaps dialogue presented in Spanish with English subtitles.

Throughout all phases of the research process we are sensitive to ethical considerations. These are especially important as we negotiate entry to the field site of the research; involve participants in our study; gather personal, emotional data that reveal the details of life; and ask participants to give considerable time to our projects. Hatch (2002) does a good job of summarizing some of the major ethical issues that researchers need to anticipate and often address in their studies. Giving back to participants for their time and efforts in our projects—reciprocity—is important, and we need to review how participants will gain from our studies. How to leave the scene of a research study—through slow withdrawal and conveying information about our departure—so that the participants do not feel abandoned is also important. We always need to be sensitive to the potential of our research to disturb the site and potentially (and often unintentionally) exploit the vulnerable populations we study, such as young children or underrepresented or marginalized groups. Along with this comes a need to be sensitive to any power imbalances our presence may establish at a site that could further marginalize the people under study. We do not want to place the participants at further risk as a result of our research. We need to anticipate how to address potential illegal activities that we see or hear, and, in some cases, report them to authorities. We need to honor who owns the account, and whether participants and leaders at our research sites will be concerned about this issue. As we work with individual participants, we need to respect them individually, such as by not stereotyping them, using their language and names, and following guidelines such as those found in the *Publication Manual of the American Psychological Association* (APA, 2001) for nondiscriminatory language. Most often our research is done within the context of a college or university setting where we need to provide evidence to institutional review boards or committees that we respect the privacy and right of participants to withdraw from the study and do not place them at risk. At this stage, too, we consciously consider ethical issues—seeking consent, avoiding the conundrum of deception, maintaining confidentiality, and protecting the anonymity of individuals with whom we speak. Weis and Fine (2000) ask us to consider our roles as insiders/outsiders to the participants; issues that we may be fearful of disclosing; how we established supportive, respectful relationships without stereotyping and using labels that participants do not embrace; whose voice will be represented in our final study; and how we will write ourselves into the study and reflect who we are as well as reflect the people we study (Weiss & Fine, 2000). We need to be sensitive to vulnerable populations, imbalanced power relations, and placing participants at risk (Hatch, 2002).

At some point we ask, "Did we get the story 'right'?" (Stake, 1995), knowing that there are no "right" stories, only multiple stories. Perhaps qualitative studies do not have endings, only questions (Wolcott, 1994b). But we seek to have our account resonate with the participants, to be an accurate reflection of what they said. So we engage in validation strategies, often using multiple strategies, which include confirming or triangulating data from several sources, having our studies reviewed and corrected by the participants, and having other researchers review our procedures.

In the end, individuals such as readers, participants, graduate committees, editorial board members for journals, and reviewers of proposals for funding will apply some criteria to assess the quality of the study. Standards for assessing the quality of qualitative research are available (Howe & Eisenhardt, 1990; Lincoln, 1995; Marshall & Rossman, 2006). Here is my short list of *characteristics of a "good" qualitative study*. You will see my emphasis on rigorous methods present in this list.

• The researcher employs rigorous data collection procedures. This means that the researcher collects multiple forms of data, adequately summarizes—perhaps in tabled form—the forms of data and detail about them, and spends adequate time in the field. It is not unusual for qualitative studies to include information about the specific amount of time in the field. I especially like to see unusual forms of qualitative data collection, such as using photographs to elicit responses, sounds, visual materials, or digital text messages.

• The researcher frames the study within the assumptions and characteristics of the qualitative approach to research. This includes fundamental characteristics such as an evolving design, the presentation of multiple realities, the researcher as an instrument of data collection, and a focus on participants' views—in short, all of the characteristics mentioned in Table 3.1.

• The researcher uses an approach to qualitative inquiry such as one of the five approaches addressed in this book. Use of a recognized approach to research enhances the rigor and sophistication of the research design. This means that the researcher identifies and defines the approach, cites studies that employ it, and follows the procedures outline in the approach. Certainly, this approach need not be "pure," and one might mix procedures from several approaches; however, for the beginning student of qualitative research, I would recommend staying within one approach, becoming comfortable with it, learning it, and keeping a study concise and straightforward. Later, especially in long and complex studies, features from several approaches may be useful.

• The researcher begins with a single focus. Although examples of qualitative research show a comparison of groups or of factors or of themes, as in case study projects or in ethnographies, I like to begin a qualitative study focused on understanding a single concept or idea (e.g., What does it mean to be a professional? A teacher? A painter? A single mother? A homeless person?). As the study progresses, it can begin incorporating the comparison (e.g., How does the case of a professional teacher differ from a professional administrator?) or relating factors (e.g., What explains why painting evokes feelings?). All too often qualitative researchers advance to the comparison or the relationship analysis without first understanding their core concept or idea.

- The study includes detailed methods, a rigorous approach to data collection, data analysis, and report writing. Rigor is seen when extensive data collection in the field occurs, or when the researcher conducts multiple levels of data analysis, from the narrow codes or themes to broader interrelated themes to more abstract dimensions. Rigor means, too, that the researcher validates the accuracy of the account using one or more of the procedures for validation, such as member checking, triangulating sources of data, or using peer or external auditors of the accounts.

- The researcher analyzes data using multiple levels of abstraction. I like to see the active work of the researcher as he or she moves from particulars to general levels of abstraction. Often, writers present their studies in stages (e.g., the multiple themes that can be combined into larger themes or perspectives) or layer their analyses from the particular to the general. The codes and themes derived from the data might show mundane, expected, and surprising ideas. Often the best qualitative studies present themes that explore the shadow side or unusual angles. I remember in one class project, the student examined how students in a distance learning class reacted to the camera focused on the class. Rather than looking at the students' reaction when the camera was on them, the researcher sought to understand what happened when the camera was *off* them. This approach led to the author taking an unusual angle, one not expected by the readers.

- The researcher writes persuasively so that the reader experiences "being there." The concept of "verisimilitude," a literary term, captures my thinking (Richardson, 1994, p. 521). The writing is clear, engaging, and full of unexpected ideas. The story and findings become believable and realistic, accurately reflecting all the complexities that exist in real life. The best qualitative studies engage the reader.

- The study reflects the history, culture, and personal experiences of the researcher. This is more than simply an autobiography, with the writer or the researcher telling about his or her background. It focuses on how individuals' culture, gender, history, and experiences shape all aspects of the qualitative project, from their choice of a question to address, to how they collect data, to how they make an interpretation of the situation. In some way—such as discussing their role, interweaving themselves into the text, or reflecting on the questions they have about the study—individuals position themselves in the qualitative study.

- The qualitative research in a good study is ethical. This involves more than simply the researcher seeking and obtaining the permission of institutional review committees or boards. It means that the researcher is aware of and addressing in the study all of the ethical issues mentioned earlier in this chapter that thread through all phases of the research study.

## The General Structure of a Plan or Proposal

Look at the diversity of final written products for qualitative research. No set format exists. But several writers suggest general topics to be included in a written plan or *proposal* for a qualitative study. I provide four examples of formats for plans or proposals for qualitative studies. In the first example, drawn from my own work (Creswell, 2003, pp. 50–51), I advance a constructionist/interpretivist form. This form (shown in Example 3.1 below) might be seen as a traditional approach to planning qualitative research, and it includes the standard introduction and procedures, including a passage in the procedures about the role of the researcher. It also incorporates anticipated ethical issues, pilot findings, and expected outcomes.

*Example 3.1 A Qualitative Constructivist/ Interpretivist Format*

Introduction

     Statement of the problem (including literature about the problem)

     Purpose of the study

     The research questions

     Delimitations and limitations

Procedures

     Characteristics of qualitative research (optional)

     Qualitative research strategy

     Role of the researcher

Data collection procedures

Data analysis procedures

Strategies for validating findings

Narrative structure

Anticipated ethical issues

Significance of the study

Preliminary pilot findings

Expected outcomes

Appendices: Interview questions, observational forms, timeline, and proposed budget

The second format provides for an advocacy perspective (Creswell, 2003, pp. 51–52). This format (as shown in Example 3.2 below) makes explicit the advocacy, transformative approach to qualitative research by stating the advocacy issue at the beginning, by emphasizing collaboration during the data collection, and by advancing the changes advocated for the group being studied.

## Example 3.2 A Qualitative Advocacy/Participatory Format

Introduction

Statement of the problem (including literature about the problem)

The advocacy/participatory issue

Purpose of the study

The research questions

Delimitations and limitations

Procedures

Characteristics of qualitative research (optional)

Qualitative research strategy

Role of the researcher

Data collection procedures (including the collaborative approaches used and sensitivity toward participants)

Data recording procedures

Data analysis procedures

Strategies for validating findings

Narrative structure of study

Anticipated ethical issues

Significance of the study

Preliminary pilot findings

Expected advocacy/participatory changes

Appendices: Interview questions, observational forms, timeline, and proposed budget

The third format, Example 3.3, is similar to the advocacy format, but it advances the use of a theoretical lens (Marshall & Rossman, 2006). Notice that this format has a section for a theoretical lens (e.g., feminist, racial, ethnic) that informs the study in the literature review, "trustworthiness" in place of what I have been calling "validation," a section for being reflexive through personal biography, and both the ethical and political considerations of the author.

## *Example 3.3 A Theoretical Lens Format*

Introduction

Overview

Type and purpose

Potential significance

Framework and general research questions

Limitations

Review of related literature

Theoretical traditions

Essays by informed experts

Related research

Design and methodology

Overall approach and rationale

Site or population selection

Data-gathering methods

Data analysis procedures

Trustworthiness

Personal biography

Ethics and political considerations

Appendices: Interview questions, observational forms, timeline, and proposed budget

In the fourth and final format, Example 3.4, Maxwell (2005) organizes the structure around a series of nine arguments that he feels need to cohere and be coherent when researchers design their qualitative proposals. I think that these nine arguments represent the most important points to include in a proposal, and Maxwell provides in his book a complete example of a qualitative dissertation proposal written by Martha G. Regan-Smith at the Harvard Graduate School of Education. My summary and adaptation of these arguments follow.

### Example 3.4 Maxwell's Nine Arguments for a Qualitative Proposal

We need to better understand . . . (the topic).

We know little about . . . (the topic).

I propose to study. . . .

The setting and participants are appropriate for this study.

The methods I plan to use will provide the data I need to answer the research questions.

Analysis will generate answers to these questions.

The findings will be validated by. . . .

The study poses no serious ethical problems.

Preliminary results support the practicability and value of the study.

These four examples speak only to designing a plan or proposal for a qualitative study. To the topics of these proposal formats, the complete study will include additional data findings, interpretations, and a discussion of the overall results, limitations of the study, and future research needs.

## Summary

The definitions for qualitative research vary, but I see it as an approach to inquiry that begins with assumptions, worldviews, possibly a theoretical lens, and the study of research problems exploring the meaning individuals or groups ascribe to a social or human problem. Researchers collect data in natural settings with a sensitivity to the people under study, and they analyze their data inductively to establish patterns or themes. The final report provides for the voices of participants, a reflexivity of the researchers, a complex description and interpretation of the problem, and a study that adds to the literature or provides a call for action. Recent introductory textbooks underscore the characteristics embedded in this definition. Given this definition, a qualitative approach is appropriate to use to study a research problem when the problem needs to be explored; when a complex, detailed understanding is needed; when the researcher wants to write in a literary, flexible style; and when the researcher seeks to understand the context or settings of participants. Qualitative research does take time, involves ambitious data analysis, results in lengthy reports, and does not have firm guidelines.

The process of designing a qualitative study emerges during inquiry, but it generally follows the pattern of scientific research. It starts with broad assumptions central to qualitative inquiry, worldview stances, and theoretical lens and a topic of inquiry. After stating a research problem or issue about this topic, the inquirer asks several open-ended research questions, gathers multiple forms of data to answer these questions, and makes sense of the data by grouping information into codes, themes or categories, and larger dimensions. The final narrative the researcher composes will have diverse formats—from a scientific type of study to narrative stories. Ethical decisions are threaded throughout the study. Several aspects will make the study a good qualitative project: rigorous data collection and analysis; the use of a qualitative approach (e.g., narrative, phenomenology, grounded theory, ethnography, case study); a single focus; a persuasive account; a reflection on the researcher's own history, culture, personal experiences, and politics; and ethical practices.

Finally, the structure of a plan or proposal for a qualitative study will vary. I include four models that differ in terms of their advocacy orientation, inclusion of personal and political considerations, and focus on the essential arguments that researchers need to address in proposals.

## Additional Readings

There are many introductory textbooks on qualitative research. At the beginning of this chapter I introduced three books that differ in their approaches: Marshall and Rossman (2006), which takes a rigorous methods approach; LeCompte and Schensul (1999), which is drawn from ethnography; and Hatch (2002), a text that was created for educators but, because of the clarity of writing and thoughts, would serve qualitative researchers well across the social and human sciences. To these books, I add the introductory text by Maxwell (2005) and my own book on research design from a qualitative, quantitative, and mixed methods approach (Creswell, 2003). The Weis and Fine (2000) book, which takes as its launching point their own study of crime and poverty, is a fascinating look at the technicalities, politics, and ethics surrounding qualitative research. Finally, rounding out my list is the Morse and Richards (2002; 2nd ed., 2007) introductory text, which takes a refreshing view of the methodological congruence of all aspects of the research process.

Creswell, J. W. (2003). *Research design: Qualitative, quantitative, and mixed methods approaches* (2nd ed.). Thousand Oaks, CA: Sage.

Hatch, J. A. (2002). *Doing qualitative research in education settings.* Albany: State University of New York Press.

LeCompte, M. D., & Schensul, J. J. (1999). *Designing and conducting ethnographic research* (Ethnographer's toolkit, Vol. 1). Walnut Creek, CA: AltaMira.

Marshall, C., & Rossman, G. B. (2006). *Designing qualitative research* (4th ed.). Thousand Oaks, CA: Sage.

Maxwell, J. (2005). *Qualitative research design: An interactive approach* (2nd ed.). Thousand Oaks, CA: Sage.

Morse, J. M., & Richards, L. (2002). *README FIRST for a user's guide to qualitative methods.* Thousand Oaks, CA: Sage.

Richards, L., & Morse, J. M. (2007). *README FIRST for a user's guide to qualitative methods* (2nd ed.). Thousand Oaks, CA: Sage.

Weis, L., & Fine, M. (2000). *Speed bumps: A study-friendly guide to qualitative research.* New York: Teachers College Press.

**Exercises**

1. Organize a two-page overview of a study you would like to conduct. At this point, you need not be concerned about the specific approach to inquiry unless you already have one selected. In your summary, include (a) the problem (or issue) you plan to study, (b) the major research question you plan to ask, (c) the data you wish to collect and analyze, (d) the significance of your study, and (e) your relationship to the topic and participants being studied. This preliminary plan will be modified later, after you have chosen an approach to inquiry.

2. For individuals new to qualitative research, examine one of the introductory texts I mentioned in the Additional Readings section and develop an outline of key ideas.

**4**

# Five Qualitative
# Approaches to Inquiry

I n this chapter, we begin our detailed exploration of narrative research,
phenomenology, grounded theory, ethnography, and case studies. For each
approach, I pose a definition, briefly trace its history, explore types of studies, introduce procedures involved in conducting a study, and indicate potential challenges in using the approach. I also review some of the similarities and
differences among the five approaches so that qualitative researchers can
decide which approach is best to use for their particular study.

## Questions for Discussion

- What are a narrative study, a phenomenology, a grounded theory, an ethnography, and a case study?
- What are the procedures and challenges to using each approach to qualitative
  research?
- What are some similarities and differences among the five approaches?

## Narrative Research

### Definition and Background

Narrative research has many forms, uses a variety of analytic practices,
and is rooted in different social and humanities disciplines (Daiute &

Lightfoot, 2004). "Narrative" might be the term assigned to any text or discourse, or, it might be text used within the context of a mode of inquiry in qualitative research (Chase, 2005), with a specific focus on the stories told by individuals (Polkinghorne, 1995). As Pinnegar and Daynes (2006) suggest, narrative can be both a method and *the phenomenon* of study. As a method, it begins with the experiences as expressed in lived and told stories of individuals. Writers have provided ways for analyzing and understanding the stories lived and told. I will define it here as a specific type of qualitative design in which "narrative is understood as a spoken or written text giving an account of an event/action or series of events/actions, chronologically connected" (Czarniawska, 2004, p. 17). The procedures for implementing this research consist of focusing on studying one or two individuals, gathering data through the collection of their stories, reporting individual experiences, and chronologically ordering (or using *life course stages*) the meaning of those experiences.

Although narrative research originated from literature, history, anthropology, sociology, sociolinguistics, and education, different fields of study have adopted their own approaches (Chase, 2005). I find a postmodern, organizational orientation in Czarniawska (2004); a human developmental perspective in Daiute and Lightfoot (2004); a *psychological approach* in Lieblich, Tuval-Mashiach, and Zilber (1998); sociological approaches in Cortazzi (1993) and Riessman (1993); and quantitative (e.g., statistical stories in event history modeling) and qualitative approaches in Elliott (2005). Interdisciplinary efforts at narrative research have also been encouraged by the *Narrative Study of Lives* annual series that began in 1993 (see, e.g., Josselson & Lieblich, 1993), and the journal *Narrative Inquiry*. With many recent books on narrative research, it is indeed a "field in the making" (Chase, 2005, p. 651). In the discussion of narrative procedures, I rely on an accessible book written for social scientists called *Narrative Inquiry* (Clandinin & Connelly, 2000) that addresses "what narrative researchers do" (p. 48).

## Types of Narrative Studies

One approach to narrative research is to differentiate types of narrative research by the analytic strategies used by authors. Polkinghorne (1995) takes this approach and distinguishes between "analysis of narratives" (p. 12), using paradigm thinking to create descriptions of themes that hold across stories or taxonomies of types of stories, and "narrative analysis," in which researchers collect descriptions of events or happenings and then configure them into a story using a plot line. Polkinghorne (1995) goes on to emphasize the second form in his writings. More recently, Chase (2005) presents an approach closely allied with Polkinghorne's "analysis of narratives." Chase suggests that researchers may use paradigmatic reasons for a narrative study, such as how individuals are enabled and constrained by social resources, socially situated in interactive performances, and how narrators develop interpretations.

A second approach is to emphasize the variety of forms found in narrative research practices (see, e.g., Casey, 1995/1996). A *biographical study* is a form of narrative study in which the researcher writes and records the experiences of another person's life. **Autobiography** is written and recorded by the individuals who are the subject of the study (Ellis, 2004). A *life history* portrays an individual's entire life, while a personal experience story is a narrative study of an individual's personal experience found in single or multiple episodes, private situations, or communal folklore (Denzin, 1989a). An *oral history* consists of gathering personal reflections of events and their causes and effects from one individual or several individuals (Plummer, 1983). Narrative studies may have a specific contextual focus, such as teachers or children in classrooms (Ollerenshaw & Creswell, 2002), or the stories told about organizations (Czarniawska, 2004). Narratives may be guided by a theoretical lens or perspective. The lens may be used to advocate for Latin Americans through using *testimonios* (Beverly, 2005), or it may be a feminist lens used to report the stories of women (see, e.g., Personal Narratives Group, 1989), a lens that shows how women's voices are muted, multiple, and contradictory (Chase, 2005).

## Procedures for Conducting Narrative Research

Using the approach taken by Clandinin and Connelly (2000) as a general procedural guide, the methods of conducting a narrative study do not follow a lock-step approach, but instead represent an informal collection of topics.

1. Determine if the research problem or question best fits narrative research. Narrative research is best for capturing the detailed stories or life experiences of a single life or the lives of a small number of individuals.

2. Select one or more individuals who have stories or life experiences to tell, and spend considerable time with them gathering their stories through multiples types of information. Clandinin and Connelly (2000) refer to the stories as "field texts." Research participants may record their stories in a journal or diary, or the researcher might observe the individuals and record fieldnotes. Researchers may also collect letters sent by the individuals; assemble

stories about the individuals from family members; gather documents such as memos or official correspondence about the individual; or obtain photographs, memory boxes (collection of items that trigger memories), and other personal-family-social *artifacts*. After examining these sources, the researcher records the individuals' life experiences.

3. Collect information about the context of these stories. Narrative researchers situate individual stories within participants' personal experiences (their jobs, their homes), their culture (racial or ethnic), and their historical contexts (time and place).

4. Analyze the participants' stories, and then "restory" them into a framework that makes sense. *Restorying* is the process of reorganizing the stories into some general type of framework. This framework may consist of gathering stories, analyzing them for key elements of the story (e.g., time, place, plot, and scene), and then rewriting the stories to place them within a chronological sequence (Ollerenshaw & Creswell, 2000). Often when individuals tell their stories, they do not present them in a chronological sequence. During the process of restorying, the researcher provides a causal link among ideas. Cortazzi (1993) suggests that the chronology of narrative research, with an emphasis on sequence, sets narrative apart from other genres of research. One aspect of the chronology is that the stories have a beginning, a middle, and an end. Similar to basic elements found in good novels, these aspects involve a predicament, conflict, or struggle; a protagonist, or main character; and a sequence with implied causality (i.e., a plot) during which the predicament is resolved in some fashion (Carter, 1993). A chronology further may consist of past, present, and future ideas (Clandinin & Connelly, 2000), based on the assumption that time has a unilinear direction (Polkinghorne, 1995). In a more general sense, the story might include other elements typically found in novels, such as time, place, and scene (Connelly & Clandinin, 1990). The plot, or story line, may also include Clandinin and Connelly's (2000) three-dimensional narrative inquiry space: the personal and social (the interaction); the past, present, and future (continuity); and the place (situation). This story line may include information about the setting or context of the participants' experiences. Beyond the chronology, researchers might detail themes that arise from the story to provide a more detailed discussion of the meaning of the story (Huber & Whelan, 1999). Thus, the qualitative data analysis may be a description of both the story and themes that emerge from it. A postmodern narrative writer, such as Czarniawska (2004), would add another element to the analysis: a deconstruction of the stories, an unmaking of them by such analytic strategies as exposing dichotomies, examining silences, and attending to disruptions and contractions.

5. Collaborate with participants by actively involving them in the research (Clandinin & Connelly, 2000). As researchers collect stories, they negotiate relationships, smooth transitions, and provide ways to be useful to the participants. In narrative research, a key theme has been the turn toward the relationship between the researcher and the researched in which both parties will learn and change in the encounter (Pinnegar & Daynes, 2006). In this process, the parties negotiate the meaning of the stories, adding a validation check to the analysis (Creswell & Miller, 2000). Within the participant's story may also be an interwoven story of the researcher gaining insight into her or his own life (see Huber & Whelan, 1999). Also, within the story may be *epiphanies* or turning points in which the story line changes direction dramatically. In the end, the narrative study tells the story of individuals unfolding in a chronology of their experiences, set within their personal, social, and *historical context,* and including the important themes in those lived experiences. "Narrative inquiry is stories lived and told," said Clandinin and Connolly (2000, p. 20).

## Challenges

Given these procedures and the characteristics of narrative research, narrative research is a challenging approach to use. The researcher needs to collect extensive information about the participant, and needs to have a clear understanding of the context of the individual's life. It takes a keen eye to identify in the source material gathered the particular stories that capture the individual's experiences. As Edel (1984) comments, it is important to uncover the "figure under the carpet" that explains the multilayered context of a life. Active collaboration with the participant is necessary, and researchers need to discuss the participant's stories as well as be reflective about their own personal and political background, which shapes how they "restory" the account. Multiple issues arise in the collecting, analyzing, and telling of individual stories. Pinnegar and Daynes (2006) raise these important questions: Who owns the story? Who can tell it? Who can change it? Whose version is convincing? What happens when narratives compete? As a community, what do stories do among us?

# Phenomenological Research

## Definition and Background

Whereas a narrative study reports the life of a *single individual,* a *phenomenological study* describes the meaning for several individuals of their *lived experiences* of a concept or a phenomenon. Phenomenologists focus on describing what all participants have in common as they experience a phenomenon (e.g., grief is universally experienced). The basic purpose of phenomenology is to reduce individual experiences with a phenomenon to a description of the universal essence (a "grasp of the very nature of the thing," van Manen, 1990, p. 177). To this end, qualitative researchers identify a phenomenon (an "object" of human experience; van Manen, 1990, p. 163). This human experience may be phenomena such as insomnia, being left out, anger, grief, or undergoing coronary artery bypass surgery (Moustakas, 1994). The inquirer then collects data from persons who have experienced the phenomenon, and develops a composite description of the essence of the experience for all of the individuals. This description consists of "what" they experienced and "how" they experienced it (Moustakas, 1994).

Beyond these procedures, phenomenology has a strong philosophical component to it. It draws heavily on the writings of the German mathematician Edmund Husserl (1859–1938) and those who expanded on his views, such as Heidegger, Sartre, and Merleau-Ponty (Spiegelberg, 1982). Phenomenology is popular in the social and health sciences, especially in sociology (Borgatta & Borgatta, 1992; Swingewood, 1991), psychology (Giorgi, 1985; Polkinghorne, 1989), nursing and the health sciences (Nieswiadomy, 1993; Oiler, 1986), and education (Tesch, 1988; van Manen, 1990). Husserl's ideas are abstract, and, as late as 1945, Merleau-Ponty (1962) still raised the question, "What is phenomenology?" In fact, Husserl was known to call any project currently under way "phenomenology" (Natanson, 1973).

Writers following in the footsteps of Husserl also seem to point to different philosophical arguments for the use of phenomenology today (contrast, for example, the philosophical basis stated in Moutakas, 1994; in Stewart and Mickunas, 1990; and in van Manen, 1990). Looking across all of these perspectives, however, we see that the philosophical assumptions rest on some common grounds: the study of the lived experiences of persons, the view that these experiences are conscious ones (van Manen, 1990), and the development of descriptions of the essences of these experiences, not explanations or analyses (Moustakas, 1994). At a broader level, Stewart and Mickunas (1990) emphasize four *philosophical perspectives* in phenomenology:

- *A return to the traditional tasks of philosophy.* By the end of the 19th century, philosophy had become limited to exploring a world by empirical means, which was called "scientism." The return to the traditional tasks of philosophy that existed before philosophy became enamored with empirical science is a return to the Greek conception of philosophy as a search for wisdom.
- *A philosophy without presuppositions.* Phenomenology's approach is to suspend all judgments about what is real—the "natural attitude"—until they are

founded on a more certain basis. This suspension is called "epoche" by Husserl.

- *The **intentionality of consciousness**.* This idea is that consciousness is always directed toward an object. Reality of an object, then, is inextricably related to one's consciousness of it. Thus, reality, according to Husserl, is not divided into subjects and objects, but into the dual Cartesian nature of both subjects and objects as they appear in consciousness.
- *The refusal of the subject-object dichotomy.* This theme flows naturally from the intentionality of consciousness. The reality of an object is only perceived within the meaning of the experience of an individual.

An individual writing a phenomenology would be remiss to not include some discussion about the philosophical presuppositions of phenomenology along with the methods in this form of inquiry. Moustakas (1994) devotes over one hundred pages to the philosophical assumptions before he turns to the methods.

## Types of Phenomenology

Two approaches to phenomenology are highlighted in this discussion: hermeneutic phenomenology (van Manen, 1990) and empirical, transcendental, or psychological phenomenology (Moustakas, 1994). Van Manen (1990) is widely cited in the health literature (Morse & Field, 1995). An educator, van Manen, has written an instructive book on *hermeneutical phenomenology* in which he describes research as oriented toward lived experience (phenomenology) and interpreting the "texts" of life (hermeneutics) (van Manen, 1990, p. 4). Although van Manen does not approach phenomenology with a set of rules or methods, he discusses phenomenology research as a dynamic interplay among six research activities. Researchers first turn to a phenomenon, an "abiding concern" (p. 31), which seriously interests them (e.g., reading, running, driving, mothering). In the process, they reflect on essential themes, what constitutes the nature of this lived experience. They write a description of the phenomenon, maintaining a strong relation to the topic of inquiry and balancing the parts of the writing to the whole. Phenomenology is not only a description, but it is also seen as an interpretive process in which the researcher makes an interpretation (i.e., the researcher "mediates" between different meanings; van Manen, 1990, p. 26) of the meaning of the lived experiences.

Moustakas's (1994) transcendental or psychological phenomenology is focused less on the interpretations of the researcher and more on a description of the experiences of participants. In addition, Moustakas focuses on one of Husserl's concepts, *epoche* (or bracketing), in which investigators set aside their experiences, as much as possible, to take a fresh perspective toward the phenomenon under examination. Hence, "transcendental" means "in which everything is perceived freshly, as if for the first time" (Moustakas, 1994, p. 34). Moustakas admits that this state is seldom perfectly achieved. However, I see researchers who embrace this idea when they begin a project by describing their own experiences with the phenomenon and bracketing out their views before proceeding with the experiences of others.

Besides bracketing, empirical, *transcendental phenomenology* draws on the *Duquesne Studies in Phenomenological Psychology* (e.g., Giorgi, 1985) and the data analysis procedures of Van Kaam (1966) and Colaizzi (1978). The procedures, illustrated by Moustakas (1994), consist of identifying a phenomenon to study, bracketing out one's experiences, and collecting data from several persons who have experienced the phenomenon. The researcher then analyzes the data by reducing the information to significant statements or quotes and combines the statements into themes. Following that, the researcher develops a *textural description* of the experiences of the persons (what participants experienced), a *structural description* of their experiences (how they experienced it in terms of the conditions, situations, or context), and a combination of the textural and structural descriptions to convey an overall **essence** of the experience.

## Procedures for Conducting Phenomenological Research

I use the psychologist Moustakas's (1994) approach because it has systematic steps in the data analysis procedure and guidelines for assembling the textual and structural descriptions. The conduct of psychological phenomenology has been addressed in a number of writings, including Dukes (1984), Tesch (1990), Giorgi (1985, 1994), Polkinghorne (1989), and, most recently, Moustakas (1994). The major procedural steps in the process would be as follows:

• The researcher determines if the research problem is best examined using a phenomenological approach. The type of problem best suited for this form of research is one in which it is important to understand several individuals' common or shared experiences of a phenomenon. It would be important to understand these common experiences in order to develop practices or policies, or to develop a deeper understanding about the features of the phenomenon.

• A phenomenon of interest to study, such as anger, professionalism, what it means to be underweight, or what it means to be a wrestler, is identified. Moustakas (1994) provides numerous examples of phenomena that have been studied.

- The researcher recognizes and specifies the broad philosophical assumptions of phenomenology. For example, one could write about the combination of objective reality and individual experiences. These lived experiences are furthermore "conscious" and directed toward an object. To fully describe how participants view the phenomenon, researchers must bracket out, as much as possible, their own experiences.

- Data are collected from the individuals who have experienced the phenomenon. Often data collection in phenomenological studies consists of in-depth interviews and multiple interviews with participants. Polkinghorne (1989) recommends that researchers interview from 5 to 25 individuals who have all experienced the phenomenon. Other forms of data may also be collected, such as observations, journals, art, poetry, music, and other forms of art. Van Manen (1990) mentions taped conversations, formally written responses, accounts of vicarious experiences of drama, films, poetry, and novels.

- The participants are asked two broad, general questions (Moustakas, 1994): What have you experienced in terms of the phenomenon? What contexts or situations have typically influenced or affected your experiences of the phenomenon? Other open-ended questions may also be asked, but these two, especially, focus attention on gathering data that will lead to a textural description and a structural description of the experiences, and ultimately provide an understanding of the common experiences of the participants.

- *Phenomenological data analysis* steps are generally similar for all psychological phenomenologists who discuss the methods (Moustakas, 1994; Polkinghorne, 1989). Building on the data from the first and second research questions, data analysts go through the data (e.g., interview transcriptions) and highlight "significant statements," sentences, or quotes that provide an understanding of how the participants experienced the phenomenon. Moustakas (1994) calls this step *horizonalization.* Next, the researcher develops *clusters of meaning* from these significant statements into themes.

- These significant statements and themes are then used to write a description of what the participants experienced (*textural description*). They are also used to write a description of the context or setting that influenced how the participants experienced the phenomenon, called *imaginative variation* or *structural description*. Moustakas (1994) adds a further step: Researchers also write about their own experiences and the context and situations that have influenced their experiences. I like to shorten Moustakas's procedures, and reflect these personal statements at the beginning of the phenomenology or include them in a methods discussion of the role of the researcher (Marshall & Rossman, 2006).

- From the structural and textural descriptions, the researcher then writes a composite description that presents the "essence" of the phenomenon, called the *essential, invariant structure (or essence)*. Primarily this passage focuses on the common experiences of the participants. For example, it means that all experiences have an underlying *structure* (grief is the same whether the loved one is a puppy, a parakeet, or a child). It is a descriptive passage, a long paragraph or two, and the reader should come away from the phenomenology with the feeling, "I understand better what it is like for someone to experience that" (Polkinghorne, 1989, p. 46).

## Challenges

A phenomenology provides a deep understanding of a phenomenon as experienced by several individuals. Knowing some common experiences can be valuable for groups such as therapists, teachers, health personnel, and policymakers. Phenomenology can involve a streamlined form of data collection by including only single or multiple interviews with participants. Using the Moustakas (1994) approach for analyzing the data helps provide a structured approach for novice researchers. On the other hand, phenomenology requires at least some understanding of the broader philosophical assumptions, and these should be identified by the researcher. The participants in the study need to be carefully chosen to be individuals who have all experienced the phenomenon in question, so that the researcher, in the end, can forge a common understanding. Bracketing personal experiences may be difficult for the researcher to implement. An interpretive approach to phenomenology would signal this as an impossibility (van Manen, 1990)—for the researcher to become separated from the text. Perhaps we need a new definition of epoche or bracketing, such as suspending our understandings in a reflective move that cultivates curiosity (LeVasseur, 2003). Thus, the researcher needs to decide how and in what way his or her personal understandings will be introduced into the study.

## Grounded Theory Research

### Definition and Background

Although a phenomenology emphasizes the meaning of an experience for a number of individuals, the intent of a *grounded theory study* is to move beyond description and to *generate or discover a theory*, an abstract analytical schema of a process (or action or interaction, Strauss & Corbin, 1998). Participants in the study would all have experienced the process, and the development of the theory might help explain practice or provide a framework for further research. A key idea is that this theory-development does not come "off the shelf," but rather is generated or "grounded" in data from participants who have experienced the process (Strauss & Corbin, 1998). Thus, grounded theory is a qualitative research design in which the inquirer generates a general explanation (a theory) of a process, action, or interaction shaped by the views of a large number of participants (Strauss & Corbin, 1998).

This qualitative design was developed in sociology in 1967 by two researchers, Barney Glaser and Anselm Strauss, who felt that theories used in research were often inappropriate and ill-suited for participants under study. They elaborated on their ideas through several books (Glaser, 1978; Glaser & Strauss, 1967; Strauss, 1987; Strauss & Corbin, 1990, 1998). In contrast to the a priori, theoretical orientations in sociology, grounded theorists held that theories should be "grounded" in data from the field, especially in the actions, interactions, and social processes of people. Thus, grounded theory provided for the generation of a theory (complete with a diagram and hypotheses) of actions, interactions, or processes through interrelating categories of information based on data collected from individuals.

Despite the initial collaboration of Glaser and Strauss that produced such works as *Awareness of Dying* (Glaser & Strauss, 1965) and *Time for Dying* (Glaser & Strauss, 1968), the two authors ultimately disagreed about the meaning and procedures of grounded theory. Glaser has criticized Strauss's approach to grounded theory as too prescribed and structured (Glaser, 1992). More recently, Charmaz (2006) has advocated for a *constructivist grounded theory*, thus introducing yet another perspective into the conversation about procedures. Through these different interpretations, grounded theory has gained popularity in fields such as sociology, nursing, education, and psychology, as well as in other social science fields.

Another recent grounded theory perspective is that of Clarke (2005) who, along with Charmaz, seeks to reclaim grounded theory from its "positivist underpinnings" (p. xxiii). Clarke, however, goes further than Charmaz, suggesting that social "situations" should form our unit of analysis in grounded theory and that three sociological modes can be useful in analyzing these situations—situational, social world/arenas, and positional cartographic maps for collecting and analyzing qualitative data. She further expands grounded theory "after the postmodern turn" (p. xxiv) and relies on postmodern perspectives (i.e., the political nature of research and interpretation, reflexivity on the part of researchers, a recognition of problems of representing information, questions of legitimacy and authority, and repositioning the researcher away from the "all knowing analyst" to the "acknowledged participant") (pp. xxvii, xxviii). Clarke frequently turns to the postmodern, poststructural writer Michael Foucault (1972) to help turn the grounded theory discourse.

## Types of Grounded Theory Studies

The two popular approaches to grounded theory are the systematic procedures of Strauss and Corbin (1990, 1998) and the constructivist approach of Charmaz (2005, 2006). In the more systematic, analytic procedures of Strauss and Corbin (1990, 1998), the investigator seeks to systematically develop a theory that explains process, action, or interaction on a topic (e.g., the process of developing a curriculum, the therapeutic benefits of sharing psychological test results with clients). The researcher typically conducts 20 to 30 interviews based on several visits "to the field" to collect interview data to saturate the categories (or find information that continues to add to them until no more can be found). A *category* represents a unit of information composed of events, happenings, and instances (Strauss & Corbin, 1990). The researcher also collects and analyzes observations and documents, but these data forms are often not used. While the researcher collects data, she or he begins analysis. My image for data collection in a grounded theory study is a "zigzag" process: out to the field to gather information, into the office to analyze the data, back to the field to gather more information, into the office, and so forth. The participants interviewed are theoretically chosen (called *theoretical sampling*) to help the researcher best form the theory. How many passes one makes to the field depends on whether the categories of information become saturated and whether the theory is elaborated in all of its complexity. This process of taking information from data collection and comparing it to emerging categories is called the *constant comparative* method of data analysis.

The researcher begins with *open coding*, coding the data for its major categories of information. From this coding, axial coding emerges in which the researcher identifies one open coding category to focus on (called the "core" phenomenon), and then goes back to the data and create categories around this core phenomenon. Strauss and Corbin (1990) prescribe the types of categories identified around the core phenomenon. They consist of **causal conditions** (what factors caused the core phenomenon), *strategies* (actions taken in response to the core phenomenon), contextual and *intervening conditions* (broad and specific situational factors that influence the strategies), and *consequences* (outcomes from using the strategies). These categories relate to and surround the core phenomenon in a visual model called the *axial coding* paradigm. The final step, then, is *selective coding*, in which the researcher takes the model and develops *propositions* (or hypotheses) that interrelate the categories in the model or assembles a story that describes the interrelationship of categories in the model. This theory, developed by the researcher, is articulated toward the end of a study and can assume several forms, such as a narrative statement (Strauss & Corbin, 1990), a visual picture (Morrow & Smith, 1995), or a series of hypotheses or propositions (Creswell & Brown, 1992).

In their discussion of grounded theory, Strauss and Corbin (1998) take the model one step further to develop a *conditional matrix*. They advance the conditional matrix as a coding device to help the researcher make connections between the macro and the micro conditions influencing the phenomenon. This matrix is a set of expanding concentric circles with labels that build outward from the individual, group, and organization to the community, region, nation, and global world. In my experience, this matrix is seldom used in grounded theory research, and researchers typically end their studies with a theory developed in selective coding, a theory that might be viewed as a substantive, low-level theory rather than an abstract, grand theory (e.g., see Creswell & Brown, 1992). Although making connections between the substantive theory and its larger implications for the community, nation, and world in the conditional matrix is important (e.g., a model of work flow in a hospital, the shortage of gloves, and the national guidelines on AIDS may all be connected; see this example provided by Strauss & Corbin, 1998), grounded theorists seldom have the data, time, or resources to employ the conditional matrix.

A second variant of grounded theory is found in the constructivist writing of Charmaz (see Charmaz, 2005, 2006). Instead of embracing the study of a single process or core category as in the Strauss and Corbin (1998) approach, Charmaz advocates for a social constructivist perspective that includes emphasizing diverse local worlds, multiple realities, and the complexities of particular worlds, views, and actions. Constructivist grounded theory, according to Charmaz (2006), lies squarely within the interpretive approach to qualitative research with flexible guidelines, a focus on theory developed that depends on the researcher's view, learning about the experience within embedded, hidden networks, situations, and relationships, and making visible hierarchies of power, communication, and opportunity. Charmaz places more emphasis on the views, values, beliefs, feelings, assumptions, and ideologies of individuals than on the methods of research, although she does describe the practices of gathering rich data, coding the data, memoing, and using theoretical sampling (Charmaz, 2006). She suggests that complex terms or jargon, diagrams, conceptual maps, and systematic approaches (such as Strauss & Corbin, 1990) detract from grounded theory and represent an attempt to gain power in their use. She advocates using active codes, such as gerund-based phrases like "recasting life." Moreover, for Charmaz, a grounded theory procedure does not minimize the role of the researcher in the process. The researcher makes decisions about the categories throughout the process, brings questions to the data, and advances personal values, experiences, and priorities. Any conclusions developed by grounded theorists are, according to Charmaz (2005), suggestive, incomplete, and inconclusive.

## Procedures for Conducting Grounded Theory Research

Although Charmaz's interpretive approach has many attractive elements (e.g., reflexivity, being flexible in structure, as discussed in Chapter 2), I rely on Strauss and Corbin (1990, 1998) to illustrate grounded theory procedures because their systematic approach is helpful to individuals learning about and applying grounded theory research.

- The researcher needs to begin by determining if grounded theory is best suited to study his or her research problem. Grounded theory is a good design to use when a theory is not available to explain a process. The literature may have models available, but they were developed and tested on samples and populations other than those of interest to the qualitative researcher. Also, theories may be present, but they are incomplete because they do not address potentially valuable variables of interest to the researcher. On the practical side, a theory may be needed to explain how people are experiencing a phenomenon, and the grounded theory developed by the researcher will provide such a general framework.

- The research questions that the inquirer asks of participants will focus on understanding how individuals experience the process and identifying the steps in the process (What was the process? How did it unfold?). After initially exploring these issues, the researcher then returns to the participants and asks more detailed questions that help to shape the axial coding phase, questions such as: What was central to the process? (the core phenomenon); What influenced or caused this phenomenon to occur? (causal conditions); What strategies were employed during the process? (strategies); What effect occurred? (consequences).

- These questions are typically asked in interviews, although other forms of data may also be collected, such as observations, documents, and audiovisual materials. The point is to gather enough information to fully develop (or *saturate*) the model. This may involve 20 to 30 interviews or 50 to 60 interviews.

- The analysis of the data proceeds in stages. In open coding, the researcher forms categories of information about the phenomenon being studied by segmenting information. Within each category, the investigator finds several *properties*, or subcategories, and looks for data to dimensionalize, or show the extreme possibilities on a continuum of, the property.

- In axial coding, the investigator assembles the data in new ways after open coding. This is presented using a *coding paradigm or logic diagram* (i.e., a visual model) in which the researcher identifies a *central phenomenon* (i.e., a central category about the phenomenon), explores *causal conditions* (i.e., categories of conditions that influence the phenomenon), specifies strategies (i.e., the actions or interactions that result from the central phenomenon), identifies the *context* and *intervening conditions* (i.e., the narrow and broad conditions that influence the strategies), and delineates the **consequences** (i.e., the outcomes of the strategies) for this phenomenon.

- In selective coding, the researcher may write a "story line" that connects the categories. Alternatively, propositions or hypotheses may be specified that state predicted relationships.

- Finally, the researcher may develop and visually portray a conditional matrix that elucidates the social, historical, and economic conditions influencing the central phenomenon. It is an optional step and one in which the qualitative inquirer thinks about the model from the smallest to the broadest perspective.

- The result of this process of data collection and analysis is a theory, a *substantive-level theory*, written by a researcher close to a specific problem or population of people. The theory emerges with help from the process of *memoing*, a process in which the researcher writes down ideas about the evolving theory throughout the process of open, axial, and selective coding. The substantive-level theory may be tested later for its empirical verification with quantitative data to determine if it can be generalized to a sample and population (see mixed methods design procedures, Creswell & Plano Clark, 2007). Alternatively, the study may end at this point with the generation of a theory as the goal of the research.

## Challenges

A grounded theory study challenges researchers for the following reasons. The investigator needs to set aside, as much as possible, theoretical ideas or notions so that the analytic, substantive theory can emerge. Despite the evolving, inductive nature of this form of qualitative inquiry, the researcher must recognize that this is a systematic approach to research with specific steps in data analysis, if approached from the Strauss and Corbin (1990) perspective. The researcher faces the difficulty of determining when categories are saturated or when the theory is sufficiently detailed. One strategy that might be used to move toward saturation is to use *discriminant sampling,* in which the researchers gathered additional information from individuals similar to those people initially interviewed to determine if the theory holds true for these additional participants. The researcher needs to recognize that the primary outcome of this study is a theory with specific components: a central phenomenon, causal conditions, strategies, conditions and context, and consequences. These are prescribed categories of information in the theory, so the Strauss and Corbin (1990, 1998) approach may not have the flexibility desired by some qualitative researchers. In this case, the Charmaz (2006) approach, which is less structured and more adaptable, may be used.

# Ethnographic Research

## Definition and Background

Although a grounded theory researcher develops a theory from examining many individuals who share in the same process, action, or interaction, the study participants are not likely to be located in the same place or interacting on so frequent a basis that they develop shared patterns of behavior, beliefs, and *language*. An ethnographer is interested in examining these shared patterns, and the unit of analysis is larger than the 20 or so individuals involved in a grounded theory study. An *ethnography* focuses on an entire cultural group. Granted, sometimes this cultural group may be small (a few teachers, a few social workers), but typically it is large, involving many people who interact over time (teachers in an entire school, a community social work group). Ethnography is a qualitative design in which the researcher describes and interprets the shared and learned patterns of values, *behaviors*, beliefs, and language of a *culture-sharing group* (Harris, 1968). As both a process and an outcome of research (Agar, 1980), ethnography is a way of studying a culture-sharing group as well as the final, written product of that research. As a process, ethnography involves extended observations of the group, most often through *participant observation*, in which the researcher is *immersed* in the day-to-day lives of the people and observes and interviews the group participants. Ethnographers study the meaning of the behavior, the language, and the interaction among members of the culture-sharing group.

Ethnography had its beginning in the comparative cultural anthropology conducted by early 20th-century anthropologists, such as Boas, Malinowski, Radcliffe-Brown, and Mead. Although these researchers initially took the natural sciences as a model for research, they differed from those using traditional scientific approaches through the firsthand collection of data concerning existing "primitive" cultures (Atkinson & Hammersley, 1994). In the 1920s and 1930s, sociologists such as Park, Dewey, and Mead at the University of Chicago adapted anthropological field methods to the study of cultural groups in the United States (Bogdan & Biklen, 1992). Recently, scientific approaches to ethnography have expanded to include "schools" or subtypes of ethnography with different theoretical orientations and aims, such as structural functionalism, symbolic interactionism, cultural and cognitive anthropology, feminism, Marxism, ethnomethodology, critical theory, cultural studies, and postmodernism (Atkinson & Hammersley, 1994). This has led to a lack of orthodoxy in ethnography and has resulted in pluralistic approaches. Many excellent books are available on ethnography, including Van Maanen (1988) on the many forms of ethnography; Wolcott (1999) on ways of "seeing" ethnography; LeCompte and Schensul (1999) on procedures of ethnography presented in a toolkit of short books; Atkinson, Coffey, and Delamont (2003) on the practices of ethnography; and Madison (2005) on critical ethnography.

## Types of Ethnographies

There are many forms of ethnography, such as a confessional ethnography, life history, autoethnography, feminist ethnography, ethnographic novels, and the visual ethnography found in photography and video, and electronic media (Denzin, 1989a; LeCompte, Millroy, & Preissle, 1992; Pink, 2001; Van Maanen, 1988). Two popular forms of ethnography will be emphasized here: the realist ethnography and the critical ethnography.

The *realist ethnography* is a traditional approach used by cultural anthropologists. Characterized by Van Maanen (1988), it reflects a particular stance taken by the researcher toward the individuals being studied. Realist ethnography is an objective account of the situation, typically written in the third-person point of view and reporting objectively on the information learned from participants at a site. In this ethnographic approach, the realist ethnographer narrates the study in a third-person dispassionate voice and reports on what is observed or heard from participants. The ethnographer remains in the background as an omniscient reporter of the "facts." The realist also reports objective data in a measured style uncontaminated by personal bias, political goals, and judgment. The researcher may provide mundane details of every-day life among the people studied. The ethnographer also uses standard cat-egories for cultural description (e.g., family life, communication networks, worklife, social networks, status systems). The ethnographer produces the participants' views through closely edited quotations and has the final word on how the culture is to be interpreted and presented.

For many researchers, ethnography today employs a "critical" approach (Carspecken & Apple, 1992; Madison, 2005; Thomas, 1993) by including in the research an advocacy perspective. This approach is in response to cur-rent society, in which the systems of power, prestige, privilege, and author-ity serve to marginalize individuals who are from different classes, races, and genders. The *critical ethnography* is a type of ethnographic research in which the authors advocate for the emancipation of groups marginalized in society (Thomas, 1993). Critical researchers typically are politically minded individuals who seek, through their research, to speak out against inequality and domination (Carspecken & Apple, 1992). For example, critical ethnog-raphers might study schools that provide privileges to certain types of students, or counseling practices that serve to overlook the needs of under-represented groups. The major components of a critical ethnography include a value-laden orientation, empowering people by giving them more author-ity, challenging the status quo, and addressing concerns about power and control. A critical ethnographer will study issues of power, empowerment, inequality, inequity, dominance, repression, hegemony, and victimization.

## Procedures for Conducting an Ethnography

As with all qualitative inquiry, there is no single way to conduct the research in an ethnography. Although current writings provide more guid-ance to this approach than ever (for example, see the excellent overview found in Wolcott, 1999), the approach taken here includes elements of both realist ethnography and critical approaches. The steps I would use to con-duct an ethnography are as follows:

• Determine if ethnography is the most appropriate design to use to study the research problem. Ethnography is appropriate if the needs are to describe how a cultural group works and to explore the beliefs, language, behaviors, and issues such as power, resistance, and dominance. The literature may be deficient in actually knowing how the group works because the group is not in the mainstream, people may not be familiar with the group, or its ways are so different that readers may not identify with the group.

- Identify and locate a culture-sharing group to study. Typically, this group is one that has been together for an extended period of time, so that their shared language, patterns of behavior, and attitudes have merged into a discernable pattern. This may also be a group that has been marginalized by society. Because ethnographers spend time talking with and observing this group, access may require finding one or more individuals in the group who will allow the researcher in—a *gatekeeper* or *key informants (or participants)*.

- Select cultural themes or issues to study about the group. This involves the *analysis of the culture-sharing group*. The themes may include such topics as enculturation, socialization, learning, cognition, domination, inequality, or child and adult development (LeCompte, Millroy, & Preissle, 1992). As discussed by Hammersley and Atkinson (1995), Wolcott (1987, 1994b), and Fetterman (1998), the ethnographer begins the study by examining people in interaction in ordinary settings and by attempting to discern pervasive patterns such as life cycles, events, and cultural themes. *Culture* is an amorphous term, not something "lying about" (Wolcott, 1987, p. 41), but something researchers attribute to a group when looking for patterns of their social world. It is inferred from the words and actions of members of the group, and it is assigned to this group by the researcher. It consists of what people do (behaviors), what they say (language), the potential tension between what they do and ought to do, and what they make and use, such as artifacts (Spradley, 1980). Such themes are diverse, as illustrated in Winthrop's (1991) *Dictionary of Concepts in Cultural Anthropology*. Fetterman (1998) discusses how ethnographers describe a *holistic* perspective of the group's history, religion, politics, economy, and environment. Within this description, cultural concepts such as the social structure, kinship, the political structure, and the social relations or *function* among members of the group may be described.

- To study cultural concepts, determine which type of ethnography to use. Perhaps how the group works needs to be described, or the critical ethnography may need to expose issues such as power, hegemony, and to advocate for certain groups. A critical ethnographer, for example, might address an inequity in society or some part of it, use the research to advocate and call for changes, and specify an issue to explore, such as inequality, dominance, oppression, or empowerment.

- Gather information where the group works and lives. This is called *fieldwork* (Wolcott, 1999). Gathering the types of information typically needed in an ethnography involves going to the research site, respecting the daily lives of individuals at the site, and collecting a wide variety of materials. Field issues of respect, *reciprocity*, deciding who owns the data, and others are central to ethnography. Ethnographers bring a sensitivity to fieldwork issues (Hammersley & Atkinson, 1995), such as attending to how they gain access, giving back or reciprocity with the participants, and being ethical in all aspects of the research, such as presenting themselves and the study. LeCompte and Schensul (1999) organize types of ethnographic data into observations, tests and measures, surveys, interviews, content analysis, interviews, elicitation methods, audiovisual methods, spatial mapping, and network research. From the many sources collected, the ethnographer analyzes the data for a *description of the culture-sharing group*, themes that emerge from the group, and an overall interpretation (Wolcott, 1994b). The researcher begins by compiling a detailed description of the culture-sharing group, focusing on a single event, on several activities, or on the group over a prolonged period of time. The ethnographer moves into a theme analysis of patterns or topics that signifies how the cultural group works and lives.

• Forge a working set of rules or patterns as the final product of this analysis. The final product is a holistic *cultural portrait* of the group that incorporates the views of the participants (*emic*) as well as the views of the researcher (*etic*). It might also advocate for the needs of the group or suggest changes in society to address needs of the group. As a result, the reader learns about the culture-sharing group from both the participants and the interpretation of the researcher. Other products may be more performance based, such as theater productions, plays, or poems.

## Challenges

Ethnography is challenging to use for the following reasons. The researcher needs to have a grounding in cultural anthropology and the meaning of a social-cultural system as well as the concepts typically explored by ethnographers. The time to collect data is extensive, involving prolonged time in the field. In many ethnographies, the narratives are written in a literary, almost storytelling approach, an approach that may limit the audience for the work and may be challenging for authors accustomed to traditional approaches to writing social and human science research. There is a possibility that the researcher will "go native" and be unable to complete the study or be compromised in the study. This is but one issue in the complex array of fieldwork issues facing ethnographers who venture into an unfamiliar cultural group or system. A sensitivity to the needs of individual studies is especially important, and the researcher needs to acknowledge his or her impact on the people and the places being studied.

# Case Study Research

## Definition and Background

The entire culture-sharing group in ethnography may be considered a case, but the intent in ethnography is to determine how the culture works rather than to understand an issue or problem using the case as a specific illustration. Thus, *case study* research involves the study of an issue explored through one or more cases within a bounded system (i.e., a setting, a context). Although Stake (2005) states that case study research is not a methodology but a choice of what is to be studied (i.e., a case within a *bounded system*), others present it as a strategy of inquiry, a methodology, or a comprehensive research strategy (Denzin & Lincoln, 2005; Merriam, 1998; Yin, 2003). I choose to view it as a methodology, a type of design in qualitative research, or an object of study, as well as a product of the inquiry. Case study research is a qualitative approach in which the investigator explores a bounded system (a *case*) or multiple bounded systems (cases) over time, through detailed, in-depth data collection involving *multiple sources of information* (e.g., observations, interviews, audiovisual material, and documents and reports), and reports a case *description* and case-based themes. For example, several programs (a *multi-site* study) or a single program (a *within-site* study) may be selected for study.

The case study approach is familiar to social scientists because of its popularity in psychology (Freud), medicine (case analysis of a problem), law (case law), and political science (case reports). Case study research has a long, distinguished history across many disciplines. Hamel, Dufour, and Fortin (1993) trace the origin of modern social science case studies through anthropology and sociology. They cite anthropologist Malinowski's study of the Trobriand Islands, French sociologist LePlay's study of families, and the case studies of the University of Chicago Department of Sociology from the 1920s and 30s through the 1950s (e.g., Thomas and Znaniecki's 1958 study of Polish peasants in Europe and America) as antecedents of qualitative case study research. Today, the case study writer has a large array of texts and approaches from which to choose. Yin (2003), for example, espouses both quantitative and qualitative approaches to case study development and discusses explanatory, exploratory, and descriptive qualitative case studies. Merriam (1998) advocates a general approach to qualitative case studies in the field of education. Stake (1995) systematically establishes procedures for case study research and cites them extensively in his example of "Harper School." Stake's most recent book on multiple case study analysis presents a step-by-step approach and provides rich illustrations of multiple case studies in the Ukraine, Slovakia, and Romania (Stake, 2006).

## Types of Case Studies

Types of qualitative case studies are distinguished by the size of the bounded case, such as whether the case involves one individual, several individuals, a group, an entire program, or an activity. They may also be distinguished in terms of the intent of the case analysis. Three variations exist in terms of intent: the single instrumental case study, the collective or multiple case study, and the *intrinsic case study*. In a single *instrumental case study* (Stake, 1995), the researcher focuses on an issue or concern, and then selects one bounded case to illustrate this issue. In a *collective case study* (or multiple case study), the one issue or concern is again selected, but the inquirer selects multiple case studies to illustrate the issue. The researcher might select for study several programs from several research sites or multiple programs within a single site. Often the inquirer purposefully selects multiple cases to show different perspectives on the issue. Yin (2003) suggests that the multiple case study design uses the logic of replication, in which the inquirer replicates the procedures for each case. As a general rule, qualitative researchers are reluctant to generalize from one case to another because the contexts of cases differ. To best generalize, however, the inquirer needs to select representative cases for inclusion in the qualitative study. The final type of case study design is an intrinsic case study in which the focus is on the case itself (e.g., evaluating a program, or studying a student having difficulty—see Stake, 1995) because the case presents an unusual or unique situation. This resembles the focus of narrative research, but the case study analytic procedures of a detailed description of the case, set within its context or surroundings, still hold true.

## Procedures for Conducting a Case Study

Several procedures are available for conducting case studies (see Merriam, 1998; Stake, 1995; Yin, 2003). This discussion will rely primarily on Stake's (1995) approach to conducting a case study.

- First, researchers determine if a case study approach is appropriate to the research problem. A case study is a good approach when the inquirer has clearly identifiable cases with boundaries and seeks to provide an in-depth understanding of the cases or a comparison of several cases.

- Researchers next need to identify their case or cases. These cases may involve an individual, several individuals, a program, an event, or an activity. In conducting case study research, I recommend that investigators first consider what type of case study is most promising and useful. The case can be single or collective, multi-sited or within-site, focused on a case or on an issue (intrinsic, instrumental) (Stake, 1995; Yin, 2003). In choosing which case to study, an array of possibilities for *purposeful sampling* is available. I prefer to select cases that show different perspectives on the problem, process, or event I want to portray (called "purposeful maximal sampling,"; Creswell, 2005), but I also may select ordinary cases, accessible cases, or unusual cases.

- The data collection in case study research is typically extensive, drawing on multiple sources of information, such as observations, interviews, documents, and audiovisual materials. For example, Yin (2003) recommends six types of information to collect: documents, archival records, interviews, direct observations, participant-observations, and physical artifacts.

- The type of analysis of these data can be a *holistic analysis* of the entire case or an *embedded analysis* of a specific aspect of the case (Yin, 2003). Through this data collection, a detailed description of the case (Stake, 1995) emerges in which the researcher details such aspects as the history of the case, the chronology of events, or a day-by-day rendering of the activities of the case. (The gunman case study in Appendix F involved tracing the campus response to a gunman for 2 weeks immediately following the near-tragedy on campus.) After this description ("relatively uncontested data"; Stake, 1995, p. 123), the researcher might focus on a few key issues (or *analysis of themes*), not for generalizing beyond the case, but for understanding the complexity of the case. One analytic strategy would be to identify issues within each case and then look for common themes that transcend the cases (Yin, 2003). This analysis is rich in the *context of the case* or setting in which the case presents itself (Merriam, 1988). When multiple cases are chosen, a typical format is to first provide a detailed description of each case and themes within the case, called a *within-case analysis,* followed by a thematic analysis across the cases, called a *cross-case analysis,* as well as *assertions* or an interpretation of the meaning of the case.

- In the final interpretive phase, the researcher reports the meaning of the case, whether that meaning comes from learning about the issue of the case (an instrumental case) or learning about an unusual situation (an intrinsic case). As Lincoln and Guba (1985) mention, this phase constitutes the "lessons learned" from the case.

## Challenges

One of the challenges inherent in qualitative case study development is that the researcher must identify his or her case. I can pose no clear solution to this challenge. The case study researcher must decide which bounded system to study, recognizing that several might be possible candidates for this selection and realizing that either the case itself or an issue, which a case or cases are selected to illustrate, is worthy of study. The researcher must consider whether to study a single case or multiple cases. The study of more than one case dilutes the overall analysis; the more cases an individual studies, the less the depth in any single case. When a researcher chooses multiple cases, the issue becomes, "How many cases?" There is not a set number of cases. Typically, however, the researcher chooses no more than four or five cases. What motivates the researcher to consider a large number of cases is the idea of "generalizability," a term that holds little meaning for most qualitative researchers (Glesne & Peshkin, 1992). Selecting the case requires that the researcher establish a rationale for his or her purposeful sampling strategy for selecting the case and for gathering information about the case. Having enough information to present an in-depth picture of the case limits the value of some case studies. In planning a case study, I have individuals develop a data collection matrix in which they specify the amount of information they are likely to collect about the case. Deciding the "boundaries" of a case—how it might be constrained in terms of time, events, and processes—may be challenging. Some case studies may not have clean beginning and ending points, and the researcher will need to set boundaries that adequately surround the case.

## The Five Approaches Compared

All five approaches have in common the general process of research that begins with a research problem and proceeds to the questions, the data, the data analysis, and the research report. They also employ similar data collection processes, including, in varying degrees, interviews, observations, documents, and audiovisual materials. Also, a couple of potential similarities among the designs should be noted. Narrative research, ethnography, and case study research may seem similar when the unit of analysis is a single individual. True, one may approach the study of a single individual from any of these three approaches; however, the types of data one would collect and analyze would differ considerably. In *narrative research*, the inquirer focuses on the stories told from the individual and arranges these stories in chronological order. In ethnography, the focus is on setting the individuals' stories within the context of their culture and culture-sharing group; in case study research, the single case is typically selected to illustrate an issue, and the researcher compiles a detailed description of the setting for the case. As Yin (2003) comments, "You would use the case study method because you deliberately wanted to cover contextual conditions—believing that they might be highly pertinent to your phenomenon of study" (p. 13). My approach is to recommend, if the researcher wants to study a single individual, the narrative approach or a single case study because ethnography is a much broader picture of the culture. Then when comparing a narrative study and a single case to study a single individual, I feel that the narrative approach is seen as more scholarly because narrative studies *tend* to focus on single individual; whereas, case studies often involve more than one case.

From these sketches of the five approaches, I can identify fundamental differences among these types of qualitative research. As shown in Table 4.1, I present several dimensions for distinguishing among the five approaches. At a most fundamental level, the five differ in what they are trying to accomplish—their foci or the primary objectives of the studies. Exploring a life is different from generating a theory or describing the behavior of a cultural group. Moreover, although overlaps exist in discipline origin, some approaches have single-disciplinary traditions (e.g., grounded theory originating in sociology, ethnography founded in anthropology or sociology) and others have broad interdisciplinary backgrounds (e.g., narrative, case study). The data collection varies in terms of emphasis (e.g., more observations in ethnography, more interviews in grounded theory) and extent of data collection (e.g., only interviews in phenomenology, multiple forms in case study research to provide the in-depth case picture). At the data analysis stage, the differences are most pronounced. Not only is the distinction one of specificity of the analysis phase (e.g., grounded theory most specific, narrative research less defined), but the number of steps to be undertaken also varies (e.g., extensive steps in phenomenology, few steps in ethnography). The result of each approach, the written report, takes shape from all the processes before it. A narrative about an individual's life forms narrative research. A description of the essence of the experience of the phenomenon becomes a phenomenology. A theory, often portrayed in a visual model, emerges in grounded theory and a holistic view of how a culture-sharing group works results in an ethnography. An in-depth study of a bounded system or a case (or several cases) becomes a case study.

Relating the dimensions of Table 4.1 to research design within the five approaches will be the focus of chapters to follow. Qualitative researchers have found it helpful to see at this point a general sketch of the overall structure of each of the five approaches. Let's examine in Table 4.2 the structure of each approach.

The outlines in Table 4.2 may be used in designing a journal-article-length study; however, because of the numerous steps in each, they also have applicability as chapters of a dissertation or a book-length work. I introduce them here because the reader, with an introductory knowledge of each approach, now can sketch the general "architecture" of a study. Certainly, this architecture will emerge and be shaped differently by the conclusion of

**Table 4.1** Contrasting Characteristics of Five Qualitative Approaches

| Characteristics | Narrative Research | Phenomenology | Grounded Theory | Ethnography | Case Study |
|---|---|---|---|---|---|
| Focus | Exploring the life of an individual | Understanding the essence of the experience | Developing a theory grounded in data from the field | Describing and interpreting a culture-sharing group | Developing an in-depth description and analysis of a case or multiple cases |
| Type of Problem Best Suited for Design | Needing to tell stories of individual experiences | Needing to describe the essence of a lived phenomenon | Grounding a theory in the views of participants | Describing and interpreting the shared patterns of culture of a group | Providing an in-depth understanding of a case or cases |
| Discipline Background | Drawing from the humanities including anthropology, literature, history, psychology, and sociology | Drawing from philosophy, psychology, and education | Drawing from sociology | Drawing from anthropology and sociology | Drawing from psychology, law, political science, medicine |
| Unit of Analysis | Studying one or more individuals | Studying several individuals that have shared the experience | Studying a process, action, or interaction involving many individuals | Studying a group that shares the same culture | Studying an event, a program, an activity, more than one individual |

| Characteristics | Narrative Research | Phenomenology | Grounded Theory | Ethnography | Case Study |
|---|---|---|---|---|---|
| Data Collection Forms | Using primarily interviews and documents | Using primarily interviews with individuals, although documents, observations, and art may also be considered | Using primarily interviews with 20–60 individuals | Using primarily observations and interviews, but perhaps collecting other sources during extended time in field | Using multiple sources, such as interviews, observations, documents, artifacts |
| Data Analysis Strategies | Analyzing data for stories, "restorying" stories, developing themes, often using a chronology | Analyzing data for significant statements, meaning units, textural and structural description, description of the "essence" | Analyzing data through open coding, axial coding, selective coding | Analyzing data through description of the culture-sharing group; themes about the group | Analyzing data through description of the case and themes of the case as well as cross-case themes |
| Written Report | Developing a narrative about the stories of an individual's life | Describing the "essence" of the experience | Generating a theory illustrated in a figure | Describing how a culture-sharing group works | Developing a detailed analysis of one or more cases |

**Table 4.2** Reporting Structures for Each Approach

| Reporting Approaches<br>General Structure of Study | Narrative | Phenomenology | Grounded Theory | Ethnography | Case Study |
|---|---|---|---|---|---|
| | • Introduction (problem, questions)<br>• Research procedures (a narrative, significance of individual, data collection, analysis outcomes)<br>• Report of stories<br>• Individuals theorize about their lives<br>• Narrative segments identified<br>• Patterns of meaning identified (events, processes, epiphanies, themes)<br>• Summary<br><br>(Adapted from Denzin, 1989a, 1989b) | • Introduction (problem, questions)<br>• Research procedures (a phenomenology and philosophical assumptions, data collection, analysis, outcomes)<br>• Significant statements<br>• Meanings of statements<br>• Themes of meanings<br>• Exhaustive description of phenomenon<br><br>(Adapted from Moustakas, 1994) | • Introduction (problem, questions)<br>• Research procedures (grounded theory, data collection, analysis, outcomes)<br>• Open coding<br>• Axial coding<br>• Selective coding and theoretical propositions and models<br>• Discussion of theory and contrasts with extant literature<br><br>(Adapted from Strauss & Corbin, 1990) | • Introduction (problem, questions)<br>• Research procedures (ethnography, data collection, analysis, outcomes)<br>• Description of culture<br>• Analysis of cultural themes<br>• Interpretation, lessons learned, questions raised<br><br>(Adapted from Wolcott, 1994b) | • Entry vignette<br>• Introduction (problem, questions, case study, data collection, analysis, outcomes)<br>• Description of the case/cases and its/their context<br>• Development of issues<br>• Detail about selected issues<br>• Assertions<br>• Closing vignette<br><br>(Adapted from Stake, 1995) |

the study, but it provides a framework for the design issue to follow. I recommend these outlines as general templates at this time. In Chapter 5, we will examine five published journal articles, with each study illustrating one of the five approaches, and explore the writing structure of each.

## Summary

In this chapter, I described each of the five approaches to qualitative research—narrative research, phenomenology, grounded theory, ethnography, and case study. I provided a definition, some history of the development of the approach, and the major forms it has assumed, and I detailed the major procedures for conducting a qualitative study. I also discussed some of the major challenges in conducting each approach. To highlight some of the differences among the approaches, I provided an overview table that contrasts the characteristics of focus, the type of research problem addressed, the discipline background, the unit of analysis, the forms of data collection, data analysis strategies, and the nature of the final, written report. I also presented outlines of the structure of each approach that might be useful in designing a study within each of the five types. In the next chapter, we will examine five studies that illustrate each approach and look more closely at the compositional structure of each type of approach.

### Additional Readings

Several readings extend this brief overview of each of the five approaches of inquiry. In Chapter 1, I presented the major books that will be used to craft discussions about each approach. Here I provide a more expanded list of references that also includes the major works.

In narrative research, I will rely on Denzin (1989a, 1989b), Czarniawska (2004), and especially Clandinin and Connelly (2000). I add to this list books on life history (Angrosino, 1989a), humanistic methods (Plummer, 1983), and a comprehensive handbook on narrative research (Clandinin, 2006).

Angrosino, M. V. (1989a). *Documents of interaction: Biography, autobiography, and life history in social science perspective.* Gainesville: University of Florida Press.

Clandinin, D. J. (Ed.). (2006). *Handbook of narrative inquiry: Mapping a methodology.* Thousand Oaks, CA: Sage.

Clandinin, D. J., & Connelly, F. M. (2000). *Narrative inquiry: Experience and story in qualitative research.* San Francisco: Jossey-Bass.

Czarniawska, B. (2004). *Narratives in social science research.* London: Sage.

Denzin, N. K. (1989a). *Interpretive biography.* Newbury Park, CA: Sage.

Denzin, N. K. (1989b). *Interpretive interactionism*. Newbury Park, CA: Sage.

Elliot, J. (2005). *Using narrative in social research: Qualitative and quantitative approaches*. London: Sage.

Plummer, K. (1983). *Documents of life: An introduction to the problems and literature of a humanistic method*. London: George Allen & Unwin.

For phenomenology, the books on phenomenological research methods by Moustakas (1994) and the hermeneutical approach by van Manen (1990) will provide a foundation for chapters to follow. Other procedural guides to examine include Giorgi (1985), Polkinghorne (1989), Van Kaam (1966), Colaizzi (1978), Spiegelberg (1982), Dukes (1984), Oiler (1986), and Tesch (1990). For basic differences between hermeneutic and empirical or transcendental phenomenology, see Lopez and Willis (2004) and for a discussion about the problems of bracketing, see LeVasseur (2003). In addition, a solid grounding in the philosophical assumptions is essential, and one might examine Husserl (1931, 1970), Merleau-Ponty (1962), Natanson (1973), and Stewart and Mickunas (1990) for this background.

Colaizzi, P. F. (1978). Psychological research as the phenomenologist views it. In R. Vaile & M. King (Eds.), *Existential phenomenological alternatives for psychology* (pp. 48–71). New York: Oxford University Press.

Dukes, S. (1984). Phenomenological methodology in the human sciences. *Journal of Religion and Health, 23,* 197–203.

Giorgi, A. (Ed.). (1985). *Phenomenology and psychological research*. Pittsburgh, PA: Duquesne University Press.

Husserl, E. (1931). *Ideas: General introduction to pure phenomenology* (D. Carr, Trans). Evanston, IL: Northwestern University Press.

Husserl, E. (1970). *The crisis of European sciences and transcendental phenomenology* (D. Carr, Trans). Evanston, IL: Northwestern University Press.

LeVasseur, J. J. (2003). The problem with bracketing in phenomenology. *Qualitative Health Research, 31*(2), 408–420.

Lopez, K. A., & Willis, D. G. (2004). Descriptive versus interpretive phenomenology: Their contributions to nursing knowledge. *Qualitative Health Research, 14*(5), 726–735.

Merleau-Ponty, M. (1962). *Phenomenology of perception* (C. Smith, Trans.). London: Routledge & Kegan Paul.

Moustakas, C. (1994). *Phenomenological research methods*. Thousand Oaks, CA: Sage.

Natanson, M. (Ed.). (1973). *Phenomenology and the social sciences*. Evanston, IL: Northwestern University Press.

Oiler, C. J. (1986). Phenomenology: The method. In P. L. Munhall & C. J. Oiler (Eds.), *Nursing research: A qualitative perspective* (pp. 69–82). Norwalk, CT: Appleton-Century-Crofts.

Polkinghorne, D. E. (1989). Phenomenological research methods. In R. S. Valle & S. Halling (Eds.), *Existential-phenomenological perspectives in psychology* (pp. 41–60). New York: Plenum.

Spiegelberg, H. (1982). *The phenomenological movement* (3rd ed.). The Hague, Netherlands: Martinus Nijhoff.

Stewart, D., & Mickunas, A. (1990). *Exploring phenomenology: A guide to the field and its literature* (2nd ed.). Athens: Ohio University Press.

Tesch, R. (1990). *Qualitative research: Analysis types and software tools.* Bristol, PA: Falmer Press.

Van Kaam, A. (1966). *Existential foundations of psychology.* Pittsburgh, PA: Duquesne University Press.

van Manen, M. (1990). *Researching lived experience: Human science for an action sensitive pedagogy.* Albany: State University of New York Press.

On grounded theory research, consult the most recent and highly readable book, Strauss and Corbin (1990), before reviewing earlier works such as Glaser and Strauss (1967), Glaser (1978), Strauss (1987), Glaser (1992), or the latest edition of Strauss and Corbin (1998). The 1990 Strauss and Corbin book provides, I believe, a better procedural guide than their 1998 book. For brief methodological overviews of grounded theory, examine Charmaz (1983), Strauss and Corbin (1994), and Chenitz and Swanson (1986). Especially helpful are Charmaz's (2006) book on grounded theory research from a constructionist's perspective and Clarke's (2005) postmodern perspective.

Charmaz, K. (1983). The grounded theory method: An explication and interpretation. In R. Emerson (Ed.), *Contemporary field research* (pp. 109–126). Boston: Little, Brown.

Charmaz, K. (2006). *Constructing grounded theory.* London: Sage.

Chenitz, W. C., & Swanson, J. M. (1986). *From practice to grounded theory: Qualitative research in nursing.* Menlo Park, CA: Addison-Wesley.

Clarke, A. E. (2005). *Situational analysis: Grounded theory after the postmodern turn.* Thousand Oaks, CA: Sage.

Glaser, B. G. (1978). *Theoretical sensitivity.* Mill Valley, CA: Sociology Press.

Glaser, B. G. (1992). *Basics of grounded theory analysis.* Mill Valley, CA: Sociology Press.

Glaser, B. G., & Strauss, A. (1967). *The discovery of grounded theory.* Chicago: Aldine.

Strauss, A. (1987). *Qualitative analysis for social scientists.* New York: Cambridge University Press.

Strauss, A., & Corbin, J. (1990). *Basics of qualitative research: Grounded theory procedures and techniques.* Newbury Park, CA: Sage.

Strauss, A., & Corbin, J. (1994). Grounded theory methodology: An overview. In N. K. Denzin & Y. S. Lincoln (Eds.), *Handbook of qualitative research* (pp. 273–285). Thousand Oaks, CA: Sage.

Strauss, A., & Corbin, J. (1998). *Basics of qualitative research: Grounded theory procedures and techniques* (2nd ed.). Newbury Park, CA: Sage.

Several recent books on ethnography will provide the foundation for the chapters to follow: Atkinson, Coffey, and Delamont (2003); the first volume in the Ethnographer's Toolkit series, *Designing and Conducting Ethnographic Research,* as well as the other six volumes in the series by LeCompte and Schensul (1999); and Wolcott (1994b, 1999). Other resources about ethnography include Spradley (1979, 1980), Fetterman (1998), and Madison (2005).

Atkinson, P., Coffey, A., & Delamont, S. (2003). *Key themes in qualitative research: Continuities and changes.* Walnut Creek, CA: AltaMira.

Fetterman, D. M. (1998). *Ethnography: Step by step* (2nd ed.). Thousand Oaks, CA: Sage.

LeCompte, M. D., & Schensul, J. J. (1999). *Designing and conducting ethnographic research* (Ethnographer's toolkit, Vol. 1). Walnut Creek, CA: AltaMira.

Madison, D. S. (2005). *Critical ethnography: Method, ethics, and performance.* Thousand Oaks, CA: Sage.

Spradley, J. P. (1979). *The ethnographic interview.* New York: Holt, Rinehart & Winston.

Spradley, J. P. (1980). *Participant observation.* New York: Holt, Rinehart & Winston.

Wolcott, H. F. (1994b). *Transforming qualitative data: Description, analysis, and interpretation.* Thousand Oaks, CA: Sage.

Wolcott, H. F. (1999). *Ethnography: A way of seeing.* Walnut Creek, CA: AltaMira.

Finally, for case study research, consult Stake (1995) or earlier books such as Lincoln and Guba (1985), Merriam (1988), and Yin (2003).

Lincoln, Y. S., & Guba, E. G. (1985). *Naturalistic inquiry.* Beverly Hills, CA: Sage.

Merriam, S. (1988). *Case study research in education: A qualitative approach.* San Francisco: Jossey-Bass.

Stake, R. (1995). *The art of case study research.* Thousand Oaks, CA: Sage.

Yin, R. K. (2003). *Case study research: Design and method* (3rd ed.). Thousand Oaks, CA: Sage.

## Exercises

1. Select one of the five approaches for a proposed study. Write a brief description of the approach, including a definition, the history, and the procedures associated with the approach. Include references to the literature.

2. Take a proposed qualitative study that you would like to conduct. Begin with presenting it as a narrative study, then shape it into a phenomenology, a grounded theory, an ethnography, and finally a case study. Discuss for each type of study the focus of the study, the types of data collection and analysis, and the final written report.

# 5

# Five Different
# Qualitative Studies

The characteristics of and steps in conducting research in the five approaches in Chapter 4 help us to understand the major characteristics of each of the five approaches. By examining published studies, we can further our understanding. In this chapter, I present several examples of qualitative research—examples that are reasonable models for a narrative study, a phenomenology, a grounded theory, an ethnography, and a case study. The entire published studies are found in Appendices B, C, D, E, and F. The best way to proceed, I believe, is to first read the entire article in the appendix, then return to my summary of the article to compare your understanding with mine. Next read my analysis of how the article illustrates a good model of the approach to research. In my analysis, I review the study and advance how it fits the characteristics of the particular approach to qualitative research taken in the study. At the conclusion of this chapter, I reflect on why one might choose one approach over another when conducting a qualitative study.

The first study, by Angrosino (1994), illustrates the broad genre of narrative research, and more specifically a biographical-type of narrative study. It is the life history of Vonnie Lee Hargrett, an individual with mental retardation. The second article, a phenomenological study by Anderson and Spencer (2002) is a study about individuals who have experienced AIDS and the images and ways they think about their disease. The third article, by Morrow and Smith (1995), is on a sensitive topic: how 11 women survived and coped with childhood sexual abuse. It is a well-constructed grounded theory study, and it provides an emotional, detailed view of the women's lives. The fourth article is an ethnographic study by Haenfler (2004) about the core values of the straight edge (sXe) movement that emerged on the East Coast of the United States from the punk subculture of the early 1980s. The sXers adopted a "clean living" ideology of abstaining for life from alcohol, tobacco, illegal drugs, and casual sex. The final article is one of my own—a qualitative case study by Asmussen and Creswell (1995)—about the reaction of people at a large Midwestern university to a student who entered a class-room in actuarial science with a machine gun and attempted to shoot at the students.

## Questions for Discussion

- What is the focus in the sample narrative study?
- What experience is examined in the sample phenomenological study?
- What concepts are the basis for a theory in the grounded theory study?
- What cultural group or people is studied in the sample ethnographic study?
- What is the "case" being examined in the case study?
- How do the five approaches differ?
- How does a researcher choose among the five for his or her particular study?

## A Narrative-Biographical Study (Angrosino, 1994; see Appendix B)

This is the story of Vonnie Lee, a 29-year-old man the author met at Opportunity House, an agency designed for the rehabilitation of adults with mental retardation and psychiatric disorders. Most of the people at the agency had criminal records. Vonnie Lee was no exception. He had experienced a troubled childhood with an absent father and an alcoholic mother who had relationships with many physically abusive men. Vonnie Lee lived mostly on the streets in the company of an older man, Lucian, who made a living by "loaning" Vonnie Lee to other men on the street. After Lucian was beaten to death, Vonnie Lee found himself in and out of psychiatric facilities until he landed at Opportunity House. When the researcher entered the story, Vonnie Lee was in transition between Opportunity House and the community through "supervised independent living." A key step in preparing individuals for this transition was to teach them how to use the public transportation system—a city bus.

The author found Vonnie Lee open to talking about his life, but within narrow strictures. Vonnie Lee's stories were almost devoid of characters and centered mainly on a description of the bus route. As Angrosino said, "He was inclined only to offer what he seemed to feel were these deeply revelatory bus itineraries" (p. 18). Following this lead, Angrosino took a bus trip with Vonnie Lee to his place of work. This bus trip held special meaning for Vonnie Lee, as he traveled for about an hour and a half to his destination with three bus transfers. Vonnie Lee had set ways; he tried to find a seat under the large red heart, the logo of the city's bus line. En route, he supplied the researcher with the details about people, places, and events of the journey. Arriving at his place of work, a plumbing supply warehouse, Vonnie Lee's supervisor commented, "It's the bus he loves, coming here on the bus" (p. 21). "Why do you like the bus so much?" asked Angrosino. Vonnie Lee exclaimed, "If I was a big shot, I'd be on the bus right now!" From this, the researcher concluded that the bus gave meaning to Vonnie Lee's life through representing both escape and empowerment, and that meaning explained why he told his life stories in the form of bus routes. Vonnie Lee's stable self-image—the bus trip—helped him survive the vicissitudes of his life.

The study ended with the researcher reflecting on the use of the metaphor as a useful framework for analyzing stories of participants in life history projects. Furthermore, the study illustrated the benefits of the "in-depth autobiographical interview methodology" for establishing the human dimension of persons with mental illness and for "contextualizing" the interview information within the ongoing life experiences of Vonnie Lee.

This article presented the biographical approach to narrative research. Written by an anthropologist, it fitted well within the cultural interpretations of anthropological life history research. Other forms of narrative research (see examples at the end of this chapter) may not contain the strong cultural issues of metaphors of self and self-images of cultural groups presented in this study. Still, this study also provided many useful "markings" of biography and narrative research:

- The author told the story of a single individual as a central focus for the study.
- The data collection consisted of "conversations" or stories: the reconstruction of life experiences through researcher participant observations.
- The individual recalled a special event of his life, an "epiphany" (e.g., the bus ride).
- The author reported detailed information about the setting or historical context of the bus trip, thus situating the epiphany within a social context.
- The author was present in the study, reflecting on his own experiences and acknowledging that the study was his interpretation of the meaning of Vonnie Lee's life.

The elements of focusing on a single individual, constructing a study out of stories and epiphanies of special events, situating them within a broader context, and evoking the presence of the author in the study all reflect the interpretive biographical form of study discussed by Denzin (1989b) and many core elements of narrative research.

# A Phenomenological Study (Anderson & Spencer, 2002; see Appendix C)

This study discusses the images or cognitive representations that AIDS patients were found to hold about their disease. The researchers explored this topic because understanding how individuals represented AIDS and their emotional response to it influenced their therapy, reduced high-risk behaviors, and enhanced their quality of life. Thus, the purpose of this study was "to explore patients' experience and cognitive representations of AIDS within the context of phenomenology" (p. 1339).

The authors introduced the study by referring to the millions of individuals infected by HIV. They advanced a framework, the Self-Regulation Model of Illness Representation, which suggested that patients were active problem solvers whose behavior was a product of their cognitive and emotional responses to a health threat. Patients formed illness representations that shaped their understanding of their diseases. It was these illness representations (e.g., images) that the researchers needed to understand more thoroughly to help patients with their therapy, behaviors, and quality of life. The authors turned to the literature on patients' experiences with AIDS. They reviewed the literature on qualitative research, noting that several phenomenological studies on such topics as coping and living with HIV had already been examined. However, how patients represented AIDS in images had not been studied.

Their design involved the study of 58 men and women with a diagnosis of AIDS. To study these individuals, they used phenomenology and the procedures advanced by Colaizzi (1978) and modified by Moustakas (1994). For over 18 months, they conducted interviews with these 58 patients, and asked them: "What is your experience with AIDS? Do you have a mental image of HIV/AIDS, or how would you describe HIV/AIDS? What feeling comes to mind? What meaning does it have in your life?" (Anderson & Spencer, 2002, pp. 1341–1342). They also asked patients to draw pictures of their disease. Although only 8 of the 58 drew pictures, the authors integrated these pictures into the data analysis. Their data analysis of these interviews consisted of the following tasks:

- reading through the written transcripts several times to obtain an overall feeling for them
- identifying significant phrases or sentences that pertained directly to the experience
- formulating meanings and clustering them into themes common to all of the participants' transcripts
- integrating the results into an in-depth, exhaustive description of the phenomenon
- validating the findings with the participants, and including participants' remarks in the final description

This analysis led to 11 major themes based on 175 significant statements. Themes such as "dreaded bodily destruction" and "devouring life" illustrated two of the themes. The results section of this study reported each of the 11 themes and provided ample quotes and perspectives to illustrate the multiple perspectives on each theme.

The study ended with a discussion in which the authors described the essence (i.e., the exhaustive description) of the patients' experiences and the coping strategies (i.e., the contexts or conditions surrounding the experience) patients used to regulate mood and disease. Finally, the authors compared their 11 themes with results reported by other authors in the literature, and they discussed the results' implications for nursing and questions for future research.

This study illustrated several aspects of a phenomenological study:

- The use of systematic data analysis procedures of significant statements, meanings, themes, and an exhaustive description of the essence of the phenomenon followed the procedures recommended by Moustakas (1994).
- The inclusion of tables illustrating the significant statements, meanings, and theme clusters showed how the authors worked from the raw data to the exhaustive description of the essence of the study in the final discussion section.
- A central phenomenon—the "cognitive representations or images" of AIDS by patients—was examined in the study.
- Rigorous data collection with 58 interviews and incorporation of patents' drawings were used.
- The study ended by describing the essence of the experience for the 58 patients and the context in which they experienced AIDS (e.g., coping mechanisms).

The authors only briefly mentioned the philosophical ideas behind phenomenology. They referred to bracketing their personal experiences and their need to explore lived experiences rather than to obtain theoretical explanations.

# A Grounded Theory Study (Morrow & Smith, 1995; see Appendix D)

This was a grounded theory study about the survival and coping strategies of 11 women who experienced childhood sexual abuse. The authors asked the following two open-ended questions. "Tell me, as much as you are comfortable sharing with me right now, what happened to you when you were sexually abused? What were the primary ways in which you survived?" Data were collected primarily through one-on-one interviews, focus group interviews, and participant observation by one of the researchers. The authors first formed categories of information and then reassembled the data through systematically relating the categories into a visual model. At the center of this model was the central phenomenon, the central category around which the theory was developed: threatening or dangerous feelings along with helplessness, powerlessness, and lack of control. Factors causing this phenomenon were cultural norms and different forms of sexual abuse. Individuals used the two strategies of avoiding being overwhelmed by feelings and managing their helplessness, powerlessness, and lack of control. These strategies were set within the context of perpetrator characteristics, sensations, and frequency as well as within larger conditions such as family dynamics, victims' ages, and rewards. The strategies were not without consequences. These women talked about consequences such as surviving, coping, healing, and hoping. The article ended by relating the theoretical model back to the literature on sexual abuse.

The authors are both distinguished qualitative researchers, and Morrow brought her expertise in counseling and psychology to the writing of the article. They presented a visual model of their substantive theory, the theory that explained the women's actions in response to feelings of threat, danger, helplessness, powerlessness, and lack of control. The authors used rigorous procedures, such as collaboration and the search for disconfirming evidence, to verify their account. In this article, they also educated the reader about grounded theory in an extensive passage on coding data into categories of information and by memoing their thoughts throughout the project. In terms of overall structure, probably because of space limitations, the study did not address all facets of grounded theory procedures, such as open coding, forming initial categories of information, developing propositions or hypotheses specifying relations among categories, and the conditional matrix. However, the authors advanced a study that models good grounded theory research:

- The authors mentioned at the beginning that their purpose was to generate a theory using a "construct-oriented" (or category) approach.

- The procedure was thoroughly discussed and systematic.
- The authors presented a visual model, a coding diagram of the theory.
- The language and feel of the article was scientific and objective while, at the same time, it addressed a sensitive topic effusively.

# An Ethnographic Study (Haenfler, 2004; see Appendix E)

This ethnography study described the core values of the straight edge (sXe) movement that emerged on the East Coast of the United States from the punk subculture of the early 1980s. The movement arose as a response to the punk subculture nihilistic tendencies of drug and alcohol abuse and promiscuous sex. The sXers adopted a "clean living" ideology of abstaining for life from alcohol, tobacco, illegal drugs, and casual sex. Involving primarily white, middle-class males from the age of 15 to 25, it has been linked inseparably with the punk genre music scene, and straight edgers made a large X on each hand before they entered punk concerts. As a study that reconceptualizes resistance to opposition, this ethnography examined how subculture group members expressed opposition individually and as a reaction to other subcultures rather than against an ambiguous "adult" culture.

The author used ethnographic methods of data collection, including participating in the movement for 14 years and attending more than 250 music shows, interviewing 28 men and women, and gathering documents from sources such as newspaper stories, music lyrics, World Wide Web pages, and sXe magazines. From these data sources, the author first provided a detailed description of the subculture (e.g., T-shirt slogans, song lyrics, and use of the symbol "X"). The description also conveyed the curious blend of conservative perspectives from religious fundamentalism and progressive influences of expressing personal values. Following this description, the author identified five themes: positivity/clean living (e.g., committed vegetarians), reserving sex for caring relationships (e.g., sex should be part of an emotional relationship based on trust), self-realization (e.g., toxins such as drugs and alcohol inhibit people from reaching their full potential), spreading the message (e.g., sXers undertook a mission to convince their peers of their values), and involvement in progressive causes (e.g., animal rights and environmental causes). The article concluded with the author conveying a broad understanding of the sXers' values. Participation in the youth subculture had meaning both individually and collectively. Also, the sXers' resistance was at the macro level when directed to a culture that marketed alcohol and tobacco to youths; at the meso level when aimed at other subcultures, such as "punks"; and at the micro level when the sXers embraced personal change, in part in defiance of family members' substance abuse or their own addictive tendencies. Resistance was seen as personal in everyday activities and in political resistance to youth culture. It short, resistance was found to be multilayered, contradictory, and personally and socially transforming.

Haenfler's ethnography nicely illustrates both core elements of an ethnographic study as well as aspects of a critical ethnography:

- It was the study of a culture-sharing group and their core values and beliefs.
- The author first described the group, then advanced five themes about the group, and ended with a broad level of abstraction beyond the themes to suggest how the subculture worked.
- The author positioned himself by describing his involvement in the subculture and his role as an observer of the group for many years.
- From a critical ethnographic perspective, the author examined the issue of resistance to opposition and studied a group of counterculture youth.
- Consistent with many critical ethnographies, the article concluded with comments about how a subculture resisted dominant culture, the complexity and multilayered (e.g., macro, meso, and micro) forms resistance took, and the personal and social transforming qualities of participating in the culture-sharing group. Unlike other critical approaches, it did not end with a call for social transformation, but the overall study stood for reexamining subculture resistance.

# A Case Study (Asmussen & Creswell, 1995; see Appendix F)

This qualitative case study describes a campus reaction to a gunman incident in which a student attempted to fire a gun at his classmates. The case study began with a detailed description of the gunman incident, a chronicle of the first 2 weeks of events following the incident, and provided details about the city, the campus, and the building in which the incident occurred. Data were collected through the multiple sources of information, such as interviews, observations, documents, and audiovisual materials. Kelly Asmussen and I did not interview the gunman or the students who were in counseling immediately following the incident, and our petition to the Institutional Review Board for Human Subjects Research had guaranteed these restrictions. From the data analysis emerged themes of denial, fear, safety, retriggering, and campus planning. Toward the end of the article we combined these narrower themes into two overarching perspectives, an organizational and a social-psychological response, and we related these to the literature, thus providing "layers" of analysis in the study and invoking broader inter-pretations of the meaning of the case. We suggested that campuses plan for their responses to campus violence, and we advanced key questions to be addressed in preparing these plans.

In this case study, we tried to follow Lincoln and Guba's (1985) case study structure—the problem, the context, the issues, and the "lessons learned." We also added our own personal perspective by presenting tables with information about the extent of our data collection and the questions necessary to be addressed in planning a campus response to an incident. The epilogue at the end of the study reflexively brought our personal experiences into the discussion without disrupting the flow of the study. With our last theme on the need for the campus to design a plan for responding to another incident, we advanced practical and useful implications of the study for per-sonnel on campuses.

Several features mark this project as a case study:

- We identified the "case" for the study, the entire campus and its response to a potentially violent crime.
- This "case" was a bounded system, bounded by time (6 months of data col-lection) and place (situated on a single campus).
- We used extensive, multiple sources of information in data collection to pro-vide the detailed in-depth picture of the campus response.
- We spent considerable time describing the context or setting for the case, situ-ating the case within a peaceful Midwestern city, a tranquil campus, a build-ing, and a classroom, along with the detailed events during a 2-week period following the incident.

## Differences Among the Approaches

A useful perspective to begin the process of differentiating among the five approaches is to assess the central purpose or focus of each approach. As shown in Figure 5.1, the focus of a narrative is on the life of an individual, and the focus of a phenomenology is a concept or phenomenon and the "essence" of the lived experiences of persons about that phenomenon. In grounded theory, the aim is to develop a theory, whereas in ethnography, it is to describe a culture-sharing group. In a case study, a specific case is exam-ined, often with the intent of examining an issue with the case illustrating the complexity of the issue. Turning to the five studies, the foci of the approaches become more evident.

The story of Vonnie Lee (Angrosino, 1994) is a case in point—one decides to write a biography or life history when the literature suggests that a single

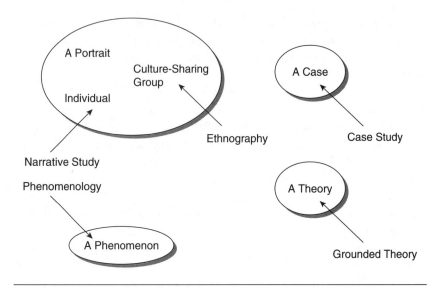

**Figure 5.1**    Differentiating Approaches by Foci

individual needs to be studied, or when an individual can illuminate a specific issue, such as the issue of being intellectually challenged. Furthermore, the researcher needs to make a case for the need to study this particular individual—someone who illustrates a problem, someone who has had a distinguished career, someone in the national spotlight, or someone who lives an ordinary life. The process of data collection involves gathering material about the person, either historically or from present-day sources, such as conversations or observations in the case of Vonnie Lee. A key consideration is whether the material is available and accessible. In the case of Vonnie Lee, Angrosino was able to win his confidence and encourage him to talk. This occurred first when Angrosino helped him with his reading assignments, and Angrosino made a mental note "to see if he would at some later time be amenable to telling me the 'story of my life'" (p. 17).

The phenomenological study, on the other hand, focuses not on the life of an individual but rather on a concept or phenomenon, such as how individuals represent their illnesses (Anderson & Spencer, 2002), and this form of study seeks to understand the meaning of experiences of individuals about this phenomenon. Furthermore, individuals are selected who have experienced the phenomenon, and they are asked to provide data, often through interviews. The researcher takes this data and, through several steps of reducing the data, ultimately develops a description of the experiences about the phenomenon that all individuals have in common—the essence of the experience.

Whereas the phenomenological project focuses on the meaning of people's experience toward a phenomenon, researchers in grounded theory have a different objective—to generate a substantive theory, such as the model about women surviving and coping with abuse in the Morrow and Smith study (1995). In the introductory passages, the authors describe the need for a "theoretical framework." Thus, grounded theorists undertake research to develop theory. The data collection method involves primarily interviewing (although other data collection procedures were used in the Morrow and Smith study). Also, the researchers use systematic procedures for analyzing and developing this theory, procedures such as open coding and axial coding, and they represent the relationship among categories with a visual model. The overall tone of this study is one of rigor and scientific credibility.

An ethnographic design is chosen when one wants to study the behaviors of a culture-sharing group, such as the sXers (Haenfler, 2004). Youths in this study were observed by the author over a prolonged period of time, and Haenfler's intent was both to provide a detailed description of the culture of the group and to identify themes about how the culture-sharing group worked. Overlaying the entire cultural portrait was a critical theory perspective of subcultural resistance, the complexity of this resistance within the group, and a contrast with other groups and adult subcultures.

Finally, a case study is chosen to study a case with clear boundaries, such as the campus in our study (Asmussen & Creswell, 1995). It is important, too, for the researcher to have contextual material available to describe the setting for the case. Also, the researcher needs to have a wide array of information about the case to provide an in-depth picture of it. In our gunman case, we went to great lengths to paint this picture for the reader through our table of information sources in the article and to illustrate our wide array of data collection procedures. With these data, we constructed a picture of the incident and the campus reaction to it through several themes.

Based now on a more thorough understanding of the five approaches, how does one choose one approach over the other? I recommend that you start with the outcome—what the approach is attempting to accomplish (e.g., the study of an individual, the examination of the meaning of experiences toward a phenomenon, the generation of a theory, the description and interpretation of a culture-sharing group, the in-depth study of a single case). In addition, other factors need also to be considered:

- *The audience question:* What approach is frequently used by gatekeepers in the field (e.g., committee members, advisers, editorial boards of journals)?
- *The background question:* What training does the researcher have in the inquiry approach?
- *The scholarly literature question:* What is needed most as contributing to the scholarly literature in the field (e.g., a study of an individual, an exploration of

the meaning of a concept, a theory, a portrait of a culture-sharing group, an in-depth case study)?

- *The personal approach question:* Are you more comfortable with a more structured approach to research or with a storytelling approach (e.g., narrative research, ethnography)? Or are you more comfortable with a firmer, more well-defined approach to research or with a flexible approach (e.g., grounded theory, case study, phenomenology)?

# Summary

This chapter examined five different short articles to illustrate good models for writing a narrative biography, a phenomenology, a grounded theory study, an ethnography, and a case study. These articles show basic characteristics of each approach and should enable readers to see differences in composing and writing varieties of qualitative studies. Choose a narrative study to examine the life experiences of a single individual when material is available and accessible and the individual is willing (assuming that he or she is living) to share stories. Choose a phenomenology to examine a phenomenon and the meaning it holds for individuals. Be prepared to interview the individuals, ground the study in philosophical tenets of phenomenology, follow set procedures, and end with the "essence" of the meaning. Choose a grounded theory study to generate or develop a theory. Gather information through interviews (primarily), and use systematic procedures of data gathering and analysis built on procedures such as open, axial, and selective coding. Although the final report will be "scientific," it can still address sensitive and emotional issues. Choose an ethnography to study the behavior of a culture-sharing group (or individual). Be prepared to observe and interview, and develop a description of the group and explore themes that emerge from studying human behaviors. Choose a case study to examine a "case," bounded in time or place, and look for contextual material about the setting of the "case." Gather extensive material from multiple sources of information to provide an in-depth picture of the "case."

These are important distinctions among the five approaches to qualitative inquiry. By studying each approach in detail, we can learn more about how to proceed and how to narrow our choice of which approach to use.

## Additional Readings

The following are published journal articles that illustrate each of the approaches of inquiry. For narrative research, I provide a range of studies that illustrate different forms of conducting a narrative study. From biography, we learn about a recovering alcohol named Freddie (Angrosino, 1989b). Two autoethnographies provide insight into the researchers' personal lives, one about a battered woman's identity (Olson, 2004) and the second about the author's personal experiences in the aftermath of her brother's death (Ellis, 1993). Women's experiences told through narratives form the central theme of Geiger's (1986) and Karen's (1990) studies about women's life histories, Nelson's (1990) oral life narrative of African American women, and Huber and Whelan's (1999) account of a teacher marginalized in her own school. Finally, I end with Smith's (1987) masterful telling of Darwin's experiences aboard the ship, the *Beagle,* a story told with multiple layers of thought about Darwin as well as the author.

Angrosino, M. V. (1989b). Freddie: The personal narrative of a recovering alcoholic— Autobiography as case history. In M. V. Angrosino, *Documents of interaction: Biography, autobiography, and life history in social science perspective* (pp. 29–41). Gainesville: University of Florida Press.

Ellis, C. (1993). "There are survivors": Telling a story of sudden death. *The Sociological Quarterly, 34,* 711–730.

Geiger, S. N. G. (1986). Women's life histories: Method and content. *Signs: Journal of Women in Culture and Society, 11,* 334–351.

Huber, J. & Whelan, K. (1999). A marginal story as a place of possibility: Negotiating self on the professional knowledge landscape. *Teaching and Teacher Education, 15,* 381–396.

Karen, C. S. (1990, April). *Personal development and the pursuit of higher education: An exploration of interrelationships in the growth of self-identity in returning women students—summary of research in progress.* Paper presented at the annual meeting of the American Educational Research Association, Boston.

Nelson, L. W. (1990). Code-switching in the oral life narratives of African-American women: Challenges to linguistic hegemony. *Journal of Education, 172,* 142–155.

Olson, L. N. (2004). The role of voice in the (re)construction of a battered woman's identity: An autoethnography of one woman's experiences of abuse. *Women's Studies in Communication, 27,* 1–33.

Smith, L. M. (1987). The voyage of the Beagle: Fieldwork lessons from Charles Darwin. *Educational Administration Quarterly, 23*(3), 5–30.

For phenomenological research journals, I have selected studies that tend to reflect the phenomenological methods discussed in Moustakas (1994) and that focus on different phenomena of interest. Brown, Sorrell, McClaren, and Creswell (2006) address experiences of individuals waiting for a liver transplant, Edwards (2006) looks at experiences of African American women with HIV/AIDS medication, Riemen (1986) studies the caring interaction between patients and nurses, and Grigsby and Megel (1995) explore the caring experiences between nurse faculty and students. In the study of mothers' experiences with deaths of wished-for babies (Lauterbach, 1993), we see the wide range of data sources that can be used in phenomenology. In Padilla's (2003) account of Clara, who sustained a head injury, we see how a phenomenology can be undertaken with the study of one individual based on extensive interviews and email messages.

Brown, J., Sorrell, J. H., McClaren, J., & Creswell, J. W. (2006). Waiting for a liver transplant. *Qualitative Health Research, 16*(1), 119–136.

Edwards, L. V. (2006). Perceived social support and HIV/AIDS medication adherence among African American women. *Qualitative Health Research, 16,* 679–691.

Grigsby, K. A., & Megel, M. E. (1995). Caring experiences of nurse educators. *Journal of Nursing Research, 34,* 411–418.

Lauterbach, S. S. (1993). In another world: A phenomenological perspective and discovery of meaning in mothers' experience with death of a wished-for baby: Doing phenomenology. In P. L. Munhall & C. O. Boyd (Eds.), *Nursing research: A qualitative perspective* (pp. 133–179). New York: National League for Nursing Press.

Padilla, R. (2003). Clara: A phenomenology of disability. *The American Journal of Occupational Therapy, 57*(4), 413–423.

Riemen, D. J. (1986). The essential structure of a caring interaction: Doing phenomenology. In P. M. Munhall & C. J. Oiler (Eds.), *Nursing research: A qualitative perspective* (pp. 85–105). Norwalk, CT: Appleton-Century-Crofts.

Our exploration of approaches continues with published grounded theory journal articles. The underlying theme of generating a theory of a process is illustrated in Conrad's (1978) study of academic change in universities; Creswell and Brown's (1992) analysis of how academic chairpersons enhance faculty research; Leipert's (2005) study of how women develop resilience in northern geographical isolated settings; Barlow and Cairns's (1997) study of women's experiences of mothering; and Kearney, Murphy, and Rosenbaum's (1994) study of mothering on crack cocaine. Also included is the constructivist grounded theory perspective of Charmaz (1994), who explores the identity dilemmas of chronically ill men.

Barlow, C. A., & Cairns, K. V. (1997). Mothering as a psychological experience: A grounded theory exploration. *Canadian Journal of Counselling, 31,* 232–247.

Charmaz, K. (1994). Identity dilemmas of chronically ill men. *The Sociological Quarterly, 35,* 269–288.

Conrad, C. F. (1978). A grounded theory of academic change. *Sociology of Education, 51,* 101–112.

Creswell, J. W., & Brown, M. L. (1992). How chairpersons enhance faculty research: A grounded theory study. *Review of Higher Education, 16*(1), 41–62.

Kearney, M. H., Murphy, S., & Rosenbaum, M. (1994). Mothering on crack cocaine: A grounded theory analysis. *Social Science Medicine, 38*(2), 351–361.

Leipert, B. D., & Reutter, L. (2005). Developing resilience: How women maintain their health in northern geographically isolated settings. *Qualitative Health Research, 15,* 49–65.

For examples of published ethnographic studies, see the different culture-sharing groups and the critical and realist lenses used in ethnographic research. Finders (1996) ethnography of adolescent females and their teen magazine, Geertz's (1973) classic notes on the Balinese cockfight, Rhoads's (1995) study of college fraternity life, and Trujillo's (1992) study of the culture of baseball all are set in different cultural settings. Wolcott's (1983) well-known study of the "sneaky kid" illustrates a realist ethnography, and our ethnographic study (Miller, Creswell, & Olander, 1998) of a soup kitchen for the homeless illustrates discussing the culture of the homeless from realist, confessional, and critical perspectives.

Finders, M. J. (1996). Queens and teen zines: Early adolescent females reading their way toward adulthood. *Anthropology & Education Quarterly, 27,* 71–89.

Geertz, C. (1973). Deep play: Notes on the Balinese cockfight. In C. Geertz (Ed.), *The interpretation of cultures: Selected essays* (pp. 412–435). New York: Basic Books.

Miller, D. L., Creswell, J. W., & Olander, L. S. (1998). Writing and retelling multiple ethnographic tales of a soup kitchen for the homeless. *Qualitative Inquiry, 4*(4), 469–491.

Rhoads, R. A. (1995). Whales tales, dog piles, and beer goggles: An ethnographic case study of fraternity life. *Anthropology and Education Quarterly, 26,* 306–323.

Trujillo, N. (1992). Interpreting (the work and the talk of) baseball. *Western Journal of Communication, 56,* 350–371.

Wolcott, H. F. (1983). Adequate schools and inadequate education: The life history of a sneaky kid. *Anthropology and Education Quarterly, 14*(1), 2–32.

Finally, for specific case study research, I suggest the published journal articles below that differ in the number of cases. The studies by Brickhous and Bodner (1992) and Rex (2000) present single case studies, while the Padula and Miller (1999) and the Hill, Vaughn, and Harrison (1995) studies examine five cases.

Brickhous, N., & Bodner, G. M. (1992). The beginning science teacher: Classroom narratives of convictions and constraints. *Journal of Research in Science Teaching, 29,* 471–485.

Hill, B., Vaughn, C., & Harrison, S. B. (1995, September/October). Living and working in two worlds: Case studies of five American Indian women teachers. *The Clearinghouse, 69*(1), 42–48.

Padula, M. A., & Miller, D. L. (1999). Understanding graduate women's reentry experiences: Case studies of four psychology doctoral students in a Midwestern university. *Psychology of Women Quarterly, 23,* 327–343.

Rex, L. A. (2000). Judy constructs a genuine question: A case for interactional inclusion. *Teaching and Teacher Education, 16,* 315–333.

## Exercises

1. Begin to sketch a qualitative study using one of the approaches. Answer the questions here that apply to the approach you are considering. For a narrative study: What individual do you plan to study? And do you have access to information about this individual's life experiences? For a phenomenology: What is the phenomenon of interest that you plan to study? And do you have access to people who have experienced it? For a grounded theory: What social science concept, action, or process do you plan to explore as the basis for your theory? For an ethnography: What cultural group or people do you plan to study? For a case study: What is the case you plan to examine?

2. Select one of the journal articles listed in the Additional Readings section. Determine the characteristics of approach being used by the author(s) and discuss why the author(s) may have used the approach.

**Exercises**

# 6

# Introducing and Focusing the Study

The design of a qualitative study begins before the researcher chooses a qualitative approach. It begins by the researcher stating the problem or issue leading to the study, formulating the central purpose of the study, and providing the research questions. However, these components need to connect or tie to the approach used in the study. It is not necessarily the case that the research problem and questions precede the design of the research. Often the logic is back and forth between these components in an integrated, consistent manner so that all parts interrelate (Morse & Richards, 2002). Thus, these introductory sections can foreshadow elements of the approach being used, or they can be written after one of the approaches (narrative, phenomenology, grounded theory, ethnography, or case study) has been selected. Regardless of the logic chosen, there are elements of writing a good qualitative research problem statement, a purpose statement, and research questions tailored to one of the approaches to qualitative research, and this chapter is devoted to conveying these elements.

## Questions for Discussion

- How can the problem statement be best written to reflect one of the approaches to qualitative research?
- How can the purpose statement be best written to convey the orientation of an approach to research?

- How can a central question be written so that it encodes and foreshadows an approach to qualitative research?
- How can subquestions be presented so that they reflect the issues being explored in an approach to qualitative research?

## The Research Problem

Qualitative studies begin with authors stating the research problem of the study. In the first few paragraphs of a design for a study, the qualitative researcher introduces the "problem" leading to the study. The term "problem" may be a misnomer, and individuals unfamiliar with writing research may struggle with this writing passage. Rather than calling this passage the "problem," it might be clearer if I call it the "need for the study." The intent of a research problem in qualitative research is to provide a rationale or need for studying a particular issue or "problem." Why is this study needed? In the following paragraphs, I consider establishing the need by considering the "source" for the problem, framing it within the literature, and encoding and foreshadowing the text for one of the five qualitative approaches to inquiry.

Research methods books (e.g., Creswell, 2005; Marshall & Rossman, 2006) advance several sources for research problems. Research problems are found in personal experience with an issue, a job-related problem, an adviser's research agenda, or the scholarly literature. It is important in qualitative research to provide a rationale or reason for studying the problem. The strongest and most scholarly rationale for a study, I believe, comes from the scholarly literature: a need exists to add to or fill a gap in the literature or to provide a voice for individuals not heard in the literature. As suggested by Barritt (1986), the rationale

> is not the discovery of new elements, as in natural scientific study, but rather the heightening of awareness for experience which has been forgotten and overlooked. By heightening awareness and creating dialogue, it is hoped research can lead to better understanding of the way things appear to someone else and through that insight lead to improvements in practice. (p. 20)

Besides dialogue and understanding, a qualitative study may fill a void in existing literature, establish a new line of thinking, or assess an issue with an understudied group or population.

Although opinions differ about the extent of literature review needed before a study begins, qualitative texts (e.g., Creswell, 2003; Marshall & Rossman, 2006) refer to the need to review the literature so that one can provide the rationale for the problem and position one's study within the ongoing literature about the topic. I have found it helpful to visually depict where my study can be positioned into the larger literature. For example, one might develop a figure—a research map (Creswell, 1994)—of existing literature and show in this figure the topics addressed in the literature and how one's proposed research fits into or extends the literature.

In addition to determining the source of the research problem and framing it within the literature, qualitative researchers need to introduce the problem in a way that the discussion foreshadows one of the five approaches to inquiry. This can be done, I believe, by mentioning how the particular choice of approach fills a need or gap in the literature about the research problem. In a problem statement for a narrative study, for example, I would expect the writer to mention how individual stories need to be told to gain personal experiences about the research problem. In a phenomenological study, I would like to hear from the author that we need to know more about a particular phenomenon and the common experiences of individuals with the phenomenon. For a grounded theory study, I would expect to learn how we need a theory that explains a process because existing theories are inadequate, nonexistent for the population, or need to be modified. In an ethnographic study, the problem statement might include thoughts about why it is important to describe and to interpret the cultural behavior of a certain group of people or how a group is marginalized and kept silent by others. For a case study, the researcher might discuss how the study of a case or cases can help inform the research problem. Thus, the need for the study, or the problem leading to it, can be related to the specific focus of one of the five approaches to research.

## The Purpose Statement

This interrelationship between design and approach continues with the purpose statement, a statement that provides the major objective or intent, or "road map," to the study. As the most important statement in an entire qualitative study, the purpose statement needs to be carefully constructed and written in clear and concise language. Unfortunately, all too many writers leave this statement implicit, causing readers extra work in interpreting and following a study. This need not be the case, so I created a "script" of this statement (Creswell, 1994, 2003), a statement containing several sentences and blanks that an individual fills in:

The purpose of this _____ (narrative, phenomenological, grounded theory, ethnographic, case) study is (was? will be?) to _____ (understand? describe? develop? discover?) the _____ (central phenomenon of the study) for _____ (the participants) at

_____ (the site). At this stage in the research, the _____ (central phenomenon) will be generally defined as _____ (a general definition of the central concept).

As I show in the script, several terms can be used to encode a passage for a specific approach to qualitative research. In the purpose statement,

- The writer identifies the specific qualitative approach used in the study by mentioning the type. The name of the approach comes first in the passage, thus foreshadowing the inquiry approach for data collection, analysis, and report writing.
- The writer encodes the passage with words that indicate the action of the researcher and the focus of the approach to research. For example, I associate certain words with qualitative research, such as "understand experiences" (useful in narrative studies), "describe" (useful in case studies, ethnographies, and phenomenologies), "meaning ascribed" (associated with phenomenologies), "develop or generate" (useful in grounded theory), and "discover" (useful in all approaches).
- I identify several words that a researcher would include in a purpose statement to encode the purpose statement for the approach chosen (see Table 6.1). These words indicate not only researchers' actions but also the foci and outcomes of the studies.
- The writer identifies the central phenomenon. The central phenomenon is the one, central concept being explored or examined in the research study. I generally recommend that qualitative researchers focus on only one concept (e.g., the campus reaction to the gunman, or the values of the sXers) at the beginning of a study. Comparing groups or looking for linkages can be included in the study as one gains experiences in the field and engages in initial exploration of the central phenomenon.
- The writer foreshadows the participants and the site for the study, whether the participants are one individual (i.e., narrative or case study), several individuals (i.e., grounded theory or phenomenology), a group (i.e., ethnography), or a site (i.e., program, event, activity, or place in a case study).
- I include a general definition for the central phenomenon. This definition may be difficult to determine with any specificity in advance. But, for example, in a narrative study, a writer might define the types of stories to be collected (e.g., life stages, childhood memories, the transition from adolescence to adulthood, attendance at an Alcoholics Anonymous meeting). In a phenomenology, the central phenomenon to be explored might be specified such as the meaning of grief, anger, or even chess playing (Aanstoos, 1985). In grounded theory, the central phenomenon might be identified as a concept central to the process being examined. In an ethnography, the writer might identify the key cultural concepts being examined such as roles, behaviors, acculturation, communication, myths, stories, or other concepts that the researcher plans to take into the

| Table 6.1 | Words to Use in Encoding the Purpose Statement | | | |
| --- | --- | --- | --- | --- |
| *Narrative* | *Phenomenology* | *Grounded Theory* | *Ethnography* | *Case Study* |
| • Narrative study | • Phenomenology | • Grounded theory | • Ethnography | • Case study |
| • Stories | • Describe | • Generate | • Culture-sharing group | • Bounded |
| • Epiphanies | • Experiences | • Develop | • Cultural behavior and language | • Single or collective case |
| • Lived experiences | • Meaning | • Propositions | • Cultural portrait | • Event, process, program, individual |
| • Chronology | • Essence | • Process | • Cultural themes | |
| | | • Substantive theory | | |

field at the beginning of the study. Finally, in a case study such as an "intrinsic" case study (Stake, 1995), the writer might define the boundaries of the case, specifying how the case is bounded in time and place. If an "instrumental" case study is desired, then the researcher might specify and define generally the issue being examined in the case.

Several examples of purpose statements follow that illustrate the encoding and foreshadowing of the five approaches to research:

## Example 6.1. A Narrative Example

From a study about the ways in which theories of narrative might be significant in the study of childbearing of 17 women:

In my research, which has involved collecting women's accounts of their experiences of becoming mothers, I am seeking to understand how women make sense of events throughout the process of childbearing, constructing these events into episodes and thereby (apparently) maintaining unity within their lives. (Miller, 2000, p. 309)

## Example 6.2. A Phenomenological Example

From a study of doctoral advisement relationships between women:

Given the intricacies of power and gender in the academy, what are doctoral advisement relationships between women advisors and women advisees really like? Because there were few studies exploring women doctoral students' experiences in the literature, a phenomenological study devoted to understanding women's lived experiences as advisees best lent itself to examining this question. (Heinrich, 1995, p. 449)

## Example 6.3. A Grounded Theory Example

From a grounded theory study of academic change in higher education:

The primary purpose of this article is to present a grounded theory of academic change that is based upon research guided by two major research questions: What are the major sources of academic change? What are the major processes through which academic change occurs? For purposes of this paper, grounded theory is defined as theory generated from data systematically obtained and analyzed through the constant comparative method. (Conrad, 1978, p. 101)

*Example 6.4. An Ethnographic Example*

From an ethnography of "ballpark" culture:

> This article examines how the work and the talk of stadium employees rein-
> force certain meanings of baseball in society, and it reveals how this work and
> talk create and maintain ballpark culture. (Trujillo, 1992, p. 351)

*Example 6.5. A Case Study Example*

From a case study using a feminist perspective to examine how men
exploit women's labor in the sport of lawn bowls at the "Roseville Club":

> Although scholars have shown that sport is fundamental in constituting and
> reproducing gender inequalities, little attention has been paid to sport and gen-
> der relations in later life. In this article we demonstrate how men exploit
> women's labor in the sport of lawn bowls, which is played predominately by
> older people. (Boyle & McKay, 1995, p. 556)

# The Research Questions

Several of these examples illustrate the interweaving of problems, research
questions, and purpose statements. For purposes of this discussion, I separate them out, although in practice some researchers combine them. But, in
many instances, the research questions are distinct and easily found in a
study. Once again, I find that these questions provide an opportunity to
encode and foreshadow an approach to inquiry.

## The Central Question

Some writers offer suggestions for writing qualitative research questions
(e.g., Creswell, 2003; Marshall & Rossman, 2006). I especially like the conceptualization of Marshall and Rossman (2006) of research questions into
four types: exploratory (e.g., to investigate phenomenon little understood),
explanatory (e.g., to explain patterns related to phenomenon), descriptive
(e.g., to describe the phenomenon), and emancipatory (e.g., to engage in
social action about the phenomenon). Qualitative research questions are
open-ended, evolving, and nondirectional; restate the purpose of the study in
more specific terms; start with a word such as "what" or "how" rather than
"why"; and are few in number (five to seven). They are posed in various

forms, from the "grand tour" (Spradley, 1979, 1980) that asks, "Tell me about yourself," to more specific questions.

I recommend that a researcher reduce her or his entire study to a single, overarching question and several subquestions. Drafting this central question often takes considerable work because of its breadth and the tendency of some to form specific questions based on traditional training. To reach the overarching question, I ask qualitative researchers to state the broadest question they could possibly pose about the research problem.

This central question can be encoded with the language of one of the five approaches to inquiry. Morse (1994) speaks directly to this issue as she reviews the types of research questions. Although she does not refer to narratives or case studies, she mentions that one finds "descriptive" questions of cultures in ethnographies, "process" questions in grounded theory studies, and "meaning" questions in phenomenological studies. For example, I searched through the five studies presented in Chapter 5 to see if I could find or imagine their central research questions.

In the life history of Vonnie Lee, Angrosino (1994) does not pose a central question, but I can infer from statements about the purpose of the study that the central question might be, "What story does Vonnie Lee have to tell?" This question implies that the individual in the narrative has a story, and that there will be some central element of interest (i.e., travel on the bus) that holds meaning for Vonnie Lee's life. In the phenomenological study of how persons living with AIDS represent and image their disease, Anderson and Spencer (2002) also did not pose a central question, but it might have been: "What meaning do 41 men and 17 women with a diagnosis of AIDS ascribe to their illness?" This central question in phenemonology implies that all of the individuals diagnosed with AIDS have something in common that provides meaning for their lives. In the grounded theory study of 11 women's survival and coping with childhood sexual abuse, Morrow and Smith (1995) do not present a central question in the introduction, but they mention several broad questions that guided their interviewing of the women: "Tell me, as much as you are comfortable sharing with me right now, what happened to you when you were sexually abused?" and "What are the primary ways in which you survived?" (p. 25). This question implies that the researchers were first interested in understanding the women's experience and then shaping it into coping strategies used to survive their abuse (as part of a theory of the process). In the ethnographic study of the sXe movement by Haenfler (2004), again no research question is advanced, but it might have been: "What are the core values of the straight edge movement, and how do the members construct and understand their subjective experiences of being a part of the subculture?" This question asks first for a description of the core values and then an understanding of experiences (that are presented as themes in the study). Finally, in our case study of a campus response to a gunman incident (Asmussen & Creswell, 1995), we asked five central guiding questions in our introduction: "What happened? Who was involved in response to the incident? What themes of response emerged during the eight-month period that followed this incident? What theoretical constructs helped us understand the campus response, and what constructs were unique to this case?" (p. 576). This example illustrates how we were interested first in simply describing their experiences and then in developing themes that represented responses of individuals on the campuses.

As these examples illustrate, authors may or may not pose a central question, although one exists in all studies. For writing journal articles, central questions may be used less than purpose statements to guide the research. However, for individuals' graduate research, such as theses or dissertations, the trend is toward writing both purpose statements and central questions.

## Subquestions

An author typically presents a small number of subquestions that follow the central question. One model for conceptualizing these subquestions is to use either issue questions or topical questions. According to Stake (1995), *issue subquestions* address the major concerns and perplexities to be resolved. The issue-oriented questions, for example,

are not simple and clean, but intricately wired to political, social, historical, and especially personal contexts. . . . Issues draw us toward observing, even teasing out the problems of the case, the conflictual outpourings, the complex backgrounds of human concern. (Stake, 1995, p. 17)

My understanding of issue-oriented subquestions is that they take the phenomenon in the central research questions and break it down into subtopics for examination. A central question such as "What does it mean to be a college professor?" would be analyzed in subquestions on topics like "What does it mean to be a college professor in the classroom? As a researcher? As an advisor?" and so forth.

Topical subquestions, on the other hand, cover the anticipated needs for information. These questions, "call for information needed for description of the case. . . . A topical outline will be used by some researchers as the primary conceptual structure and by others as subordinate to the issue structure" (Stake, 1995, p. 25). I view topical subquestions as questions that advance the procedural steps in the process of research, steps that are typically conducted within one of the approaches to research (see Chapter 4 for the procedures of each approach). To be more descriptive, I would change the name from "topical" to **"procedural" subquestions**. For example, in grounded theory, the steps involve identifying a central phenomenon, the causal conditions, the intervening conditions, and the strategies and consequences. By writing procedural subquestions, authors can mirror the procedures they intend to use in one of the five approaches to inquiry and foreshadow their choice of approach.

Several illustrations in the following examples represent both issue and procedural subquestions.

In writing a biographical narrative, Denzin (1989b) suggests that research questions follow an interpretive format and be formulated into a single statement, beginning with how, not why, and starting with one's own personal history and building on other information. From his own studies, Denzin illustrates types of issue questions: "How is emotion, as a form of consciousness, lived, experienced, articulated and felt?"; "How do ordinary men and women live and experience the alcoholic self active alcoholism produces?" (p. 50).

Then, one could pose procedural subquestions that relate to the manner or procedure of narrative research. For example, these procedural questions might be:

- What are the experiences in this individual's life?
- What are the stories that can be told from these experiences?
- What are some "turning points" in the stories?
- What are some theories that relate to this individual's life?

In an example of a phenomenological study, Riemen (1986) poses this central question in her nursing-caring interaction study: "What is essential for the experience to be described by the client as being a caring interaction?" (p. 91). By adding a set of procedural questions related to the procedures in phenomenology, one emerges with subquestions. For example, following Moustakas's (1994) procedures, she might have asked the following procedural subquestions related to phenomenology:

- What statements describe these experiences?
- What themes emerge from these experiences?
- What are the contexts of and thoughts about the experiences?
- What is the overall essence of the experience?

To illustrate both issue and procedural subquestions in a study, Gritz (1995, p. 4) studied "teacher professionalism" as it was understood by practicing elementary classroom teachers in her phenomenology study. She posed the following central question and two sets of subquestions, one issue oriented and the other procedural.

## Central question

- What does it mean (to practitioners) to be a professional teacher?

## Issue subquestions

- What do professional teachers do?
- What don't professional teachers do?
- What does a person do who exemplifies the term "teacher professionalism"?
- What is difficult or easy about being a professional educator?
- How or when did you first become aware of being a professional?

## Procedural subquestions

- What are the structural meanings of teacher professionalism?
- What are the underlying themes and contexts that account for this view of teacher professionalism?
- What are the universal structures that precipitate feelings and thoughts about "teacher professionalism"?
- What are the invariant structural themes that facilitate a description of "teacher professionalism" as it is experienced by practicing elementary classroom teachers?

For a grounded theory study, the procedural subquestions might be posed as aspects of the coding steps, such as open coding, axial coding, selective coding, and the development of propositions:

- What are the general categories to emerge in a first review of the data? (open coding)
- What is the phenomenon of interest?
- What caused the phenomenon of interest? What contextual and intervening conditions influenced it? What strategies or outcomes resulted from it? What were the consequences of these strategies? (axial coding)

For example, in Mastera's (1995) dissertation proposal, she advances a study of the process of revising the general education curriculum in three private baccalaureate colleges. Her plan calls for both issue and procedural questions. The issue questions that guided her study were "What is the theory that explains the change process in the revision of general education curricula on three college campuses?" and "How does the chief academic officer participate in the process on each campus?" She then poses several procedural subquestions specifically related to open and axial coding:

- How did the process unfold?
- What were the major events or benchmarks in the process?
- What were the obstacles to change?
- Who were the important participants? How did they participate in the process?
- What were the outcomes?

In another study, Valerio (1995) uses procedural sub-questions directly rounded theory questions directly related to the steps in grounded theory data analysis:

> The overarching question for my grounded theory research study is: What theory explains why teenage girls become pregnant? The sub-questions follow the paradigm for developing a theoretical model. The questions seek to explore each of the interview coding steps and include: What are the general categories to emerge in open coding? What central phenomenon emerges? What are its causal conditions? What specific interaction issues and larger conditions have been influential? What are the resulting associated strategies and outcomes? (p. 3)

In an ethnography, one might present procedural subquestions that relate to (a) a description of the context, (b) an analysis of the major themes, and (c) the interpretation of cultural behavior (Wolcott, 1994b). Using Spradley's (1979, 1980) approach to ethnography, these procedural subquestions might reflect the 12 steps in his "decision research sequence." They might be as follows:

- What is the social situation to be studied?
- How does one go about observing this situation?
- What is recorded about this situation?
- What is observed about this situation?
- What cultural domains emerge from studying this situation?
- What more specific, focused observations can be made?
- What taxonomy emerges from these focused observations?
- Looking more selectively, what observations can be made?
- What components emerge from these observations?
- What themes emerge?
- What is the emerging cultural inventory?
- How does one write the ethnography?

In using good research question format for our gunman case study (Asmussen & Creswell, 1995), I would redraft the questions presented in the article. To foreshadow the case of a single campus and individuals on it,

I would pose the central question—"What was the campus response to the gunman incident at the Midwestern university?"—and then I would present the issue subquestions guiding my study (although we presented these questions more as central questions, as already noted):

1. What happened?

2. Who was involved in response to the incident?

3. What themes of response emerged during the 8-month period that followed this incident?

4. What theoretical constructs helped us understand the campus response?

5. What constructs were unique to this case? (p. 576)

Then, I would present the procedural subquestions:

1. How might the campus (case), and the events following the incident, be described? (description of the case)

2. What themes emerge from gathering information about the case? (analysis of the case materials)

3. How would I interpret these themes within larger social and psychological theories? (lessons learned from the case surrounded by the literature)

These illustrations show that, in a qualitative study, one can write subquestions that address issues on the topic being explored and use terms that encode the work within an approach. Also, procedural subquestions can be used that foreshadow the steps in the procedures of data collection, analysis, and narrative format construction.

## Summary

In this chapter, I addressed three topics related to introducing and focusing a qualitative study: the problem statement, the purpose statement, and the research questions. Although I discussed general features of designing each section in a qualitative study, I related the topics to the five approaches advanced in this book. The problem statement should indicate the source of the issue leading to the study, be framed in terms of existing literature, and be related to one of the approaches to research using words that convey the approach. The purpose statement also should include terms that encode the statement for a specific approach. Including comments about the site or people to be studied foreshadows the approach as well. The research questions continue this encoding within an approach for the central question, the overarching question being addressed in the study. Following the central question are subquestions, and I expand a model presented by Stake (1995) that groups subquestions into two sets: issue subquestions, which divide the central phenomenon into subtopics of study, and procedural subquestions, which convey the steps in the research within an approach. Procedural subquestions foreshadow how the researcher will be presenting and analyzing the information.

## Additional Readings

For writing problem statements in general, examine Marshall and Rossman (2006). For several basic principles in writing purpose statements, explore Creswell (2003) and the references mentioned in my chapter on writing purpose statements. For a good overview of writing research questions, I recommend Miles and Huberman (1994). Also, in standard qualitative texts, most authors address qualitative research questions (e.g., Hatch, 2002; Maxwell, 2005). I particularly like the conceptualization of issue and topical (procedural) questions by Stake (1995). Also, the reader should examine qualitative journal articles and reports to find good illustrations of problem statements, purpose statements, and research questions.

Creswell, J. W. (2003). *Research design: Qualitative, quantitative, and mixed methods approaches* (2nd ed.). Thousand Oaks, CA: Sage.

Marshall, C., & Rossman, G. B. (2006). *Designing qualitative research* (4th ed.). Thousand Oaks, CA: Sage.

Maxwell, J. (2005). *Qualitative research design: An interactive approach* (2nd ed.). Thousand Oaks, CA: Sage.

Miles, M. B., & Huberman, A. M. (1994). *Qualitative data analysis: A sourcebook of new methods* (2nd ed.). Thousand Oaks, CA: Sage.

Stake, R. (1995). *The art of case study research*. Thousand Oaks, CA: Sage.

## Exercises

1. Consider how you would write about the research problem or issue in your study. State the issue in a couple of sentences, then discuss the research literature that will provide evidence for a need for studying the problem. Finally, within the context of one of the five approaches to research, what rationale exists for studying the problem that reflects your approach to research?

Exercises

2. For the study you are designing, write a central question for your approach to research using the guidelines in this chapter for writing a good central question and using the words that encode the question within your approach to research.

3. In this chapter, I have presented a model for writing the subquestions in an issue and procedural format. Write five to seven issue-oriented subquestions and five to seven procedural subquestions in your approach to inquiry for your proposed study.

**Exercises**

# 7

# Data Collection

D
ata collection offers one more instance for assessing research design within each approach to inquiry. However, before exploring this idea, I find it useful to visualize the phases of data collection common to all approaches. A "circle" of interrelated activities best displays this process, a process of engaging in activities that include but go beyond collecting data.

I begin this chapter by presenting this circle of activities, briefly introducing each activity. These activities are locating a site or an individual, gaining access and making rapport, sampling purposefully, collecting data, recording information, exploring field issues, and storing data. Then I explore how these activities differ in the five approaches to inquiry, and I end with a few summary comments about comparing the data collection activities across the five approaches.

## Questions for Discussion

- What are the steps in the overall data collection process of qualitative research?
- What are typical access and rapport issues?
- How does one select people or places to study?
- What type of information typically is collected?
- How is information recorded?
- What are common issues in collecting data?
- How is information typically stored?
- How are the five approaches both similar and different during data collection?

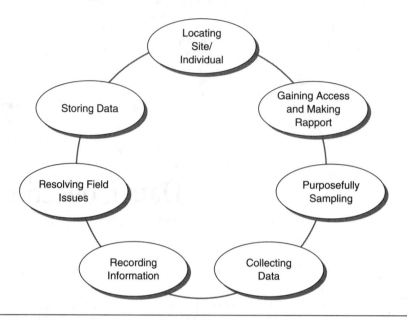

**Figure 7.1**    Data Collection Activities

## The Data Collection Circle

I visualize data collection as a series of interrelated activities aimed at gathering good information to answer emerging research questions. As shown in Figure 7.1, a qualitative researcher engages in a series of activities in the process of collecting data. Although I start with locating a site or an individual to study, an investigator may begin at another entry point in the circle. Most importantly, I want the researcher to consider the multiple phases in collecting data, phases that extend beyond the typical reference point of conducting interviews or making observations.

An important step in the process is to find people or places to study and to gain access to and establish rapport with participants so that they will provide good data. A closely interrelated step in the process involves determining a strategy for the purposeful sampling of individuals or sites. This is not a probability sample that will enable a researcher to determine statistical inferences to a population; rather, it is purposeful sample that will intentionally sample a group of people that can best inform the researcher about the research problem under examination. Thus, the researcher needs to determine which type of purposeful sampling will be best to use.

Once the inquirer selects the sites or people, decisions need to be made about the most appropriate data collection approaches. Increasingly, a qualitative researcher has more choices regarding data collection, such as e-mail messages and online data gathering, and typically the researcher will collect data from more than one source. To collect this information, the researcher develops protocols or written forms for recording the information and needs to develop some forms for recording the data, such as interview or observational protocols. Also, the researcher needs to anticipate issues of data collection, called "field issues," which may be a problem, such as having inadequate data, needing to prematurely leave the field or site, or contributing to lost information. Finally, a qualitative researcher must decide how he or she will store data so that they can easily be found and protected from damage or loss.

I now turn to each of these data collection activities, and I address each for general procedures and within each approach to inquiry. As shown in Table 7.1, these activities are both different and similar across the five approaches to inquiry.

## The Site or Individual

In a narrative study, one needs to find one or more individuals to study, individuals who are accessible, willing to provide information, and distinctive for their accomplishments and ordinariness or who shed light on a specific phenomenon or issue being explored. Plummer (1983) recommends two sources of individuals to study. The pragmatic approach is where individuals are met on a chance encounter, emerge from a wider study, or are volunteers. Alternatively, one might identify a "marginal person" who lives in conflicting cultures, a "great person" who impacts the age in which he or she lives, or an "ordinary person" who provides an example of a large population. An alternative perspective is available from Gergen (1994), who suggests that narratives "come into existence" (p. 280) not as a product of an individual, but as a facet of relationships, as a part of culture, as reflected in social roles such as gender and age. Thus, to ask which individuals will participate is not to focus on the right question. Instead, narrative researchers need to focus on the stories to emerge, recognizing that all people have stories to tell. Also instructive in considering the individual in narrative research is to consider whether first-order or second-order narratives are the focus of inquiry (Elliot, 2005). In first-order narratives, individuals tell stories about themselves and their own experiences, while in second-order narratives, researchers construct a narrative about other people's experiences (e.g., biography) or present a collective story that represents the lives of many.

In a phenomenological study, the participants may be located at a single site, although they need not be. Most importantly, they must be individuals who have all experienced the phenomenon being explored and can articulate

**Table 7.1** Data Collection Activities by Five Approaches

| Data Collection Activity | Narrative | Phenomenology | Grounded Theory | Ethnography | Case Study |
|---|---|---|---|---|---|
| What is traditionally studied? (sites or individuals) | Single individual, accessible and distinctive | Multiple individuals who have experienced the phenomenon | Multiple individuals who have responded to an action or participated in a process about a central phenomenon | Members of a culture-sharing group or individuals representative of the group | A bounded system. such as a process, an activity, an event, a program, or multiple individuals |
| What are typical access and rapport issues? (access and rapport) | Gaining permission from individuals, obtaining access to information in archives | Finding people who have experienced the phenomenon | Locating a homogeneous sample | Gaining access through the gatekeeper, gaining the confidence of informants | Gaining access through the gatekeeper, gaining the confidence of participants |
| How does one select a site or individuals to study? (purposeful sampling strategies) | Several strategies, depending on the person (e.g., convenient, politically important, typical, a critical case) | Finding individuals who have experienced the phenomenon, a "criterion" sample | Finding a homogeneous sample, a "theory-based" sample, a "theoretical" sample | Finding a cultural group to which one is a "stranger," a "representative" sample | Finding a "case" or "cases," an "atypical" case, or a "maximum variation" or "extreme" case |

| Data Collection Activity | Narrative | Phenomenology | Grounded Theory | Ethnography | Case Study |
|---|---|---|---|---|---|
| What type of information typically is collected? (forms of data) | Documents and archival material, open-ended interviews, subject journaling, participant observation, casual chatting | Interviews with 5 to 25 people (Polkinghorne, 1989) | Primarily interviews with 20 to 30 people to achieve detail in the theory | Participant observations, interviews, artifacts, and documents | Extensive forms, such as documents and records, interviews, observation, and physical artifacts |
| How is information recorded? (recording information) | Notes, interview protocol | Interviews, often multiple interviews with the same individuals | Interview protocol, memoing | Fieldnotes, interview and observational protocols | Fieldnotes, interview and observational protocols |
| What are common data collection issues? (field issues) | Access to materials, authenticity of account and materials | Bracketing one's experiences, logistics of interviewing | Interviewing issues (e.g., logistics, openness) | Field issues (e.g., reflexivity, reactivity, reciprocality, "going native," divulging private information, deception) | Interviewing and observing issues |
| How is information typically stored? (storing data) | File folders, computer files | Transcriptions, computer files | Transcriptions, computer files | Fieldnotes, transcriptions, computer files | Fieldnotes, transcriptions, computer files |

their lived experiences. The more diverse the characteristics of the individuals, the more difficult it will be for the researcher to find common experiences, themes, and the overall essence of the experience for all participants. In a grounded theory study, the individuals may not be located at a single site; in fact, if they are dispersed, they can provide important contextual information useful in developing categories in the axial coding phase of research. They need to be individuals who have participated in the process or action the researcher is studying in the grounded theory study. For example, in Creswell and Brown (1992), we interviewed 32 department chairpersons located across the United States who had mentored faculty in their departments. In an ethnographic study, a single site, in which an intact culture-sharing group has developed shared values, beliefs, and assumptions, is often important. The researcher needs to identify a group (or an individual or individuals representative of a group) to study, preferably one to which the inquirer is a "stranger" (Agar, 1986) and can gain access. For a case study, the researcher needs to select a site or sites to study, such as programs, events, processes, activities, individuals, or several individuals. Although Stake (1995) refers to an individual as an appropriate "case," I turn to the narrative biographical approach or the life history approach in studying a single individual. However, the study of multiple individuals, each defined as a case and considered a collective case study, is acceptable practice.

A question that students often ask is whether they can study their own organization, place of work, or themselves. Such a study may raise issues of power and risk to the researcher, the participants, and to the site. To study one's own workplace, for example, raises questions about whether good data can be collected when the act of data collection may introduce a power imbalance between the researcher and the individuals being studied. Although studying one's own "backyard" is often convenient and eliminates many obstacles to collecting data, researchers can jeopardize their jobs if they report unfavorable data or if participants disclose private information that might negatively influence the organization or workplace. A hallmark of all good qualitative research is the report of multiple perspectives that range over the entire spectrum of perspectives (see the section in Chapter 3 on the characteristics of qualitative research). I am not alone in sounding this cautionary note about studying one's own organization or workplace. Glesne and Peshkin (1992) question research that examines "your own *backyard*—within your own institution or agency, or among friends or colleagues" (p. 21), and they suggest that such information is "dangerous knowledge" that is political and risky for an "inside" investigator. When it becomes important to study one's own organization or workplace, I typically recommend that multiple strategies of validation (see Chapter 10) be used to ensure that the account is accurate and insightful.

Studying yourself can be a different matter. There is an approach that has gained prominence in qualitative research—autoethnography—an approach championed by Ellis (2004) and others. For example, Ellis's (1993) story of the experiences of her brother's sudden death illustrates the power of personal emotion and providing cultural perspectives around one's own experiences. I recommend that individuals wanting to study themselves and their own experiences turn to autoethnography or biographical memoir for scholarly procedures in how to conduct their studies.

## Access and Rapport

Gaining access to sites and individuals also involves several steps. Regardless of the approach to inquiry, permissions need to be sought from a human subjects review board, a process in which campus committees review research studies for their potential harmful impact on and risk to participants. This process involves submitting to the board a proposal that details the procedures in the project. Most qualitative studies are exempt from a lengthy review (e.g., the expedited or full review), but studies involving individuals as minors (i.e., 18 years or under) or studies of high-risk, sensitive populations (e.g., HIV-positive individuals) require a thorough review, a process involving detailed, lengthy applications and an extended time for review. Because many review boards are more familiar with the quantitative approaches to social and human science research than they are with qualitative approaches, the qualitative project description may need to conform to some of the standard procedures and language of quantitative research (e.g., research questions, results) as well as provide information about the protection of human subjects. To the review board, it might be argued, qualitative interviews, if unstructured, may actually provide participants considerable control over the interview process (Corbin & Morse, 2003). It is helpful to examine a sample consent form that participants need to review and sign in a qualitative study. An example is shown in Figure 7.2.

This consent form often requires that specific elements be included, such as:

- the right of participants to voluntarily withdraw from the study at any time
- the central purpose of the study and the procedures to be used in data collection
- comments about protecting the confidentiality of the respondents
- a statement about known risks associated with participation in the study
- the expected benefits to accrue to the participants in the study
- the signature of the participant as well as the researcher

For a narrative study, inquirers gain information from individuals by obtaining their permission to participate in the study. Study participants should be

---

"Experiences in Learning Qualitative Research: A Qualitative Case Study"

Dear Participant,

The following information is provided for you to decide whether you wish to participate in the present study. You should be aware that you are free to decide not to participate or to withdraw at any time without affecting your relationship with this department, the instructor, or the University of Nebraska-Lincoln.

The purpose of this study is to understand the process of learning qualitative research in a doctoral-level college course. The procedure will be a single, holistic case study design. At this stage in the research, process will be generally defined as perceptions of the course and making sense out of qualitative research at different phases in the course.

Data will be collected at three points—at the beginning of the course, at the midpoint, and at the end of the course. Data collection will involve documents (journal entries made by students and the instructor, student evaluations of the class and the research procedure), audio-visual material (a videotape of the class), interviews (transcripts of interviews between students), and classroom observation fieldnotes (made by students and the instructor). Individuals involved in the data collection will be the instructor and the students in the class.

Do not hesitate to ask any questions about the study either before participating or during the time that you are participating. We would be happy to share our findings with you after the research is completed. However, your name will not be associated with the research findings in any way, and your identity as a participant will be known only to the researchers.

There are no known risks and/or discomforts associated with this study. The expected benefits associated with your participation are the information about the experiences in learning qualitative research, the opportunity to participate in a qualitative research study, and co-authorship for those students who participate in the detailed analysis of the data. If submitted for publication, a byline will indicate the participation of all students in the class.

Please sign your consent with full knowledge of the nature and purpose of the procedures. A copy of this consent form will be given to you to keep.

_____                    _____
Signature of Participant                                              Date

John W. Creswell, Ed. Psy., UNL, Principal Investigator

---

**Figure 7.2**    Sample Human Subjects Consent-to-Participate Form

appraised of the motivation of the researcher for their selection, granted anonymity (if they desire it), and told by the researcher about the purpose of the study. This disclosure helps build rapport. Access to biographical documents and archives requires permission and perhaps travel to distant libraries.

In a phenomenological study in which the sample includes individuals who have experienced the phenomenon, it is also important to obtain participants' written permission to be studied. In the Anderson and Spencer (2002) study of the patients' images of AIDS, 58 men and women participated in the project at three sites dedicated to persons with HIV/AIDS: a hospital clinic, a long-term care facility, and a residence. These were all individuals with a diagnosis of AIDS, 18 years of age or older, able to communicate in English, and with a Mini-Mental Status score above 22. In such a study, it was important to obtain permission to have access to the vulnerable individuals participating in the study.

In a grounded theory study, the participants need to provide permission to be studied, while the researcher should have established rapport with the participants so that they will disclose detailed perspectives about responding to an action or process. The grounded theorist starts with a homogeneous sample, individuals who have commonly experienced the action or process. In an ethnography, access typically begins with a "gatekeeper," an individual who is a member of or has insider status with a cultural group. This gatekeeper is the initial contact for the researcher and leads the researcher to other participants (Hammersley & Atkinson, 1995). Approaching this gatekeeper and the cultural system slowly is wise advice for "strangers" studying the culture. For both ethnographies and case studies, gatekeepers require information about the studies that often includes answers from the researchers to the following questions, as Bogdan and Biklen (1992) suggest:

- Why was the site chosen for study?
- What will be done at the site during the research study? How much time will be spent at the site by the researchers?
- Will the researcher's presence be disruptive?
- How will the results be reported?
- What will the gatekeeper, the participants, and the site gain from the study? (reciprocity)

## Purposeful Sampling Strategy

The concept of purposeful sampling is used in qualitative research. This means that the inquirer selects individuals and sites for study because they can purposefully inform an understanding of the research problem and central phenomenon in the study. Decisions need to be made about who or what should be sampled, what form the sampling will take, and how many people or sites need to be sampled. Further, the researchers need to decide if the sampling will be consistent with the information needed by one of the five approaches to inquiry.

I will begin with some general remarks about sampling and then turn to sampling within each of the five approaches. The decision about who or what should be sampled can benefit from the conceptualization of Marshall and Rossman (2006), who provide an example of sampling four aspects: events, settings, actors, and artifacts. They also note that sampling can change during a study and researchers need to be flexible, but despite this, plan ahead as much as possible for their sampling strategy. I like to think as well in terms of levels of sampling in qualitative research. Researchers can sample at the site level, at the event or process level, and at the participant level. In a good plan for a qualitative study, one or more of these levels might be present and they each need to be identified.

On the question of what form the sampling will take, we need to note that there are several qualitative sampling strategies available (see Table 7.2 for a list of possibilities). These strategies have names and definitions, and they can be described in research reports. Also, researchers might use one or more of the strategies in a single study. Looking down the list, maximum variation is listed first because it is a popular approach in qualitative studies. This approach consists of determining in advance some criteria that differentiate the sites or participants, and then selecting sites or participants that are quite different on the criteria. This approach is often selected because when a researcher maximizes differences at the beginning of the study, it increases the likelihood that the findings will reflect differences or different perspectives—an ideal in qualitative research. Other sampling strategies frequently used are critical cases, which provide specific information about a problem, and convenience cases, which represent sites or individuals from which the researcher can access and easily collect data.

The size question is an equally important decision to sampling strategy in the data collection process. One general guideline in qualitative research is not only to study a few sites or individuals but also to collect extensive detail about each site or individual studied. The intent in qualitative research is not to generalize the information (except in some forms of case study research), but to elucidate the particular, the specific (Pinnegar & Daynes, 2006). Beyond these general suggestions, each of the five approaches to research raises specific size considerations.

In narrative research, I have found many examples with one or two individuals, unless a larger pool of participants is used to develop a collective story (Huber & Whelan, 1999). In phenomenology, I have seen the number of participants range from 1 (Dukes, 1984) up to 325 (Polkinghorne, 1989). Dukes (1984) recommends studying 3 to 10 subjects, and in one phenomenology, Riemen (1986) studied 10 individuals. In grounded theory, I recommend including 20 to 30 individuals in order to develop a well-saturated

**Table 7.2**    Typology of Sampling Strategies in Qualitative Inquiry

| Type of Sampling | Purpose |
|---|---|
| Maximum variation | Documents diverse variations and identifies important common patterns |
| Homogeneous | Focuses, reduces, simplifies, and facilitates group interviewing |
| Critical case | Permits logical generalization and maximum application of information to other cases |
| Theory based | Find examples of a theoretical construct and thereby elaborate on and examine it |
| Confirming and disconfirming cases | Elaborate on initial analysis, seek exceptions, looking for variation |
| Snowball or chain | Identifies cases of interest from people who know people who know what cases are information-rich |
| Extreme or deviant case | Learn from highly unusual manifestations of the phenomenon of interest |
| Typical case | Highlights what is normal or average |
| Intensity | Information-rich cases that manifest the phenomenon intensely but not extremely |
| Politically important | Attracts desired attention or avoids attracting undesired attention |
| Random purposeful | Adds credibility to sample when potential purposeful sample is too large |
| Stratified purposeful | Illustrates subgroups and facilitates comparisons |
| Criterion | All cases that meet some criterion; useful for quality assurance |
| Opportunistic | Follow new leads; taking advantage of the unexpected |
| Combination or mixed | Triangulation, flexibility; meets multiple interests and needs |
| Convenience | Saves time, money, and effort, but at the expense of information and credibility |

SOURCE: Miles & Huberman (1994, p. 28). Reprinted with permission from Miles, M. B., & Huberman, A. M. (1994). *Qualitative data analysis: A sourcebook of new methods* (2nd ed.). Thousand Oaks, CA: Sage.

theory, but this number may be much larger (Charmaz, 2006). In ethnography, I like well-defined studies of single culture-sharing groups, with numerous artifacts, interviews, and observations collected until the workings of the cultural-group are clear. For case study research, I would not include more than 4 or 5 case studies in a single study. This number should provide ample opportunity to identify themes of the cases as well as conduct cross-case theme analysis.

In a narrative study, the researcher reflects more on who to sample—the individual may be convenient to study because she or he is available, a politically important individual who attracts attention or is marginalized, or a typical, ordinary person. All of the individuals need to have stories to tell about their lived experiences. Inquirers may select several options, depending on whether the person is marginal, great, or ordinary (Plummer, 1983). Vonnie Lee, who consented to participate and provided insightful information about individuals with mental retardation (Angrosino, 1994), was convenient to study but also was a critical case to illustrate the types of challenges surrounding the issues of mental retardation in our society.

I have found, however, a much more narrow range of sampling strategies for a phenomenological study. It is essential that all participants have experience of the phenomenon being studied. Criterion sampling works well when all individuals studied represent people who have experienced the phenomenon. In a grounded theory study, the researcher chooses participants who can contribute to the development of the theory. Strauss and Corbin (1998) refer to theoretical sampling, which is a process of sampling individuals that can contribute to building the opening and axial coding of the theory. This begins with selecting and studying a homogeneous sample of individuals (e.g., all women who have experienced childhood abuse) and then, after initially developing the theory, selecting and studying a heterogeneous sample (e.g., types of support groups other than women who have experienced childhood abuse). The rationale for studying this heterogeneous sample is to confirm or disconfirm the conditions, both contextual and intervening, under which the model holds.

In ethnography, once the investigator selects a site with a cultural group, the next decision is who and what will be studied. Thus, within-culture sampling proceeds, and several authors offer suggestions for this procedure. Fetterman (1998) recommends proceeding with the "big net approach" (p. 32), where at first the researcher mingles with everyone. Ethnographers rely on their judgment to select members of the subculture or unit based on their research questions. They take advantage of opportunities (i.e., opportunistic sampling; Miles & Huberman, 1994) or establish criteria for studying select individuals (criterion sampling). The criteria for selecting who and what to study, according to Hammersley and Atkinson (1995), are based on gaining some perspective on chronological time in the social life of the group, people representative of the culture-sharing group in terms of demographics, and the contexts that lead to different forms of behavior.

In a case study, I prefer to select unusual cases in collective case studies and employ maximum variation as a sampling strategy to represent diverse cases and to fully describe multiple perspectives about the cases. Extreme and deviant cases may comprise my collective case study, such as the study of the unusual gunman incident on the university campus (Asmussen & Creswell, 1995).

## Forms of Data

New forms of qualitative data continually emerge in the literature (see Creswell, 2003), but all forms might be grouped into four basic types of information: observations (ranging from nonparticipant to participant), interviews (ranging from close-ended to open-ended), documents (ranging from private to public), and audiovisual materials (including materials such as photographs, compact disks, and videotapes). Over the years, I have kept an evolving list of data types, as shown in Figure 7.3.

I organize my list into the four basic types, although some forms may not be easily placed into one category or the other. In recent years, new forms of data have emerged, such as journaling in narrative story writing, using text from e-mail messages, and observing through examining videotapes and photographs. Stewart and Williams (2005) discuss using online focus groups for social research. They reviewed both synchronous (real-time) and asynchronous (non-real-time) applications highlighting new developments such as virtual reality applications as well as advantages (participants can be questioned over long periods of time, larger numbers can be managed, and more heated and open exchanges occur). Problems arise with online focus groups, such as obtaining complete informed consent, recruiting individuals to participate, and choosing times to convene given different international time zones.

Despite problems in innovative data collection such as these, I encourage individuals designing qualitative projects to include new and creative data collection methods that will encourage readers and editors to examine their studies. Researchers need to consider visual ethnography (Pink, 2001), or the possibilities of narrative research to include living stories, metaphorical visual narratives, and digital archives (see Clandinin, 2006). I like the technique of "photo elicitation" in which participants are shown pictures (their own or those taken by the researcher) and asked by the researcher to discuss

**Observations**

- Gather fieldnotes by conducting an observation as a participant.
- Gather fieldnotes by conducting an observation as an observer.
- Gather fieldnotes by spending more time as a participant than as an observer.
- Gather fieldnotes by spending more time as an observer than as a participant.
- Gather fieldnotes first by observing as an "outsider" and then by moving into the setting and observing as an "insider."

**Interviews**

- Conduct an unstructured, open-ended interview and take interview notes.
- Conduct an unstructured, open-ended interview, audiotape the interview, and transcribe the interview.
- Conduct a semistructured interview, audiotape the interview, and transcribe the interview.
- Conduct a focus group interview, audiotape the interview, and transcribe the interview.
- Conduct different types of interviews: e-mail, face-to-face, focus group, online focus group, telephone interviews.

**Documents**

- Keep a journal during the research study.
- Have a participant keep a journal or diary during the research study.
- Collect personal letters from participants.
- Analyze public documents (e.g., official memos, minutes, records, archival material).
- Examine autobiographies and biographies.
- Have informants take photographs or videotapes (i.e., photo elicitation).
- Conduct chart audits.
- Review medical records.

**Audiovisual materials**

- Examine physical trace evidence (e.g., footprints in the snow).
- Videotape or film a social situation or an individual or group.
- Examine photographs or videotapes.
- Collect sounds (e.g., musical sounds, a child's laughter, car horns honking).
- Collect e-mail or electronic messages.
- Gather phone text messages.
- Examine possessions or ritual objects.

---

**Figure 7.3**     A Compendium of Data Collection Approaches in Qualitative Research

the contents of the pictures (Denzin & Lincoln, 1994). Ziller (1990), for example, handed one loaded Polaroid camera each to 40 male and 40 female 4th graders in Florida and West Germany and asked them to take pictures of images that represented war and peace.

The particular approach to research often directs a qualitative researchers' attention toward preferred approaches to data collection, although these preferred approaches cannot be seen as rigid guidelines. For a narrative study, Czarniawska (2004) mentioned three ways to collect data for stories: recording spontaneous incidents of storytelling, eliciting stories through interviews, and asking for stories through such mediums as the Internet. Clandinin and Connelly (2000) suggest collecting field texts through a wide array of sources, autobiography, journal, researcher fieldnotes, letters, conversations, interviews, stories of families, documents, photographs, and personal-family-social artifacts. For a phenomenological study, the process of collecting information involves primarily in-depth interviews (see, e.g., the discussion about the long interview in McCracken, 1988) with as many as 10 individuals. The important point is to describe the meaning of the phenomenon for a small number of individuals who have experienced it. Often multiple interviews are conducted with the each of the research participants. Besides interviewing and self-reflection, Polkinghorne (1989) advocates gathering information from depictions of the experience outside the context of the research projects, such as descriptions drawn from novelists, poets, painters, and choreographers. I recommend Lauterbach (1993), the study of wished-for babies from mothers, as an especially rich example of phenomenological research using diverse forms of data collection.

Interviews play a central role in the data collection in a grounded theory study. In the study Brown and I conducted with academic chairpersons (Creswell & Brown, 1992), each of our interviews with 33 individuals lasted approximately an hour. Other data forms besides interviewing, such as participant observation, researcher reflection or journaling (memoing), participant journaling, and focus groups, may be used to help develop the theory (see how Morrow and Smith, 1995, use these forms in their study of women's childhood abuse). However, in my experience, these multiple data forms often play a secondary role to interviewing in grounded theory studies.

In an ethnographic study, the investigator collects descriptions of behavior through observations, interviewing, documents, and artifacts (Hammersley & Atkinson, 1995; Spradley, 1980), although observing and interviewing appear to be the most popular forms of ethnographic data collection. Ethnography has the distinction among the five approaches, I believe, of advocating the use of quantitative surveys and tests and measures as part of data collection. For example, examine the wide array of forms of data in ethnography as advanced by LeCompte and Schensul (1999). They reviewed ethnographic data collection techniques of observation, tests and repeated measures, sample surveys, interviews, content analysis of secondary or visual data, elicitation methods, audiovisual information, spatial mapping, and network research. Participant observation, for example, offers possibilities for the researcher on a continuum from being a complete outsider to being a complete insider (Jorgensen, 1989). The approach of changing one's role from that of an outsider to that of an insider through the course of the ethnographic study is well documented in field research (Jorgensen, 1989). Wolcott's (1994b) study of the Principal Selection Committee illustrates an outsider perspective, as he observed and recorded events in the process of selecting a principal for a school without becoming an active participant in the committee's conversations and activities.

Like ethnography, case study data collection involves a wide array of procedures as the researcher builds an in-depth picture of the case. I am reminded of the multiple forms of data collection recommended by Yin (2003) in his book about case studies. He refers to six forms: documents, archival records, interviews, direct observation, participant observation, and physical artifacts. Because of the extensive data collection in the gunman case study, Asmussen and I present a matrix of information sources for the reader (Asmussen & Creswell, 1995). This matrix contains four types of data (interviews, observations, documents, and audiovisual materials) in the columns and specific forms of information (e.g., students at large, central administration) in the rows. Our intent was to convey through this matrix the depth and multiple forms of data collection, thus inferring the complexity of our case. The use of a matrix, which is especially applicable in an information-rich case study, might serve the inquirer equally well in all approaches of inquiry.

Of all the data collection sources in Figure 7.3, interviewing and observing deserve special attention because they are frequently used in all five of the approaches to research. Entire books are available on these two topics (e.g., Kvale, 1996, on interviewing; Spradley, 1980, on observing), thus I highlight only basic procedures that I recommend to prospective interviewers and observers.

## Interviewing

One might view interviewing as a series of steps in a procedure:

- Identify interviewees based on one of the purposeful sampling procedures mentioned in the preceding discussion (see Miles & Huberman, 1994).

- Determine what type of interview is practical and will net the most useful information to answer research questions. Assess the types available, such as a telephone interview, a focus group interview, or a one-on-one interview. A telephone interview provides the best source of information when the researcher does not have direct access to individuals. The drawbacks of this approach are that the researcher cannot see the informal communication and the phone expenses. Focus groups are advantageous when the interaction among interviewees will likely yield the best information, when interviewees are similar and cooperative with each other, when time to collect information is limited, and when individuals interviewed one-on-one may be hesitant to provide information (Krueger, 1994; Morgan, 1988; Stewart & Shamdasani, 1990). With this approach, however, care must be taken to encourage all participants to talk and to monitor individuals who may dominate the conversation. For one-on-one interviewing, the researcher needs individuals who are not hesitant to speak and share ideas, and needs to determine a setting in which this is possible. The less articulate, shy interviewee may present the researcher with a challenge and less than adequate data.

- Use adequate recording procedures when conducting one-on-one or focus group interviews. I recommend equipment such as a lapel mike for both the interviewer and interviewee or an adequate mike sensitive to the acoustics of the room.

- Design and use an interview protocol, a form about four or five pages in length, with approximately five open-ended questions and ample space between the questions to write responses to the interviewee's comments (see the sample protocol in Figure 7.4 below). How are questions developed? The questions are a narrowing of the central question and subquestions in the research study. These might be seen as the core of the interview protocol, bounded on the front end by questions to invite the interviewee to open up and talk and located at the end by questions about "Who should I talk to in order to learn more?" or comments thanking the participants for their time for the interview.

- Refine the interview questions and the procedures further through pilot testing. Sampson (2004), in an ethnographic study of boat pilots aboard cargo vessels, recommends the use of a pilot test to refine and develop research instruments, assess the degrees of observer bias, frame questions, collect background information, and adapt research procedures. During her pilot testing, Sampson participated at the site, kept detailed fieldnotes, and conducted detailed tape-recorded, confidential interviews. In case study research, Yin (2003) also recommends a pilot test to refine data collection plans and develop relevant lines of questions. These pilot cases are selected on the basis of convenience, access, and geographic proximity.

- Determine the place for conducting the interview. Find, if possible, a quiet location free from distractions. Ascertain if the physical setting lends itself to audiotaping, a necessity, I believe, in accurately recording information.

- After arriving at the interview site, obtain consent from the interviewee to participate in the study. Have the interviewee complete a consent form for the human relations review board. Go over the purpose of the study, the amount of time that will be needed to complete the interview, and plans for using the results from the interview (offer a copy of the report or an abstract of it to the interviewee).

- During the interview, stay to the questions, complete the interview within the time specified (if possible), be respectful and courteous, and offer few questions and advice. This last point may be the most important, and it is a reminder of how a good interviewer is a good listener rather than a frequent speaker during an interview. Also, record information on the interview protocol in the event that the audio-recording does not work. Recognize that quickly inscribed notes may be incomplete and partial because of the difficulty of asking questions and writing answers at the same time.

## Observing

Observing in a setting is a special skill that requires addressing issues such as the potential *deception* of the people being interviewed, impression management, and the potential marginality of the researcher in a strange setting (Hammersley & Atkinson, 1995). Like interviewing, I also see observing as a series of steps:

- Select a site to be observed. Obtain the required permissions needed to gain access to the site.
- At the site, identify who or what to observe, when, and for how long. A gatekeeper helps in this process.
- Determine, initially, a role to be assumed as an observer. This role can range from that of a complete participant (going native) to that of a complete observer. I especially like the procedure of being an outsider initially, followed by becoming an insider over time.
- Design an observational protocol as a method for recording notes in the field. Include in this protocol both descriptive and reflective notes (i.e., notes about your experiences, hunches, and learnings).
- Record aspects such as portraits of the informant, the physical setting, particular events and activities, and your own reactions (Bogdan & Biklen, 1992).
- During the observation, have someone introduce you if you are an outsider, be passive and friendly, and start with limited objectives in the first few sessions

of observation. The early observational sessions may be times in which to take few notes and simply observe.

- After observing, slowly withdraw from the site, thanking the participants and informing them of the use of the data and their accessibility to the study.

# Recording Procedures

In discussing observation and interviewing procedures, I mention the use of a protocol, a predesigned form used to record information collected during an observation or interview. The interview protocol enables a person to take notes during the interview about the responses of the interviewee. It also helps a researcher organize thoughts on items such as headings, information about starting the interview, concluding ideas, information on ending the interview, and thanking the respondent. In Figure 7.4, I provide the interview protocol used in the gunman case study (Asmussen & Creswell, 1995).

Besides the five open-ended questions in the study, this form contains several features I recommend. The instructions for using the interview protocol are as follows:

- Use a header to record essential information about the project and as a reminder to go over the purpose of the study with the interviewee. This heading might also include information about confidentiality and address aspects included in the consent form.
- Place space between the questions in the protocol form. Recognize that an individual may not always respond directly to the questions being asked. For example, a researcher may ask Question 2, but the interviewee's response may be to Question 4. Be prepared to write notes on all of the questions as the interviewee speaks.
- Memorize the questions and their order to minimize losing eye contact with the participant. Provide appropriate verbal transitions from one question to the next.
- Write out the closing comments that thank the individual for the interview and request follow-up information, if needed, from them.

During an observation, use an observational protocol to record information. As shown in Figure 7.5, this protocol contains notes taken by one of my students on a class visit by Harry Wolcott. I provide only one page of the protocol, but this is sufficient for one to see what it includes. It has a header giving information about the observational session, and then includes a "descriptive notes" section for recording a description of activities. The section with a box around it in the "descriptive notes" column indicates the observer's

Interview Protocol Project: University Reaction to a Terrorist Incident

Time of interview:

Date:

Place:

Interviewer:

Interviewee:

Position of interviewee:

(Briefly describe the project)

**Questions:**

1.  What has been your role in the incident?

2.  What has happened since the event that you have been involved in?

3.  What has been the impact on the university community of this incident?

4.  What larger ramifications, if any, exist from the incident?

5.  To whom should we talk to find out more about campus reaction to the incident?

(Thank the individual for participating in this interview. Assure him or her of confidentiality of responses and potential future interviews.)

**Figure 7.4**    Sample Interview Protocol

| Length of Activity: 90 Minutes | |
|---|---|
| Descriptive Notes | Reflective Notes |
| General: What are the experiences of graduate students as they learn qualitative research in the classroom? | |
| See classroom layout and comments about physical setting at the bottom of this page. | *Overhead with flaps: I wonder if the back of the room was able to read it.* |
| Approximately 5:17 p.m., Dr. Creswell enters the filled room, introduces Dr. Wolcott. Class members seem relieved. | *Overhead projector not plugged in at the beginning of the class: I wonder if this was a distraction (when it took extra time to plug it in).* |
| Dr. Creswell gives brief background of guest, concentrating on his international experiences; features a comment about the educational ethnography "The Man in the Principal's Office." | *Lateness of the arrival of Drs. Creswell and Wolcott: Students seemed a bit anxious. Maybe it had to do with the change in starting time to 5 p.m. (some may have had 6:30 classes or appointments to get to).* |
| Dr. Wolcott begins by telling the class he now writes out educational ethnography and highlights this primary occupation by mentioning two books: *Transferring Qualitative Data* and *The Art of Fieldwork*. | *Drs. Creswell and Wolcott seem to have a good rapport between them, judging from many short exchanges that they had.* |
| While Dr. Wolcott begins his presentation by apologizing for his weary voice (due to talking all day, apparently), Dr. Creswell leaves the classroom to retrieve the guest's overhead transparencies. | |
| Seemed to be three parts to this activity: (1) the speaker's challenge to the class of detecting pure ethnographical methodologies, (2) the speaker's presentation of the "tree" that portrays various strategies and substrategies for qualitative research in education, and (3) the relaxed "elder statesman" fielding class questions, primarily about students' potential research projects and prior studies Dr. Wolcott had written. |  SKETCH OF CLASSROOM |
| The first question was "How do you look at qualitative research?" followed by "How does ethnography fit in?" | |

**Figure 7.5**   Sample Observational Protocol Length of Activity: 90 Minutes

attempt to summarize, in chronological fashion, the flow of activities in the classroom. This can be useful information for developing a chronology of the ways the activities unfolded during the class session. There is also a "reflective notes" a section for notes about the process, reflections on activities, and summary conclusions about activities for later theme development. A line down the center of the page divides descriptive notes from reflective notes. A visual sketch of the setting and a label for it provide additional useful information.

Whether a researcher uses an observational or interview protocol, the essential process is recording information or, as Lofland and Lofland (1995) state it, "logging data" (p. 66). This process involves recording information through various forms, such as observational fieldnotes, interview write-ups, mapping, census taking, photographing, sound recording, and documents. An informal process may occur in recording information composed of initial "jottings" (Emerson, Fretz, & Shaw, 1995), daily logs or summaries, and descriptive summaries (see Sanjek, 1990, for examples of fieldnotes). These forms of recording information are popular in narrative research, ethnographies, and case studies.

# Field Issues

Researchers engaged in studies within all five approaches face issues in the field when gathering data that need to be anticipated. During the last several years, the number of books and articles on field issues has expanded considerably as interpretive issues (see Chapter 2) have been widely discussed. Beginning researchers are often overwhelmed by the amount of time needed to collect qualitative data and the richness of the data encountered. As a practical recommendation, I suggest that beginners start with limited data collection and engage in a pilot project to gain some initial experiences (Sampson, 2004). This limited data collection might consist of one or two interviews or observations, so that researchers can estimate the time needed to collect data.

One way to think about and anticipate the types of issues that may arise during data collection is to view the issues as they relate to several aspects of data collection, such as entry and access, the types of information collected, and potential ethical issues.

## Access to the Organization

Gaining access to organizations, sites, and individuals to study has its own challenges. Convincing individuals to participate in the study, building trust and credibility at the field site, and getting people from a site to respond are all important access challenges. Factors related to considering the appropriateness of a site need to be considered as well (see Weis & Fine, 2000). For example, researchers may choose a site that is one in which they have a vested interest (e.g., employed at the site, studying superiors or subordinates at the site) that would limit ability to develop diverse perspectives on coding data or developing themes. A researcher's own particular "stance" within the group may keep him or her from acknowledging all dimensions of the experiences. The researchers may hear or see something uncomfortable when they collect data. In addition, participants' may be fearful that their issues will be exposed to people outside their community, and this may make them unwilling to accept the researcher's interpretation of the situation.

Also related to access is the issue of working with an institutional review board that may not be familiar with unstructured interviews in qualitative research and the risks associated with these interviews (Corbin & Morse, 2003). Weis and Fine (2000) raise the important question of whether the response of the institutional review board to a project influences the researcher's telling of the narrative story.

## Observations

The types of challenges experienced during observations will closely relate to the role of the inquirer in observation, such as whether the researcher assumes a participant, nonparticipant, or middle-ground position. There are challenges as well with the mechanics of observing, such as remembering to take fieldnotes, recording quotes accurately for inclusion in fieldnotes, determining the best timing for moving from a nonparticipant to a participant (if this role change is desired), and keeping from being overwhelmed at the site with information, and learning how to funnel the observations from the broad picture to a narrower one in time. Participant observation has attracted several commentaries by writers (Labaree, 2002; Ezeh, 2003). Labaree (2002), who was a participant in an academic senate on a campus, notes the advantages of this role but also discusses the dilemmas of entering the field, disclosing oneself to the participants, sharing relationships with other individuals, and attempting to disengage from the site. Ezeh (2003), a Nigerian, studied the Orring, a little-known minority ethnic group in Nigeria. Although his initial contact with the group was supportive, the more the researcher became integrated into the host community, the more he experienced human relations problems, such as being accused of spying, pressured to be more generous in his material gifts, and suspected of trysts with women. Ezeh concluded that being of the same nationality was no guarantee of a lack of challenges at the site.

## Interviews

Challenges in qualitative interviewing often focus on the mechanics of conducting the interview. Roulston, deMarrais, and Lewis (2003) chronicle the challenges in interviewing by postgraduate students during a 15-day intensive course. These challenges related to unexpected participant behaviors and students' ability to create good instructions, phrase and negotiate questions, deal with sensitive issues, and do transcriptions. Suoninen and Jokinen (2005), from the field of social work, ask whether the phrasing of our interview questions leads to subtle persuasive questions, responses, or explanations.

Undoubtedly, conducting interviews is taxing, especially for inexperienced researchers engaged in studies that require extensive interviewing, such as phenomenology, grounded theory, and case study research. Equipment issues loom large as a problem in interviewing, and both recording equipment and transcribing equipment need to be organized in advance of the interview. The process of questioning during an interview (e.g., saying "little," handling "emotional outbursts," using "ice-breakers") includes problems that an interviewer must address. Many inexperienced researchers express surprise at the difficulty of conducting interviews and the lengthy process involved in transcribing audiotapes from the interviews. In addition, in phenomenological interviews, asking appropriate questions and relying on participants to discuss the meaning of their experiences require patience and skill on the part of the researcher.

Recent discussions about qualitative interviewing highlight the importance of reflecting about the relationship that exists between the interviewer and interviewee (Kvale, 2006; Nunkoosing, 2005; Weis & Fine, 2000). Kvale (2006), for example, questions the warm, caring, and empowering dialogues in interviews, and states that the interview is actually a hierarchical relationship with an asymmetrical power distribution between the interviewer and interviewee. Kvale discusses the interview as being "ruled" by the interviewer, enacting a one-way dialogue, serving the interviewer, containing hidden agendas, leading to the interviewer's monopoly over interpretation, enacting "counter control" by the interviewee who does not answer or deflects questions, and leading to a false security when the researcher checks the account (i.e., member checking, as discussed in Chapter 10 of this book) with the participants. Nunkoosing (2005) extends the discussion by reflecting on the problems of power and resistance, distinguishing truth from authenticity, the impossibility of consent, and projection of the interviewers' own self (their status, race, culture, and gender). Weiss and Fine (2000) raise additional questions for consideration: Are your interviewees able to articulate the forces that interrupt or suppress or oppress them? Do they erase their history, approaches, and cultural identity? Do they choose not to expose their history or go on record about the difficult aspects of their lives? These questions and the points raised about the nature of the interviewer-interviewee relationship cannot be easily answered with pragmatic decisions that encompass all interview situations. They do, however, sensitize us to important challenges in qualitative interviewing that need to be anticipated.

## Documents and Audiovisual Materials

In document research, the issues involve locating materials, often at sites far away, and obtaining permission to use the materials. For biographers, the primary form of data collection might be archival research from documents. When researchers ask participants in a study to keep journals, additional field issues emerge. Journaling is a popular data collection process in case studies and narrative research. What instructions should be given to individuals prior to writing in their journals? Are all participants equally comfortable with journaling? Is it appropriate, for example, with small children who express themselves well verbally but have limited writing skills? The researcher also may have difficulty reading the handwriting of participants who journal. Recording on videotape raises issues for the qualitative researcher such as keeping disturbing room sounds to a minimum, deciding on the best location for the camera, and determining whether to provide close-up shots or distant shots.

## Ethical Issues

Regardless of the approach to qualitative inquiry, a qualitative researcher faces many ethical issues that surface during data collection in the field and in analysis and dissemination of qualitative reports. Lipson (1994) groups ethical issues into informed consent procedures; deception or covert activities; confidentiality toward participants, sponsors, and colleagues; benefits of research to participants over risks; and participant requests that go beyond social norms. The criteria of the American Anthropological Association (see Glesne & Peshkin, 1992) reflect appropriate standards. A researcher protects the anonymity of the informants, for example, by assigning numbers or aliases to individuals. A researcher develops case studies of individuals that represent a composite picture rather than an individual picture. Furthermore, to gain support from participants, a qualitative researcher conveys to participants that they are participating in a study, explains the purpose of the study, and does not engage in deception about the nature of the study. What if the study is on a sensitive topic and the participants decline to be involved if they are aware of the topic? This issue of disclosure of the researcher, widely discussed in cultural anthropology (e.g., Hammersley & Atkinson, 1995), is handled by the researcher by presenting *general* information, not specific information about the study. Another issue likely to develop is participants sharing information "off the record." Although in most instances this information is deleted from analysis by the researcher, the issue becomes problematic when the information, if reported, harms individuals. I am reminded of a researcher who studied incarcerated Native Americans and learned about a potential "breakout" during one of the interviews. This researcher concluded that it would be a breach of faith with the participants if she reported the matter, and she kept quiet. Fortunately, the breakout did not occur. A final ethical issue is whether the researcher shares personal experiences with participants in an interview setting such as in a case study, phenomenology, or ethnography. This sharing minimizes the "bracketing" that is essential to construct the meaning of participants in phenomenology and reduces information shared by participants in case studies and ethnographies.

## Storing Data

I am surprised at how little attention is given in books and articles to storing qualitative data. The approach to storage will reflect the type of information collected, which varies by approach to inquiry. In writing a narrative life history, the researcher needs to develop a filing system for the "wad of hand-written notes or a tape" (Plummer, 1983, p. 98). Davidson's (1996) suggestions about backing up information collected and noting changes made to the database is sound advice for all types of research studies. With extensive use of computers in qualitative research, more attention will likely be given to how qualitative data are organized and stored, whether the data are fieldnotes, transcripts, or rough jottings. With extremely large databases being used by some qualitative researchers, this aspect assumes major importance.

Some principles about data storage and handling that are especially well suited for qualitative research include the following:

- Always develop backup copies of computer files (Davidson, 1996).
- Use high-quality tapes for audio-recording information during interviews. Also, make sure that the size of the tapes fits the transcriber's machine.
- Develop a master list of types of information gathered.

- Protect the anonymity of participants by masking their names in the data.
- Develop a data collection matrix as a visual means of locating and identifying information for a study.

## Five Approaches Compared

Returning again to Table 7.1, there are both differences and similarities among the activities of data collection for the five approaches to inquiry. Turning to differences, certain approaches seem more directed toward specific types of data collection than others. For case studies and narrative studies, the researcher uses multiple forms of data to build the in-depth case or the storied experiences. For grounded theory studies and phenomenological projects, inquirers rely primarily on interviews as data. Ethnographers highlight the importance of participant observation and interviews, but, as noted earlier, they may use many different sources of information. Unquestionably, some mixing of forms occurs, but in general these patterns of collection by approach hold true.

Second, the unit of analysis for data collection varies. Narrative researchers, phenomenologists, and ground theorists study individuals; case study researchers examine groups of individuals participating in an event or activity or an organization; and ethnographers study entire cultural systems or some subcultures of the systems.

Third, I found the amount of discussion about field issues to vary among the five approaches. Ethnographers have written extensively about field issues (e.g., Hammersley & Atkinson, 1995)—even more so, it seems, than those in other approaches to qualitative research. This may reflect historical concerns about imbalanced power relationships, imposing objective, external standards on participants, and failures to be sensitive to marginalized groups. Narrative researchers are less specific about field issues, although their concerns are mounting about how to conduct the interview (Elliot, 2005). Across all approaches, ethical issues are widely discussed.

Fourth, the approaches vary in their intrusiveness of data collection. Conducting interviews seems less intrusive in phenomenological projects and grounded theory studies than in the high level of access needed in personal narratives, the prolonged stays in the field in ethnographies, and the immersion into programs or events in case studies.

These differences do not lessen some important similarities that need to be observed. All qualitative studies sponsored by public institutions need to be approved by a human subjects review board. Also, the use of interviews and observations is central to many of the approaches. Furthermore, the recording devices, such as observational and interview protocols, can be similar regardless of approach (although specific questions on each protocol will reflect the language of the approach). Finally, the issue of data storage of information is closely related to the form of data collection, and the basic objective of researchers, regardless of approach, is to develop some filing and storing system for organized retrieval of information.

## Summary

In this chapter, I addressed several components of the data collection process. The researcher attends to locating a site or person to study; gaining access to and building rapport at the site or with the individual; sampling purposefully using one or more of the many approaches to sampling in qualitative research; collecting information through many forms, such as interviews, observations, documents, and audiovisual materials and newer forms emerging in the literature; establishing approaches for recording information such as the use of interview or observational protocols; anticipating and addressing field issues ranging from access to ethical concerns; and developing a system for storing and handling the databases. The five approaches to inquiry differ in the diversity of information collected, the unit of study being examined, the extent of field issues discussed in the literature, and the intrusiveness of the data collection effort. Researchers, regardless of approach, need approval from review boards, engage in similar data collection of interviews and observations, and use similar recording protocols and forms for storing data.

### Additional Readings

For a discussion about purposeful sampling strategies, I recommend Miles and Huberman (1994) and Creswell (2005).

Miles, M. B., & Huberman, A. M. (1994). *Qualitative data analysis: A sourcebook of new methods* (2nd ed.). Thousand Oaks, CA: Sage.
Creswell, J. W. (2005). *Educational research: Planning, conducting, and evaluating quantitative and qualitative research* (2nd ed.). Upper Saddle River, NJ: Pearson Education.

For interviewing, I direct researchers to Gubrium and Holstein (2003), Kvale (1996), McCracken (1988), and Rubin and Rubin (1995).

Gubrium, J. F., & Holstein, J. A. (2003). *Postmodern interviewing*. Thousand Oaks, CA: Sage.
Kvale, S. (1996). *InterViews: An introduction to qualitative research interviewing*. Thousand Oaks, CA: Sage.

McCracken, G. (1988). *The long interview*. Newbury Park, CA: Sage.

Rubin, H. J., & Rubin, I. S. (1995). *Qualitative interviewing*. Thousand Oaks, CA: Sage.

For discussions about making observations and taking fieldnotes, I suggest several writers: Bernard (1994), Bogdewic (1992), Emerson, Fritz, and Shaw (1995), Hammersley and Atkinson (1995), Jorgensen (1989), and Sanjek (1990).

Bernard, H. R. (1994). *Research methods in anthropology: Qualitative and quantitative approaches* (2nd ed.). Thousand Oaks, CA: Sage.

Bogdewic, S. P. (1992). Participant observation. In B. F. Crabtree & W. L. Miller (Eds.), *Doing qualitative research* (pp. 45–69). Newbury Park, CA: Sage.

Emerson, R. M., Fretz, R. I., & Shaw, L. L. (1995). *Writing ethnographic fieldnotes*. Chicago: University of Chicago Press.

Hammersley, M., & Atkinson, P. (1995). *Ethnography: Principles in practice* (2nd ed.). New York: Routledge.

Jorgensen, D. L. (1989). *Participant observation: A methodology for human studies*. Newbury Park, CA: Sage.

Sanjek, R. (1990). *Fieldnotes: The makings of anthropology*. Ithaca, NY: Cornell University Press.

For information about the issues and use of documents, see:

Prior, L. (2003). *Using documents in social research*. London: Sage.

For a discussion of field relations and issues, see the books by Hammersley and Atkinson (1995) and Lofland and Lofland (1995) and the two articles on interviewing by Kvale (2006) and Nunkoosing ((2005).

Hammersley, M., & Atkinson, P. (1995). *Ethnography: Principles in practice* (2nd ed.). New York: Routledge.

Lofland, J., & Lofland, L. H. (1995). *Analyzing social settings: A guide to qualitative observation and analysis* (3rd ed.). Belmont, CA: Wadsworth.

Kvale, S. (2006). Dominance through interviews and dialogues. *Qualitative Inquiry, 12*, 480–500.

Nunkoosing, K. (2005). The problems with interviews. *Qualitative Health Research, 15*, 698–706.

**Exercises**

1. Gain some experience in collecting data for your project. Conduct either an interview or an observation and record the information on a protocol form. After this experience, identify issues that posed challenges in data collection.

2. It is helpful to design the data collection activities for a project. Examine Figure 7.1 for the seven activities. Develop a matrix that describes data collection for all seven activities for your project.

# 8

# Data Analysis and Representation

$A$ nalyzing text and multiple other forms of data presents a challenging task for qualitative researchers. Deciding how to represent the data in tables, matrices, and narrative form adds to the challenge. In this chapter, I first discuss several general procedures for qualitative data analysis and then detail the analysis procedures often used in each of the five approaches to inquiry.

I begin by summarizing three general approaches to analysis provided by leading authors. I then present a visual model—a data analysis spiral—that I find useful to conceptualize a larger picture of all steps in the data analysis process in qualitative research. I use this spiral as a conceptualization to further explore each of the five approaches to inquiry, and I examine specific data analysis procedures within each approach and compare these procedures. I end with the use of computers in qualitative analysis and introduce four software programs—Atlas.ti, NVivo, HyperRESEARCH, and Maxqda—and discuss the common features of using software programs in data analysis as well as templates for coding data within each of the five approaches.

## Questions for Discussion

- What are common data analysis strategies used in qualitative research?
- How might the overall data analysis process be conceptualized in qualitative research?

147

- What are specific data analysis procedures used within each of the approaches to inquiry, and how do they differ?
- What are the procedures available in qualitative computer analysis programs, and how would these procedures differ by approach to qualitative inquiry?

## Three Analysis Strategies

Data analysis in qualitative research consists of preparing and organizing the data (i.e., text data as in transcripts, or image data as in photographs) for analysis, then reducing the data into themes through a process of coding and condensing the codes, and finally representing the data in figures, tables, or a discussion. Across many books on qualitative research, this is the general process that researchers use. Undoubtedly, there will be some variations in this approach. Beyond these steps, the five approaches to inquiry have additional analysis steps. Before examining the specific analysis steps in the five approaches, it is helpful to have in mind the general analysis procedures.

Table 8.1 presents typical general analysis procedures as illustrated through the writings of three qualitative researchers. I have chosen these three authors because they represent different perspectives. Madison (2005) presents a perspective taken from critical ethnography, Huberman and Miles (1994) adopt a systematic approach to analysis, and Wolcott (1994b) uses a more traditional approach to research from ethnography and case study analysis. These three sources, advocate many similar processes, as well as a few different processes, in the analytic phase of qualitative research.

All three authors comment on the central steps of coding the data (reducing the data into meaningful segments and assigning names for the segments), combining the codes into broader categories or themes, and displaying and making comparisons in the data graphs, tables, and charts. These are the core elements of qualitative data analysis.

Beyond these elements, the authors present different phases in the data analysis process. Huberman and Miles (1994), for example, provide more detailed steps in the process, such as writing marginal notes, drafting summaries of fieldnotes, and noting relationships among the categories. Madison (2005), however, introduces the need to create a point of view— a stance that signals the theoretical perspective (e.g., critical, feminist) taken in the study. This point of view is central to the analysis in critical, theoretically oriented qualitative studies. Wolcott (1994b), on the other hand, discusses the importance of forming a description from the data, as well as relating the description to the literature and cultural themes in cultural anthropology.

**Table 8.1**    General Data Analysis Strategies, by Authors

| Analytic Strategy | Madison (2005) | Huberman & Miles (1994) | Wolcott (1994b) |
|---|---|---|---|
| Sketching ideas | | Write margin notes in fieldnotes | Highlight certain information in description |
| Taking notes | | Write reflective passages in notes | |
| Summarizing fieldnotes | | Draft a summary sheet on fieldnotes | |
| Working with words | | Make metaphors | |
| Identifying codes | Do abstract coding or concrete coding | Write codes, memos | |
| Reducing codes to themes | Identify salient themes or patterns | Note patterns and themes | Identify patterned regularities |
| Counting frequency of codes | | Count frequency of codes | |
| Relating categories | | Factor, note relations among variables, build a logical chain of evidence | |
| Relating categories to analytic framework in literature | | | Contextualize in framework from literature |
| Creating a point of view | For scenes, audience, readers | | |
| Displaying the data | Create a graph or picture of the framework | Make contrasts and comparisons | Display findings in tables, charts, diagrams, and figures; compare cases; compare with a standard |

## The Data Analysis Spiral

Data analysis is not off-the-shelf; rather, it is custom-built, revised, and "choreographed" (Huberman & Miles, 1994). The processes of data collection, data analysis, and report writing are not distinct steps in the process—they are interrelated and often go on simultaneously in a research project. Qualitative researchers often "learn by doing" (Dey, 1993, p. 6) data analysis. This leads critics to claim that qualitative research is largely intuitive, soft, and relativistic or that qualitative data analysts fall back on the three "I's"—"insight, intuition, and impression" (Dey, 1995, p. 78). Undeniably, qualitative researchers preserve the unusual and serendipitous, and writers craft each study differently, using analytic procedures that evolve in the field. But given this perspective, I believe that the analysis process conforms to a general contour.

The contour is best represented in a spiral image, a data analysis spiral. As shown in Figure 8.1, to analyze qualitative data, the researcher engages in the process of moving in analytic circles rather than using a fixed linear approach. One enters with data of text or images (e.g., photographs, videotapes) and exits with an account or a narrative. In between, the researcher touches on several facets of analysis and circles around and around.

Data management, the first loop in the spiral, begins the process. At an early stage in the analysis process, researchers organize their data into file folders, index cards, or computer files. Besides organizing files, researchers convert their files to appropriate text units (e.g., a word, a sentence, an entire story) for analysis either by hand or by computer. Materials must be easily located in large databases of text (or images). As Patton (1980) says,

> The data generated by qualitative methods are voluminous. I have found no way of preparing students for the sheer massive volumes of information with which they will find themselves confronted when data collection has ended. Sitting down to make sense out of pages of interviews and whole files of field notes can be overwhelming. (p. 297)

Computer programs help with this phase of analysis, and their role in this process will be addressed later in this chapter.

Following the organization of the data, researchers continue analysis by getting a sense of the whole database. Agar (1980), for example, suggests that researchers " . . . read the transcripts in their entirety several times. Immerse yourself in the details, trying to get a sense of the interview as a whole before breaking it into parts" (p. 103). Writing memos in the margins of fieldnotes or transcripts or under photographs helps in this initial process

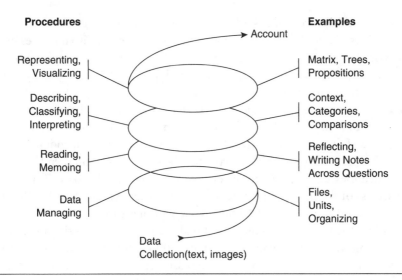

**Figure 8.1**    The Data Analysis Spiral

of exploring a database. These memos are short phrases, ideas, or key concepts that occur to the reader.

We used this procedure in our gunman case study (Asmussen & Creswell, 1995) (see Appendix F). We scanned all of our databases to identify major organizing ideas. Looking over our fieldnotes from observations, interview data, physical trace evidence, and audio and visual images, we disregarded predetermined questions so we could "hear" what interviewees said. We reflected on the larger thoughts presented in the data and formed initial categories. These categories were few in number (about 10), and we looked for multiple forms of evidence to support each. Moreover, we found evidence that portrayed multiple perspectives about each category (Stake, 1995).

This process consists of moving from the reading and memoing loop into the spiral to the describing, classifying, and interpreting loop. In this loop, code or category (and these two terms will be used interchangeably) formation represents the heart of qualitative data analysis. Here researchers describe in detail, develop themes or dimensions through some classification system, and provide an interpretation in light of their own views or views of perspectives in the literature. Authors employ descriptive detail, classification, or interpretation or some combination of these analysis procedures. Detailed description means that authors describe what they see. This detail is provided *in situ*, that is, within the context of the setting of the person, place, or event. Description becomes a good place to start in a qualitative study (after reading and managing data), and it plays a central role in ethnographic and case studies.

During this process of describing, classifying and interpreting, qualitative researchers develop codes or categories and to sort text or visual images into categories. I think about "winnowing" the data here; not all information is used in a qualitative study, and some may be discarded (Wolcott, 1994b). Researchers develop a short list of tentative codes (e.g., 12 or so) that match text segments, regardless of the length of the database. Beginning researchers tend to develop elaborate lists of codes when they review their databases. I proceed differently. I begin with a short list, "lean coding" I call it—five or six categories with shorthand labels or codes—and then I expand the categories as I continue to review and re-review my database. Typically, regardless of the size of the database, I do not develop more than 25–30 categories of information, and I find myself working to reduce and combine them into the five or six themes that I will use in the end to write my narrative. Those researchers who end up with 100 or 200 categories—and it is easy to find this many in a complex database—struggle to reduce the picture to the five or six themes that they must end with for most publications.

Several issues are important to address in this coding process. The first is whether qualitative researchers should count codes. Huberman and Miles (1994), for example, suggest that investigators make preliminary counts of data codes and determine how frequently codes appear in the database. Some (but not all) qualitative researchers feel comfortable counting and reporting the number of times the codes appear in their databases. It does provide an indicator of frequency of occurrence, something typically associated with quantitative research or systematic approaches to qualitative research. In my own work, I may look at the number of passages associated with each code as an indicator of participant interest in a code, but I do not report counts in my articles (see Asmussen & Creswell, 1995). This is because counting conveys a quantitative orientation of magnitude and frequency contrary to qualitative research. In addition, a count conveys that all codes should be given equal emphasis and it disregards that the passages coded may actually represent contradictory views.

Another issue is the use of pre-existing or *a priori* codes that guide my coding process. Again, we have a mixed reaction to the use of this procedure. Marshall and Rossman (2006) and Crabtree and Miller (1992) discuss a continuum of coding strategies that range from "prefigured" categories to "emergent" categories (p. 151). Using "prefigured" codes or categories (often from a theoretical model or the literature) are popular in the health sciences (Crabtree & Miller, 1992), but they do serve to limit the analysis to the "prefigured" codes rather than opening up the codes to reflect the views of participants in a traditional qualitative way. If a "prefigured" coding scheme is used in analysis, I typically encourage the researchers to be open to additional codes emerging during the analysis.

Another issue is the question as to the origin of the code names or labels. Code labels emerge from several sources. They might be *in vivo codes*, names that are the exact words used by participants. They might also be code names drawn from the social or health sciences (e.g., coping strategies), or names the researcher composes that seem to best describe the information. In the process of data analysis, I encourage qualitative researchers to look for code segments that can be used to describe information and develop themes. These codes can:

- represent information that researchers expect to find before the study;
- represent surprising information that researchers did not expect to find;
- and represents information that is conceptually interesting or unusual to researchers (and potentially participants and audiences)

Moving beyond coding, classifying pertains to taking the text or qualitative information apart, and looking for categories, themes, or dimensions of information. As a popular form of analysis, classification involves identifying five to seven general themes. These themes, in turn, I view as a "family" of themes with children, or subthemes, and even grandchildren, sub–subthemes represented by segments of data. It is difficult, especially in a large database, to reduce the information down into five or seven "families," but my process involves winnowing the data, reducing them to a small, manageable set of themes to write into my final narrative.

A related topic is the types of information a qualitative researcher codes and develops into themes. The researcher might look for stories (as in narrative research), individual experiences and the context of those experiences (in phenomenology), processes, actions, or interactions (in grounded theory), cultural themes and how the culture-sharing group works that can be described or categorized (in ethnography), or a detailed description of the particular case or cases (in case study research). Another way of thinking about the types of information would be to use a deconstructive stance, a stance focused on issues of desire and power (Czarniawska, 2004). Czarniawska identifies the data analysis strategies used in deconstruction, adapted from Martin (1990, p. 355), that helps focus attention on types of information to analyze from qualitative data in all approaches:

- Dismantling a dichotomy, exposing it as a false distinction (e.g., public/private, nature/culture, etc.)
- Examining silences—what is not said (e.g., noting who or what is excluded by the use of pronouns such as "we")
- Attending to disruptions and contradictions; places where a texts fails to make sense or does not continue

- Focusing on the element that is most alien or peculiar in the text—to find the limits of what is conceivable or permissible
- Interpreting metaphors as a rich source of multiple meanings
- Analyzing double entendres that may point to an unconscious subtext, often sexual in content
- Separating group-specific and more general sources of bias by 'reconstructing' the text with substitution of its main elements

Researchers engage in interpreting the data when they conduct qualitative research. Interpretation involves making sense of the data, the "lessons learned," as described by Lincoln and Guba (1985). Several forms exist, such as interpretation based on hunches, insights, and intuition. Interpretation also might be within a social science construct or idea or a combination of personal views as contrasted with a social science construct or idea. In the process of interpretation, researchers step back and form larger meanings of what is going on in the situations or sites. For postmodern and interpretive researchers, these interpretations are seen as tentative, inconclusive, and questioning.

In the final phase of the spiral, researchers present the data, a packaging of what was found in text, tabular, or figure form. For example, creating a visual image of the information, a researcher may present a comparison table (see Spradley, 1980) or a matrix—for example, a 2 × 2 table that compares men and women in terms of one of the themes or categories in the study (see Miles & Huberman, 1994). The cells contain text, not numbers. A hierarchical tree diagram represents another form of presentation. This shows different levels of abstraction, with the boxes in the top of the tree representing the most abstract information and those at the bottom representing the least abstract themes. Figure 8.2 illustrates the levels of abstraction that we used in the gunman case (Asmussen & Creswell, 1995). Although I have presented this figure at conferences, we did not include it in the published journal article version of the study. This illustration shows inductive analysis that begins with the raw data consisting of multiple sources of information and then broadens to several specific themes (e.g., safety, denial) and on to the most general themes represented by the two perspectives of social-psychological and psychological factors.

Hypotheses or propositions that specify the relationship among categories of information also represent information. In grounded theory, for example, investigators advance propositions that interrelate the causes of a phenomenon with its context and strategies. Finally, authors present metaphors to analyze the data, literary devices in which something borrowed from one domain applies to another (Hammersley & Atkinson, 1995). Qualitative writers may compose entire studies shaped by analyses of metaphors.

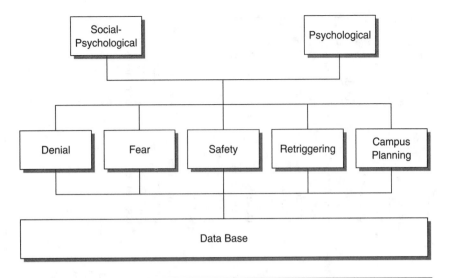

**Figure 8.2**    Layers of Analysis in Gunman Case

SOURCE: Asmussen & Creswell (1995).

At this point, the researcher might obtain feedback on the initial summaries by taking information back to informants, a procedure to be discussed in Chapter 10 as a key validation step in research.

## Analysis Within Approaches to Inquiry

Beyond the general spiral analysis processes, I can now relate the procedures to each of the five approaches to inquiry and highlight specific differences in analysis and representing data. My organizing framework for this discussion is found in Table 8.2. I address each approach and discuss specific analysis and representing characteristics. At the end of this discussion, I return to significant differences and similarities among the five approaches.

### Narrative Research Analysis and Representation

The data collected in a narrative study need to be analyzed for the story they have to tell, a chronology of unfolding events, and turning points or epiphanies. Within this broad sketch of analysis, several options exist for the narrative researcher. Using a story in science education told by four fourth graders in one elementary school, Ollerenshaw and I (Ollerenshaw &

**Table 8.2**   Data Analysis and Representation, by Research Approaches

| Data Analysis and Representation | Narrative | Phenomenology | Grounded Theory Study | Ethnography | Case Study |
|---|---|---|---|---|---|
| Data managing | • Create and organize files for data | • Create and organize files for data | • Create and organize files for data | • Create and organize files for data | • Create and organize files for data |
| Reading, memoing | • Read through text, make margin notes, form initial codes | • Read through text, make margin notes, form initial codes | • Read through text, make margin notes, form initial codes | • Read through text, make margin notes, form initial codes | • Read through text, make margin notes, form initial codes |
| Describing | • Describe the story or objective set of experiences and place it in a chronology | • Describe personal experiences through epoche <br> • Describe the essence of the phenomenon | • Describe open coding categories | • Describe the social setting, actors, events; draw picture of setting | • Describe the case and its context |
| Classifying | • Identify stories <br> • Locate epiphanies <br> • Identify contextual materials | • Develop significant statements <br> • Group statements into meaning units | • Select one open coding category for central phenomenon in process <br> • Engage in axial coding—causal condition, context, intervening conditions, strategies, consequences | • Analyze data for themes and patterned regularities | • Use categorical aggregation to establish themes or patterns |

| Data Analysis and Representation | Narrative | Phenomenology | Grounded Theory Study | Ethnography | Case Study |
|---|---|---|---|---|---|
| Interpreting | • Interpret the larger meaning of the story | • Develop a textural description, "What happened"<br>• Develop a structural description, "How" the phenomenon was experienced<br>• Develop the "essence" | • Engage in selective coding and interrelate the categories to develop "story" or propositions<br>• Develop a conditional matrix | • Interpret and make sense of the findings – how the culture "works" | • Use direct interpretation<br>• Develop naturalistic generalizations |
| Representing, visualizing | • Present narration focusing on processes, theories, and unique and general features of the life | • Present narration of the "essence" of the experience; in tables, figures, or discussion | • Present a visual model or theory<br>• Present propositions | • Present narrative presentation augmented by tables, figures, and sketches | • Present in-depth picture of the case (or cases) using narrative, tables, and figures |

Creswell, 2002) related two approaches to narrative analysis. The first approach was an analytic process advanced by Yussen and Ozcan (1997) that involved analyzing text data for five elements of plot structure (i.e., characters, setting, problem, actions, and resolution). The second approach was found in the three-dimensional space approach of Clandinin and Connelly (2000) that involved analyzing the data for three elements: interaction (personal and social), continuity (past, present, and future), and situation (physical places or the storyteller's places). In these approaches, we saw common elements of narrative analysis: collecting stories of personal experiences in the form of field texts such as interviews or conversations, retelling the stories based on narrative elements (e.g., three-dimensional space approach or the five elements of plot), rewriting the stories into a chronological sequence, and incorporating the setting or place of the participants' experiences.

These common elements are found in biographical narrative writing. Denzin (1989b) suggests that a researcher begin biographical analysis by identifying an objective set of experiences in the subject's life. Having the individual journal a sketch of his or her life may be a good beginning point for analysis. In this sketch, the researcher looks for life-course stages or experiences (e.g., childhood, marriage, employment) to develop a *chronology* of the individual's life. Stories and epiphanies will emerge from the individual's journal or from interviews. The researcher looks in the database (typically interviews or documents) for concrete, contextual biographical materials. During the interview, the researcher prompts the participant to expand on various sections of the stories and asks the interviewee to theorize about his or her life. These theories may relate to career models, processes in the life course, models of the social world, relational models of biography, and natural history models of the life course. Then, narrative segments and categories within the interview-story are isolated by the researcher, and larger patterns and meanings are determined. Finally, the individual's biography is reconstructed, and the researcher identifies factors that have shaped the life. This leads to the writing of an analytic abstraction of the case that highlights (a) the processes in the individual's life, (b) the different theories that relate to these life experiences, and (c) the unique and general features of the life.

In the life history of Vonnie Lee (Angrosino, 1994), for example, the reader finds many of these forms of analysis in the chronology of the bus trip, the specific stories such as the logo on the bus, and the theorizing (at least by the author) about the meaning of the bus trip as a metaphor for Vonnie Lee's experiences in life as an individual with mental retardation.

# Phenomenological Analysis and Representation

The suggestions for narrative analysis present a general template for qualitative researchers. In contrast, in phenomenology, there have been specific, structured methods of analysis advanced, especially by Moustakas (1994). Moustakas reviews several approaches in his book, but I see his modification of the Stevick-Colaizzi-Keen method as providing the most practical, useful approach. My approach, a simplified version of the Stevick-Colaizzi-Keen method discussed by Moustakas (1994), is as follows:

- First describe personal experiences with the phenomenon under study. The researcher begins with a full description of his or her own experience of the phenomenon. This is an attempt to set aside the researcher's personal experiences (which cannot be done entirely) so that the focus can be directed to the participants in the study.
- Develop a list of significant statements. The researcher then finds statements (in the interviews or other data sources) about how individuals are experiencing the topic, lists these significant statements (horizonalization of the data) and treats each statement as having equal worth, and works to develop a list of nonrepetitive, nonoverlapping statements.
- Take the significant statements and then group them into larger units of information, called "meaning units" or themes.
- Write a description of "what" the participants in the study experienced with the phenomenon. This is called a "textural description" of the experience—what happened—and includes verbatim examples.
- Next write a description of "how" the experience happened. This is called "structural description," and the inquirer reflects on the setting and context in which the phenomenon was experienced. For example, in a phenomenological study of the smoking behavior of high school students (McVea, Harter, McEntarffer, & Creswell, 1999), my colleagues and I provided a structural description about where the phenomenon of smoking occurred, such as in the parking lot, outside the school, by student lockers, in remote locations at the school, and so forth.
- Finally, write a composite description of the phenomenon incorporating both the textural and structural descriptions. This passage is the "essence" of the experience and represents the culminating aspect of a phenomenological study. It is typically a long paragraph that tells the reader "what" the participants experienced with the phenomenon and "how" they experienced it (i.e., the context).

The phenomenological study by Riemen (1986) tends to follow this general analytic approach. In Riemen's study of caring by patients and their nurses, she presented significant statements of caring and noncaring interactions for both males and females. Furthermore, Riemen formulated meaning statements from these significant statements and presented them in tables. Finally, Riemen advanced two "exhaustive" descriptions for the essence of the experience—two short paragraphs—and sets them apart by enclosing them in tables. In the phenomenological study of individuals with AIDS by Anderson and Spencer (2002; see Appendix C) reviewed in Chapter 5, the authors used Colaizzi's (1978) method of analysis, one of the approaches mentioned by Moustakas (1994). This approach followed the general guideline of analyzing the data for significant phrases, developing meanings and clustering them into themes, and presenting an exhaustive description of the phenomenon.

## Grounded Theory Analysis and Representation

Similar to phenomenology, grounded theory uses detailed procedures for analysis. It consists of three phases of coding—open, axial, and selective—as advanced by Strauss and Corbin (1990, 1998). Grounded theory provides a procedure for developing categories of information (open coding), interconnecting the categories (axial coding), building a "story" that connects the categories (selective coding), and ending with a discursive set of theoretical propositions (Strauss & Corbin, 1990).

In the open coding phase, the researcher examines the text (e.g., transcripts, fieldnotes, documents) for salient categories of information supported by the text. Using the constant comparative approach, the researcher attempts to "saturate" the categories—to look for instances that represent the category and to continue looking (and interviewing) until the new information obtained does not further provide insight into the category. These categories are composed of subcategories, called "properties," that represent multiple perspectives about the categories. Properties, in turn, are dimensionalized and presented on a continuum. Overall, this is the process of reducing the database to a small set of themes or categories that characterize the process or action being explored in the grounded theory study.

Once an initial set of categories has been developed, the researcher identifies a single category from the open coding list as the central phenomenon of interest. The open coding category selected for this purpose is typically one which is extensively discussed by the participants or one of particular conceptual interest because it seems central to the process being studied in the grounded theory project. The inquirer selects this one open coding category (a central phenomenon), positions it as the central feature of the theory, and then returns to the database (or collects additional data) to understand the categories that relate to this central phenomenon. Specifically, the researcher engages in the coding process called axial coding in which the database is reviewed (or new data are collected) to provide insight into specific coding categories that relate or explain the central phenomenon. These are causal conditions that influence the central phenomenon, the strategies for addressing the phenomenon, the context and intervening conditions that shape the strategies, and the consequences of undertaking the strategies. Information from this coding phase are then organized into a figure, a coding paradigm, that presents a theoretical model of the process under study. In this way, a theory is built or generated. From this theory, the inquirer generates propositions (or hypotheses) or statements that interrelate the categories in the coding paradigm. This is called selective coding. Finally, at the broadest level of analysis, the researcher can create a conditional matrix. This matrix is an analytical aid—a diagram—that helps the researcher visualize the wide range of conditions and consequences (e.g., society, world) related to the central phenomenon (Strauss & Corbin, 1990). Seldom have I found the conditional matrix actually used in studies.

The specific form for presenting the theory differs. In our study of department chairs, Brown and I present it as hypotheses (Creswell & Brown, 1992), and in their study of coping strategies of sexually abused women, Morrow and Smith (1995) advance a visual model.

The grounded theory study of survival and coping from childhood abuse by Morrow and Smith (1995) (see Appendix D) reflects several of these phases of data analysis. Although they referred to open coding, they did not present the results of this analysis, probably because of the space limitations of the journal. They did present results of the axial coding by discussing causal conditions that influenced the central phenomenon—threatening or dangerous feelings as well as helplessness, powerlessness, and lack of control. They specified two groups of strategies the women in the study used and indicated the narrower context in which these strategies occurred as well as the broader intervening conditions such as family dynamics and the victim's age. Morrow and Smith detailed the consequences of using strategies such as coping, healing, and empowerment. They also presented these categories in a visual model, called a "theoretical model for surviving and coping with childhood sexual abuse" (p. 27).

## Ethnographic Analysis and Representation

For ethnographic research, I recommend the three aspects of data analysis advanced by Wolcott (1994b): description, analysis, and interpretation of the culture-sharing group. Wolcott (1990b) believes that a good starting point for writing an ethnography is to describe the culture-sharing group and setting:

> Description is the foundation upon which qualitative research is built. . . . Here you become the storyteller, inviting the reader to see through your eyes what you have seen. . . . Start by presenting a straightforward description of the setting and events. No footnotes, no intrusive analysis—just the facts, carefully presented and interestingly related at an appropriate level of detail. (p. 28)

From an interpretive perspective, the researcher may only present one set of facts; other facts and interpretations await the reading of the ethnography by the participants and others. But this description may be analyzed by presenting information in chronological order. The writer describes through progressively focusing the description or chronicling a "day in the life" of the group or individual. Finally, other techniques involve focusing on a critical or key event, developing a "story" complete with a plot and characters, writing it as a "mystery," examining groups in interaction, following an analytical framework, or showing different perspectives through the views of informants.

Analysis for Wolcott (1994b) is a sorting procedure—"the quantitative side of qualitative research" (p. 26). This involves highlighting specific material introduced in the descriptive phase or displaying findings through tables, charts, diagrams, and figures. The researcher also analyzes through using systematic procedures such as those advanced by Spradley (1979, 1980), who calls for building taxonomies, generating comparison tables, and developing semantic tables. Perhaps the most popular analysis procedure, also mentioned by Wolcott (1994b), is the search for patterned regularities in the data. Other forms of analysis consist of comparing the cultural group to others, evaluating the group in terms of standards, and drawing connections between the culture-sharing group and larger theoretical frameworks. Other analysis steps include critiquing the research process and proposing a redesign for the study.

Making an ethnographic interpretation of the culture-sharing group is a data transformation step as well. Here the researcher goes beyond the database and probes "what is to be made of them" (Wolcott, 1994b, p. 36). The researcher speculates outrageous, comparative interpretations that raise doubts or questions for the reader. The researcher draws inferences from the data or turns to theory to provide structure for his or her interpretations. The researcher also personalizes the interpretation: "This is what I make of it" or "This is how the research experience affected me" (p. 44). Finally, the investigator forges an interpretation through expressions such as poetry, fiction, or performance.

The ethnography presented in Chapter 4 by Haenfler (2004) (and presented in Appendix E) applies a critical perspective to these analytic procedures of ethnography. He provides a detailed description of the straight edge core values of resistance to other cultures and then discussed five themes related to these core values (e.g., positive, clean living). Then the conclusion to the article includes broad interpretations of the group's core values, such as the individualized and collective meanings for participation in the subculture. However, Haenfler began the methods discussion with a self-disclosing, positioning statement about his background and participation in the straight edge movement. This positioning is also presented as a chronology of his experiences from 1989 to 2001.

## Case Study Analysis and Representation

For a case study, as in ethnography, analysis consists of making a detailed description of the case and its setting. If the case presents a chronology of events, I then recommend analyzing the multiple sources of data to determine evidence for each step or phase in the evolution of the case. Moreover, the setting is particularly important. In our gunman case (Asmussen & Creswell, 1995)(see Appendix F), Asmussen and I analyzed the information to determine how the incident fit into the setting—in our situation, a tranquil, peaceful Midwestern community.

In addition, Stake (1995) advocates four forms of data analysis and interpretation in case study research. In categorical aggregation, the researcher seeks a collection of instances from the data, hoping that issue-relevant meanings will emerge. In *direct interpretation*, on the other hand, the case study researcher looks at a single instance and draws meaning from it without looking for multiple instances. It is a process of pulling the data apart and putting them back together in more meaningful ways. Also, the researcher establishes *patterns* and looks for a correspondence between two or more categories. This correspondence might take the form of a table, possibly a 2 × 2 table, showing the relationship between two categories. Yin (2003) advances the cross-case synthesis as an analytic technique when the researcher studies two or more cases. He suggests that a word table can be created to display the data from individual cases according to some uniform framework. The implication of this is that the researcher can then look for similarities and differences among the cases. Finally, the researcher develops *naturalistic generalizations* from analyzing the data, generalizations that people can learn from the case either for themselves or to apply to a population of cases.

To these analysis steps I would add description of the case, a detailed view of aspects about the case—the "facts." In our gunman case study (Asmussen

& Creswell, 1995), we describe the events following the incident for 2 weeks, highlighting the major players, the sites, and the activities. We then aggregate the data into about 20 categories (categorical aggregation) and collapse them into five themes. In the final section of the study, we develop generalizations about the case in terms of the themes and how they compare and contrast with published literature on campus violence.

## Comparing the Five Approaches

Returning to Table 8.2, data analysis and representation in the five approaches have several common and distinctive features. Across all five approaches, the researcher typically begins by creating and organizing files of information. Next, the process of a general reading and memoing of information occurs to develop a sense of the data and to begin the process of making sense of them. Then, all approaches have a phase of description, with the exception of grounded theory, in which the inquirer seeks to begin building toward a theory of the action or process.

However, several important differences exist in the five approaches. Grounded theory and phenomenology have the most detailed, explicated procedure for data analysis. Ethnography and case studies have analysis procedures that are common, and narrative research represents the least structured procedure. Also, the terms used in the phase of classifying show distinct language among these approaches (see Appendix A for a glossary of terms); what is called open coding in grounded theory is similar to the first stage of identifying significant statements in phenomenology and to categorical aggregation in case study research. The researcher needs to become familiar with the definition of these terms of analysis and employ them correctly in the chosen approach to inquiry. Finally, the presentation of the data, in turn, reflects the data analysis steps, and it varies from a narration in narrative to tabled statements, meanings, and description in phenomenology to a visual model or theory in grounded theory.

## Computer Use in Qualitative Data Analysis

Qualitative computer programs have been available since the late 1980s, and they have become more refined and helpful in computerizing the process of analyzing text and image data (see Weitzman and Miles, 1995, for a review of 24 programs). The process used for qualitative data analysis is the same for hand coding or using a computer: The inquirer identifies a text segment or image segment, assigns a code label, and then searches through the database for all text segments that have the same code label. In this process the researcher, not the computer program, does the coding and categorizing.

## Advantages and Disadvantages

The computer program simply provides a means for storing the data and easily accessing the codes provided by the researcher. I feel that computer programs are most helpful with large databases, such as 500 or more pages of text. Although using a computer may not be of interest to all qualitative researchers, there are advantages to using them:

- A computer program provides an organized storage file system so that the researcher can quickly and easily locate material and store it in one place. This aspect becomes especially important in locating entire cases or cases with specific characteristics.
- A computer program helps a researcher locate material easily, whether this material is an idea, a statement, a phrase, or a word. No longer do we need to "cut and paste" material onto file cards and sort and resort the cards according to themes. No longer do we need to develop an elaborate "color code" system for text related to themes or topics. The search for text can be easily accomplished with a computer program. Once researchers identify categories in grounded theory, or themes in case studies, the names of the categories can be searched using the computer program for other instances when the names occur in the database.
- A computer program encourages a researcher to look closely at the data, even line by line, and think about the meaning of each sentence and idea. Sometimes, without a program, the researcher is likely to casually read through the text files or transcripts and not analyze each idea carefully.
- The concept mapping feature of computer programs enables the researcher to visualize the relationship among codes and themes by drawing a visual model.
- A computer program allows the researcher to easily retrieve memos associated with codes, themes, or documents.

The disadvantages of using computer programs go beyond their cost:

- Using a computer program requires that the researcher learn how to run the program. This is sometimes a daunting task that is above and beyond learning required for understanding the procedures of qualitative research. Granted, some people learn computer programs more easily than do others, and prior experience with programs shortens the learning time.
- A computer program may, to some individuals, put a machine between the researcher and the actual data. This causes an uncomfortable distance between the researcher and his or her data.

- Although individuals may believe that the data are fixed or set by the program (Kelle, 1995), the categories and organization of the data may be changed by the software user. Some individuals may find changing the categories or moving information around less desirable than others and find that the computer program slows down or inhibits this process.
- Instructions for using computer programs vary in their ease of use and accessibility. Many documents for computer programs do not provide information about how to use the program to generate a qualitative study, or one of the five approaches to research discussed in this book.
- A computer program may not have the features or capability that researchers need, so researchers can shop comparatively to find a program that meets their needs.

## A Sampling of Computer Programs

There are many computer programs available for analysis; some have been developed by individuals on campuses and some are available for commercial purchase. I highlight four commercial programs that are popular and that I have examined closely (see Creswell, 2005; Creswell & Maietta, 2002)—Atlas.ti, NVivo, Maxqda, and HyperRESEARCH. I have intentionally left out the version numbers because the developers are continually upgrading the programs. Although the first three programs to be reviewed are PC-based, HyperRESEARCH is the only program available for the Macintosh or the PC. To use the other programs on a Macintosh, the user must run virtual PC.

### Atlas.ti (http://www.atlasti.com)

This PC, Windows-based program enables you to organize your text, graphic, audio, and visual data files, along with your coding, memos, and findings, into a project. Further, you can code, annotate, and compare segments of information. You can drag and drop codes within an interactive margin screen. You can rapidly search, retrieve, and browse all data segments and notes relevant to an idea and, importantly, build unique visual networks that allow you to connect visually selected passages, memos, and codes in a concept map. Data can be exported to SPSS, HTML, XML, and CSV. Less computer memory is needed for this program as compared with other programs because it directly links data files to a project. This program also allows for a group of researchers to work on the same project and make comparisons of how each researcher coded the data. A demonstration software package is available to test out this program, which is described by and available from Scientific Software Development in Germany.

## QSR NVivo (http://www.qsrinternational.com/)

NVivo is the latest version of software from QSR International. NVivo combines the features of the popular software program N6 (or Nud.ist) and NVivo 2.0. It is available for Windows PC only. NVivo helps analyze, manage, shape, and analyze qualitative data. Its streamlined look makes it easy to use. It provides security by storing the database and files together in a single file, it enables a researcher to use multiple languages, it has a merge function for team research, and it enables the researcher to easily manipulate the data and conduct searches. Further, it can display graphically the codes and categories. A good overview of the evolution of the software from N3 to Nvivo is available from Bazeley (2002). NVivo is distributed by QSR International in Australia. A demonstration copy is available to see and try out the features of this software program.

## HyperRESEARCH (http://www.researchware.com/)

This program is available for the Windows or Macintosh platform. It is an easy-to-use qualitative software package enabling you to code and retrieve, build theories, and conduct analyses of the data. Now with advanced multimedia capabilities, HyperRESEARCH allows the researcher to work with text, graphics, audio, and video sources—making it a valuable research analysis tool. HyperRESEARCH is a solid code-and-retrieve data analysis program, with additional theory-building features provided by the Hypothesis Tester. This program also allows the researcher to draw visual diagrams, and it now has a module that can be added, called "Hyper-Transcriber" that will allow researchers to create a transfer of video and audio data. This program, developed by Research Ware, is available in the United States.

## MAXqda (http://www.maxqda.com/)

MAXqda is a PC-based software program that helps the researcher to systematically evaluate and interpret qualitative texts. It is also a powerful tool for developing theories and testing theoretical conclusions. The main menu has four windows: the data, the code or category system, the text being analyzed, and the results of basic and complex searches. It uses a hierarchical code system, and the researcher can attach a weight score to a text segment to indicate the relevance of the segment. Memos can be easily written and stored as different types of memos (e.g., theory memos or methodological memos). Data can be exported to statistical programs, such as SPSS or Excel, and the software can import Excel or SPSS programs as well. It is easily used by multiple coders on research teams. Images and video segments can also be stored and coded in this program. MAXqda is distributed by VERBI Software in Germany. A demonstration program is available to learn more about the unique features of this program.

## Use of Computer Software Programs With the Five Approaches

After reviewing all of these computer programs, I see several ways that they can facilitate qualitative data analysis:

- Computer programs help store and organize qualitative data. The programs provide a convenient way to store qualitative data. Data are stored in document files (files converted from a word processing program to DOS, ASCII, or rich-text in some programs). These document files consist of information from one discrete unit of information such as a transcript from one interview, one set of observational notes, or one article scanned from a newspaper. For all five of the approaches to qualitative inquiry, the document could be one interview, one observation, or one document.

- Computer programs help locate text or image segments associated with a code or theme. When using a computer program, the researcher goes through the text or images one line or image at a time and asks, "What is the person saying (or doing) in this passage?" Then the researcher assigns a code label using the words of the participant, social or human science terms, or composes a term that seems to relate to the situation. After reviewing many pages or images, the researcher can use the search function of the program to locate all the text or image segments that fit a code label. In this way, the researcher can easily see how participants are discussing the code in a similar or different way.

- Computer programs help locate common passages or segments that relate to two or more code labels. The search process can be extended to include two or more code labels. For example, the code label "two-parent family" might be combined with "females" to yield text segments in which women are discussing a "two-parent family." Alternatively, "two-parent family" might be combined with "males" to generate text segments in which men talk about the "two-parent family." One helpful code label is "quotes," and researchers can assign interesting quotes to use in a qualitative report into this code label and easily retrieve useful quotes for a report. Computer programs also enable the user to search for specific words to see how frequently they occur in the texts; in this way, specific words might be elevated to the status of code labels or possible themes based on the frequency of their use. In another usage, a code label may be created for the "title" in the study, and the information in the label might change as the author revises the title in the process of conducting the study.

- Computer programs help make comparisons among code labels. If the researcher makes both of these requests about females and males in the prior example, data then exist for making comparisons among the responses of females and males on their views about the "two-parent family." The computer program thus enables a researcher to interrogate the database about the interrelationship among codes or categories.

- Computer programs help the researcher to conceptualize different levels of abstraction in qualitative data analysis. The process of qualitative data analysis, as discussed earlier in this chapter, starts with the researcher analyzing the raw data (e.g., interviews), forming the raw data into codes, and then combining the codes into broader themes. These themes can be and often are "headings" used in a qualitative study. The software programs provide a means for organizing codes hierarchically so that smaller units, such as codes, can be placed under larger units, such as themes. In NVivo, the concept of children and parent codes illustrates two levels of abstraction. In this way, the computer program helps the researcher to build levels of analysis and see the relationship between the raw data and the broader themes.

- Computer programs provide a visual picture of codes and themes. Many computer programs contain the feature of concept mapping so that the user can generate a visual diagram of the codes and themes and their interrelationships. These codes and themes can be continually moved around and reorganized under new categories of information as the project progresses.

- Computer programs provide the capability to write memos and store them as codes. In this way, the researcher can begin to create the qualitative report during data analysis or simply record insights as they emerge during the data analysis.

- With computer programs, the researcher can create a template for coding data within each of the five approaches. The researcher can establish a preset list of codes that match the data analysis procedure within the approach of choice. Then, as data are reviewed during computer analysis, the researcher can identify information that fits into the codes or write memos that become codes. As shown in Figure 8.3 through Figure 8.7, I created templates for coding within each approach that fit the general structure in analyzing data with the approach. I developed these codes as a hierarchical picture, but they could be drawn as circles or in a less linear fashion. Hierarchical organization of codes is the approach often used in the concept-mapping feature of software programs.

In narrative research (see Figure 8.3), I created codes that relate to the story, such as the chronology, the plot or the three-dimensional space model, and the themes that might arise from the story. The analysis might proceed using the plot structure approach or the three-dimensional model, but

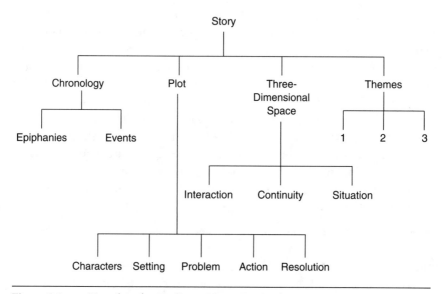

**Figure 8.3**    Template for Coding a Narrative Study

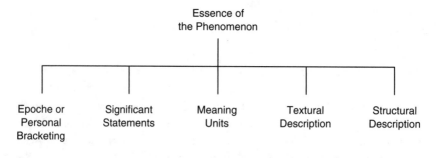

**Figure 8.4**    Template for Coding a Phenomological Study

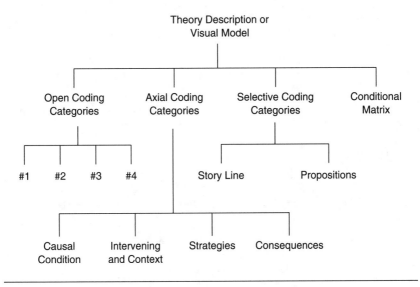

**Figure 8.5**    Template for Coding a Grounded Theory Study

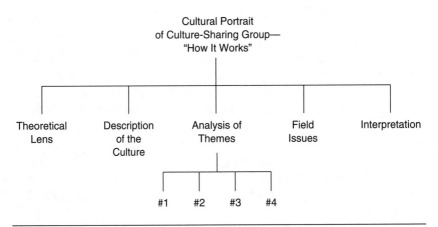

**Figure 8.6**    Template for Coding an Ethnography

I placed both in the figure to provide the most options for analysis. The researcher will not know what approach to use until he or she actually starts the data analysis process. The code "story" might be used by the researcher to actually begin writing out the story based on the elements analyzed.

In the template for coding a phenomenological study (see Figure 8.4), I used the categories mentioned earlier in data analysis. I placed codes for epoche or bracketing (if this is used), significant statements, meaning

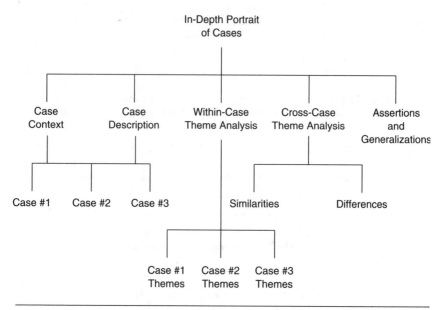

**Figure 8.7**    Template for Coding a Case Study (Using a Multiple or Collective Case Approach)

units, and textural and structural descriptions (which both might be written as memos). The code at the top, "essence of the phenomenon," is written as a memo about the "essence" that will become the "essence" description in the final written report. In the template for coding a grounded theory study (see Figure 8.5), I included the three major coding phases: open coding, axial coding, and selective coding. I also included a code for the conditional matrix if that feature is used by the grounded theorist. The code at the top, "theory description or visual model," can be used by the researcher to actually create a visual model of the process that is linked to this code.

In the template for coding an ethnography (see Figure 8.6), I included a code that might be a memo or reference to text about the theoretical lens used in the ethnography, codes on the description of the culture and an analysis of themes, a code on field issues, and a code on interpretation. The code at the top, "cultural portrait of culture-sharing group—'how it works,'" can be a code in which the ethnographer writes a memo summarizing the major cultural rules that pertain to the group. Finally, in the template for coding a case study (see Figure 8.7), I chose a multiple case study to illustrate the precode specification. For each case, codes exist for the context and description of the case. Also, I advanced codes for themes within each case, and for themes that are similar and difference in cross-case analysis. Finally, I included codes for assertions and generalizations across all cases.

## How to Choose Among the Computer Programs

With different programs available, decisions need to be made about the proper choice of a qualitative software program. Basically, all of the programs provide similar features, and some have more features than others. Many of the programs have a demonstration copy available at their Web sites so that you can examine and try out the program. Also, other researchers can be approached who have used the program and you can determine their views of the software. In 2002, I wrote a chapter with Maietta (Creswell & Maietta, 2002) in which we assessed several computer programs using eight criteria. As shown in Figure 8.8, the criteria for selecting a program were the ease of using the program; the type of data it accepted; its capability to read and review text; its provision of memo-writing functions; its processes of categorization; its analysis features, such as concept mapping; the ability of the program to input quantitative data; and its support for multiple researchers and merging different databases. These criteria can be used to identify a computer program that will meet a researcher's needs.

## Summary

This chapter presented data analysis and representation. I began with a review of data analysis procedures advanced by three authors and noted the common features of coding, developing themes, and providing a visual diagram of the data. I also noted some of the differences among their approaches. Then I advanced a spiral of analysis that captured the general process. This spiral contained aspects of data management; reading and memoing; describing, classifying, and interpreting; and representing and visualizing data. I next introduced each of the five approaches to inquiry and discussed how they had unique data analysis steps beyond the concept of the spiral. Finally, I described how computer programs aid in the analysis and representation of data; discussed four programs, common features of using computer software, and templates for coding each of the five approaches to inquiry; and ended with information about criteria for choosing a computer software program.

- Ease of Integration in Using the Program
  Is it Windows or Macintosh compatible?
  Is it easy to use in getting started?
  Can you easily work through a document?

- Type of Data the Program Will Accept
  Will it handle text data?
  Will it handle multimedia (image) data?

- Reading and Reviewing Text
  Can it highlight and connect quotations?
  Can it search for specific text passages?

- Memo Writing
  Does it have the capability for you to add notes or memos?
  Can you easily access the memos you write?

- Categorization
  Can you develop codes?
  Can you easily apply codes to text or images?
  Can you easily display codes?
  Can you easily review and make changes in the codes?

- Analysis Inventory and Assessment
  Can you sort for specific codes?
  Can you combine codes in a search?
  Can you develop a concept map with the codes?
  Can you make demographic comparisons with the codes?

- Quantitative Data
  Can you import a quantitative database (e.g., SPSS)?
  Can you export a word or image qualitative database to a quantitative program?

- Merging Project
  Can two or more researchers analyze the data and can these analyses be merged?

**Figure 8.8**    Features to Consider When Comparing Qualitative Data Analysis Software

SOURCE: Adapted from Creswell & Maietta (2002), Qualitative research. In D. C. Miller & N. J. Salkind (Eds.), *Handbook of social research* (pp. 143–184). Thousand Oaks, CA: Sage. Used with permission.

## Additional Readings

The classical book on qualitative data analysis is Miles and Huberman (1994), now in its second edition. Also, I recommend books that address the process of conducting qualitative research, such as Marshall and Rossman (2006).

Miles, M. B., & Huberman, A. M. (1994). *Qualitative data analysis: A sourcebook of new methods* (2nd ed.). Thousand Oaks, CA: Sage.
Marshall, C., & Rossman, G. B. (2006). *Designing qualitative research* (4th ed.). Thousand Oaks, CA: Sage.

Specific data analysis strategies for each of the five approaches to inquiry are available in Clandinin and Connelly (2000), Czarniawska (2004), and Denzin (1989a) for narrative research; Moustakas (1994) for phenomenology; Stake (1995) for case studies; Strauss and Corbin (1990) for grounded theory; and Wolcott (1994b) for ethnography.

Clandinin, D. J., & Connelly, F. M. (2000). *Narrative inquiry: Experience and story in qualitative research.* San Francisco: Jossey-Bass.
Czarniawska, B. (2004). *Narratives in social science research.* Thousand Oaks, CA: Sage.
Denzin, N. K. (1989a). *Interpretive biography.* Newbury Park, CA: Sage.
Moustakas, C. (1994). *Phenomenological research methods.* Thousand Oaks, CA: Sage.
Stake, R. (1995). *The art of case study research.* Thousand Oaks, CA: Sage.
Strauss, A., & Corbin, J. (1990). *Basics of qualitative research: Grounded theory procedures and techniques.* Newbury Park, CA: Sage.
Wolcott, H. F. (1994b). *Transforming qualitative data: Description, analysis, and interpretation.* Thousand Oaks, CA: Sage.

For a review of computer programs available for analyzing text data, I recommend Creswell and Maieta (2002), Kelle (1995), and Weitzman and Miles (1995).

Creswell, J. W., & Maietta, R. C. (2002). Qualitative research. In D. C. Miller & N. J. Salkind (Eds.), *Handbook of social research* (pp. 143–184). Thousand Oaks, CA: Sage.
Kelle, E. (Ed.). (1995). *Computer-aided qualitative data analysis.* Thousand Oaks, CA: Sage.
Weitzman, E. A., & Miles, M. B. (1995). *Computer programs for qualitative data analysis.* Thousand Oaks, CA: Sage.

**Exercises**

**Exercises**

1. Analyze data from your data collection in the Exercises in Chapter 7. Analyze the data using the steps or phases for your approaches to inquiry. Present a summary of findings.

2. Plan the data analysis steps for your project. Using Table 8.2 as a guide, discuss how you plan to describe, classify, and interpret your information.

3. Gain some experience using a computer software program. Select one of the computer programs mentioned in this chapter, go to its Web site, and find the demonstration program. Try out the program.

**Exercises**

# 9

# Writing a Qualitative Study

W riting and composing the narrative report brings the entire study together. Borrowing a term from Strauss and Corbin (1990), I am fascinated by the "architecture" of a study, how it is composed and organized by writers. I also like Strauss and Corbin's (1990) suggestion that writers use a "spatial metaphor" (p. 231) to visualize their full reports or studies. To consider a study "spatially," they ask the following questions. Is coming away with an idea like walking slowly around a statue, studying it from a variety of interrelated views? Like walking downhill step by step? Like walking through the rooms of a house?

In this chapter, I assess the general architecture of a qualitative study, and then I invite the reader to enter specific rooms of the study to see how they are composed. In this process, I begin with four rhetorical issues in the rendering of a study regardless of approach: reflexivity and representation, audience, encoding, and quotes. Then I take each of the five approaches to inquiry and assess two rhetorical structures: the overall structure (i.e., overall organization of the report or study) and the embedded structure (i.e., specific narrative devices and techniques that the writer uses in the report). I return once again to the five examples of studies in Chapter 5 to illustrate overall and embedded structures. Finally, I compare the narrative structures for the five approaches in terms of four dimensions. In this chapter I will not address the use of grammar and syntax and will refer readers to books that provide a detailed treatment of these subjects (e.g., Creswell, 2003).

## Questions for Discussion

- What are several broad rhetorical issues associated with writing a qualitative study? What are the rhetorical structures for writing a study within each of the five approaches of inquiry?
- What are the embedded rhetorical structures for writing a study within each of the five approaches of inquiry?
- How do the narrative structures for the five approaches differ?

## Several Rhetorical Issues

Unquestionably, the narrative forms are extensive in qualitative research. In reviewing the forms, Glesne and Peshkin (1992) note that narratives in "storytelling" modes blur the lines between fiction, journalism, and scholarly studies. Other forms engage the reader through a chronological approach as events unfold slowly over time, whether the subject is a study of a culture-sharing group, the narrative story of the life of an individual, or the evolution of a program or an organization. Another technique is to narrow and expand the focus, evoking the metaphor of a camera lens that pans out, zooms in, then zooms out again. Some reports rely heavily on description of events, whereas others advance a small number of "themes" or perspectives. A narrative might capture a "typical day in the life" of an individual or a group. Some reports are heavily oriented toward theory, whereas others, such as Stake's (1995) "Harper School," employ little literature and theory. Since the publication of Clifford and Marcus's (1986) edited volume *Writing Culture* in ethnography, qualitative writing has been shaped by a need for researchers to be self-disclosing about their role in the writing, the impact of it on participants, and how information conveyed is read by audiences. Researcher reflexivity and representations is the first issue to which we turn.

## Reflexivity and Representations in Writing

Qualitative researchers today are much more self-disclosing about their qualitative writings than they were a few years ago. No longer is it acceptable to be the omniscient, distanced qualitative writer. As Laurel Richardson wrote, researchers "do not have to try to play God, writing as disembodied omniscient narrators claiming universal and atemporal general knowledge" (Richardson & St. Pierre, 2005, p. 961). Through these omniscient narrators, postmodern thinkers "deconstruct" the narrative, challenging text as contested terrain that cannot be understood without references to ideas being concealed by the author and contexts within the author's life (Agger, 1991). This theme was espoused by Denzin (1989a) in his "interpretive" approach to biographical writing. As a response, qualitative research today acknowledges the impact of the writing on the researcher, on the participants, and on the reader.

How we write is a reflection of our own interpretation based on the cultural, social, gender, class, and personal politics that we bring to research. All writing is "positioned" and within a stance. All researchers shape the writing that emerges, and qualitative researchers need to accept this interpretation and be open about it in their writings. According to Richardson (1994), the best writing acknowledges its own "undecidability" forthrightly, that all writing has "subtexts" that "situate" or "position" the material within a particular historical and locally specific time and place. In this perspective, no writing has "privileged status" (Richardson, 1994, p. 518) or superiority over other writings. Indeed, writings are co-constructions, representations of interactive processes between researchers and the researched (Gilgun, 2005).

Also, there is increased concern about the impact of the writing on the participants. How will they see the write up? Will they be marginalized because of it? Will they be offended? Will they hide their true feelings and perspectives? Have the participants reviewed the material, and interpreted, challenged, and dissented from the interpretation (Weis & Fine, 2000)? Perhaps researchers' writing objectively, in a scientific way, has the impact of silencing the participants, and silencing the researchers as well (Czarniawska, 2004). Gilgun (2005) makes the point that this silence is contradictory to qualitative research that seeks to hear all voices and perspectives.

Also, the writing has an impact on the reader, who also makes an interpretation of the account and may form an entirely different interpretation than the author or the participants. Should the researcher be afraid that certain people will see the final report? Can the researcher give any kind of definitive account when it is the reader who makes the ultimate interpretation of the events? Indeed, the writing may be a performance, and the standard writing of qualitative research into text has expanded to include split-page writings, theater, poetry, photography, music, collage, drawing, sculpture, quilting, stained glass, and dance (Gilgun, 2005). Language may "kill" whatever it touches, and qualitative researchers understand that it is impossible to truly "say" something (van Manen, 2006).

Weis and Fine (2000) discuss a "set of self-reflective points of critical consciousness around the questions of how to represent responsibility" in qualitative writings (p. 33). There are questions that can be formed from their major points and should be considered by all qualitative researchers about their writings:

- Should I write about what people say or recognize that sometimes they cannot remember or choose not to remember?
- What are my political reflexivities that need to come into my report?
- Has my writing connected the voices and stories of individuals back to the set of historic, structural, and economic relations in which they are situated?
- How far should I go in theorizing the words of participants?
- Have I considered how my words could be used for progressive, conservative, and repressive social policies?
- Have I backed into the passive voice and decoupled my responsibility from my interpretation?
- To what extent has my analysis (and writing) offered an alternative to common sense or the dominant discourse?

## Audience for Our Writings

A basic axiom holds that all writers write for an audience. As Clandinin and Connelly (2000) say, "A sense of an audience peering over the writer's shoulder needs to pervade the writing and the written text" (p. 149). Thus, writers consciously think about their audience or multiple audiences for their studies (Richardson, 1990, 1994). Tierney (1995), for example, identifies four potential audiences: colleagues, those involved in the interviews and observations, policy makers, and the general public. In short, how the findings are presented depends on the audience with whom one is communicating (Giorgi, 1985). For example, because Fischer and Wertz (1979) disseminated information about their phenomenological study at public forums, they produced several expressions of their findings, all responding to different audiences. They used a general structure, four paragraphs in length, an approach that they admitted lost its richness and concreteness. Another form consisted of case synopses, each reporting the experiences of one individual and each two and a half pages in length.

## Encoding Our Writings

A closely related topic is recognizing the importance of language in shaping our qualitative texts. The words we use encode our report, revealing ourselves and how we perceive the needs of our audiences. Earlier, in Chapter 6, I presented encoding the problem, purpose, and research questions; now I consider encoding the entire narrative report. Richardson's (1990) study of women in affairs with married men illustrates how a writer can shape a work for a trade audience, an academic audience, or a moral/political audience. For a trade audience, she encoded her work with literary devices such as

> jazzy titles, attractive covers, lack of specialized jargon, marginalization of methodology, common-world metaphors and images, and book blurbs and prefatory material about the "lay" interest in the material. (Richardson, 1990, p. 32)

For the moral/political audience, she encoded through devices such as,

> in-group words in the title, for example, woman/women/feminist in feminist writing; the moral or activist "credentials" of the author, for example, the author's role in particular social movements; references to moral and activist authorities; empowerment metaphors, and book blurbs and prefatory material about how this work relates to real people's lives. (Richardson, 1990, pp. 32–33)

Finally, for the academic publications (e.g., journals, conference papers, academic books), she marked it by a

> prominent display of academic credentials of author, references, footnotes, methodology sections, use of familiar academic metaphors and images (such as "exchange theory," "roles," and "stratification"), and book blurbs and prefatory material about the science or scholarship involved. (Richardson, 1990, p. 32)

Although I emphasize academic writing here, researchers encode qualitative studies for audiences other than academics. For example, in the social and human sciences, policy makers may be a primary audience, and this necessitates writing with less methods, more parsimony, and a focus on practice and results.

Richardson's (1990) ideas triggered my own thoughts about how one might encode a qualitative narrative. Such encoding might include the following:

- An overall structure that does not conform to the standard quantitative introduction, methods, results, and discussion format. Instead, the methods might be called "procedures," and the results might be called "findings." In fact, the researcher might phrase the heading in the words of participants in the study as they discuss "denial," "retriggering," and so forth, as I did in the gunman case. (Asmussen & Creswell, 1995)

- A writing style that is personal, familiar, perhaps "up-close," highly readable, friendly, and applied for a broad audience. Our structures hope for a "persuasive" effect (Czarniaswka, 2004, p. 124). Readers should find the material interesting and memorable, the "grab" in writing. (Gilgun, 2005)
- A level of detail that makes the work come alive—*verisimilitude* comes to mind (Richardson, 1994, p. 521)—a criterion for a good literary study where the writing seems "real" and "alive," transporting the reader directly into the world of the study, whether this world is the cultural setting of youths' resistance to both the counterculture and the dominant culture (Haenfler, 2004) or women expressing emotion about their abusive childhoods (Morrow & Smith, 1995). Still, we must recognize that the writing is only a representation of what we see or understand.

## Quotes in Our Writings

In addition to encoding text with the language of qualitative research, authors bring in the voice of participants in the study. Writers use ample quotes, and I find Richardson's (1990) discussion about three types of quotes most useful. The first consists of short eye-catching quotations. These are easy to read, take up little space, and stand out from the narrator's text and are indented to signify different perspectives. For example, in the phenomenological study of how persons live with AIDS, Anderson and Spencer (2002) used paragraph-long quotes from men and women in the study to convey the "magic of not thinking" theme:

> It's a sickness, but in my mind I don't think that I got it. Because if you think about having HIV, it comes down more on you. It's more like a mind game. To try and stay alive is that you don't even think about it. It's not in the mind. (p. 1347)

Dialogue, a variation of quotes, may be used, such as in the Principal Selection Committee study by Wolcott (1994a) in which he states conversation between candidates (e.g., "Mr. Fifth") and the interviewing principals.

The second approach consists of embedded quotes, briefly quoted phrases within the analyst's narrative. These quotes, according to Richardson (1990), prepare a reader for a shift in emphasis or display a point and allow the writer (and reader) to move on. Asmussen and I used embedded quotes extensively in our gunman study (Asmussen & Creswell, 1995) because they consume little space and provide specific concrete evidence, in the informants' words, to support a theme. Embedded quotes also are used extensively in the childhood sexual abuse grounded theory study by Morrow and Smith (1995).

A third type of quote is the longer quotation, used to convey more complex understandings. These are difficult to use because of space limitations in publications and because longer quotes may contain many ideas and so the reader needs to be guided both "into" the quote and "out of" the quote to focus his or her attention on the controlling idea that the writer would like the reader to see. In the Vonnie Lee biography, Angrosino (1994) states several long quotes to provide complete answers to questions posed to Vonnie Lee and to develop for the reader a sense of Vonnie Lee's voice, questions such as "Why do you like the bus so much?" (p. 21).

In addition to these rhetorical issues, the writer needs to address how he or she is going to compose the overall narrative structure of the report and use embedded structures within the report to provide a narrative within the approach of choice. I offer Table 9.1 as a guide to the discussion to follow, in which I list many overall and embedded structural approaches as they apply to the five approaches of inquiry.

## Narrative Research Structure

As I read about the writing of studies in narrative research, I find authors unwilling to prescribe a structure or specific writing strategies (Clandinin & Connelly, 2000; Czarniawska, 2004). Instead, I find the authors suggesting maximum flexibility in structure (see Ely, 2006), but emphasizing core elements that might go into the narrative study.

### Overall Rhetorical Structure

Narrative researchers encourage individuals to write narrative studies that experiment with form (Clandinin & Connelly, 2000). Researchers can come to their narrative form by first looking to their own preferences in reading (e.g., memoirs, novels), reading other narrative dissertations and books, and viewing the narrative study as back-and-forth writing, as a process (Clandinin & Connelly, 2000). Within these general guidelines, Clandinin and Connelly (2000) review two doctoral dissertations that employ narrative research. The two have different narrative structures: one provides narratives of a chronology of the lives of three women; the other adopts a more classical approach to a dissertation including an introduction, a literature review, and a methodology. For this second example, the remaining chapters then go into a series of letters that tell the stories of the author's experiences with the participants. Reading through these two examples, I am struck by how they both reflect the three-dimensional inquiry space that Clandinin

**Table 9.1**    Overall and Embedded Rhetorical Structures and the Five Approaches

| Approach to Inquiry | Overall Rhetorical Structures | Embedded Rhetorical Structures |
| --- | --- | --- |
| Narrative | • Flexible and evolving as a process (Clandinin & Connelly, 2000)<br>• Three-dimensional space inquiry model (Clandinin & Connelly, 2000)<br>• Chronology to the stories (Clandinin & Connelly, 2000) | • Epiphany (Denzin, 1989b)<br>• Theme, key event, or plot (Czarniawska, 2004; Smith, 1994)<br>• Metaphors and transitions (Clandinin & Connelly, 2000; Lomask, 1986)<br>• Progressive-regressive method or zooming in and out (Czarniawska, 2004; Denzin, 1989b) |
| Phenomenology | • Chapters in a "research manuscript" (Moustakas, 1994)<br>• The "research report" (Polkinghorne, 1989) | • Figure or table for essence (Grigsby & Megel, 1995)<br>• Discussion about philosophy (Harper, 1981)<br>• Creative closing (Moustakas, 1994) |
| Grounded theory | • Components of a grounded theory study (May, 1986)<br>• Parameters of a grounded theory study (Strauss & Corbin, 1990) | • Extent of analysis (Chenitz & Swanson, 1986)<br>• Form of propositions (Strauss & Corbin, 1990)<br>• Use of visual diagram (Morrow & Smith, 1995) |
| Ethnography | • Types of tales (Van Maanen, 1988)<br>• Description, analysis, and interpretation (Wolcott, 1994b)<br>• "Thematic narrative" (Emerson, Fretz, & Shaw, 1995) | • Tropes (Hammersley & Atkinson, 1995)<br>• "Thick" description (Denzin, 1989b)<br>• Dialogue (Nelson, 1990)<br>• Scenes (Emerson, Fretz, & Shaw, 1995)<br>• Literary devices (Richardson, 1990) |
| Case study | • Report format with vignettes (Stake, 1995)<br>• Substantive case report format (Lincoln & Guba, 1985)<br>• Types of cases (Yin, 2003) | • Funnel approach (Asmussen & Creswell, 1995)<br>• Amount of description (Merriam, 1988) |

and Connelly (2000) discuss. This space, as mentioned earlier, is a text that looks backward and forward, looks inward and outward, and situates the experiences within place. For example, the dissertation of He, cited by Clandinin, is a study about the lives of two participants and the author in China and Canada. The story

> looks backward to the past for her and her two participants and forward to the puzzle of who they are and who they are becoming in their new land. She looks inward to her personal reasons for doing this study and outward to the social significance of the work. She paints landscapes of China and Canada and the in-between places where she images herself to reside. (Clandinin & Connelly, 2000, p. 156)

Later in Clandinin and Connelly (2000), there is a story about Jean Clandinin's advising students about the narrative form of their studies. This form again relates to the three-dimensional space model:

> When they came to Jean for conversations about their emerging texts, she found herself responding not so much with comments about preestablished and accepted forms but with response that raised questions situated within the three-dimensional narrative inquiry space. (Clandinin & Connelly, 2000, p. 165)

Notice in this passage how Clandinin "raised questions" rather than told the student how to proceed, and how she returned to the larger rhetorical structure of the three-dimensional inquiry space model as a framework for thinking about the writing of a narrative study. This framework also suggests a chronology to the narrative report.

## Embedded Rhetorical Structure

Assuming that the larger writing structure proceeds with experimentation and flexibility, the writing structure at the more micro level relates to several elements of writing strategies that authors might use in composing a narrative study. These were drawn from Czarniawska (2004) and Clandinin and Connelly (2000):

- The writing of a narrative needs to not silence some of the voices, and it ultimately gives more space to certain voices than others (Czarniawska, 2004).
- There can be a spatial element to the writing, such as in the *progressive-regressive method* (Denzin, 1989b) whereby the biographer begins with a key event in the participant's life and then works forward and backward from that event, such as in Denzin's study of alcoholics. Alternatively, there can be a

"zooming in" and "zooming out," such as describing a large context to a concrete field of study (e.g., a site) and then telescoping out again (Czarniawska, 2004).

- The writing may emphasize the "key event" or the "epiphany," defined as interactional moments and experiences that mark people's lives (Denzin, 1989b). He distinguishes four types: the major event that touches the fabric of the individual's life; the cumulative or representative events, experiences that continue for some time; the minor epiphany, which represents a moment in an individual's life; and episodes or relived epiphanies, which involve reliving the experience. Czarniawska (2004) introduces the key element of the plot or the emplotment, a means of introducing structure that allows making sense of the events reported.

- Themes can be reported in narrative writing. Smith (1994) recommends finding a theme to guide the development of the life to be written. This theme emerges from preliminary knowledge or a review of the entire life, although researchers often experience difficulty in distinguishing the major theme from lesser or minor themes. Clandinin and Connelly (2000) refer to writing research texts at the reductionistic boundary, an approach consisting of a "reduction downward" (p. 143) to themes in which the researcher looks for common threads or elements across participants.

- Other narrative rhetorical devices include the use of transitions, at which biographers excel. Lomask (1986) refers to these as built into the narratives in natural chronological linkages. Writers insert them through words or phrases, questions (which Lomask calls being "lazy"), and time-and-place shifts moving the action forward or backward. In addition to transitions, biographers employ foreshadowing, the frequent use of narrative hints of things to come or of events or themes to be developed later. Narrative researchers also use metaphors, and Clandinin and Connelly (2000) used the metaphor of a soup (i.e., with description of people, places, and things; arguments for understandings, and richly textured narratives of people situated in place, time, scene, and plot) within containers (i.e., dissertation, journal article) to describe their narrative texts.

Angrosino's (1994) study of Vonnie Lee, for example, illustrates many narrative writing structures. At the outset, we are told that "explorations in life history and metaphor" (p. 14) is the type of biographical writing. Although difficult to classify according to Clifford's (1970) taxonomy of biographies, it has elements of the artistic and scholarly biography where Angrosino retells Vonnie Lee's stories within the scholarly context of Vonnie Lee's life, his bus ride, and the thematic meanings of this bus ride. Certainly, Angrosino focuses on a key event, perhaps a minor epiphany in Vonnie Lee's life of the bus ride. When Angrosino joins Vonnie Lee on his bus ride (and hears Vonnie Lee's stories), the reader gains a sense of movement from one bus stop to another until they reach Vonnie Lee's place of employment. The transitions of this journey are natural, and his struggles in life are foreshadowed early in the story through the recapitulation of his abusive early life. This bus journey, on several levels, becomes a metaphor for Vonnie Lee's life of empowerment and stability.

# Phenomenological Structure

Those who write about phenomenology (e.g., Moustakas, 1994) provide more extensive attention to overall writing structures than to embedded ones. However, as in all forms of qualitative research, one can learn much from a careful study of research reports in journal article, monograph, or book form.

## Overall Rhetorical Structure

The highly structured approach to analysis by Moustakas (1994) presents a detailed form for composing a phenomenological study. The analysis steps—identifying significant statements, creating meaning units, clustering themes, advancing textural and structural descriptions, and making a composite description of textural and structural descriptions into an exhaustive description of the essential invariant structure (or essence) of the experience—provide a clearly articulated procedure for organizing a report (Moustakas, 1994). In my experience, individuals are quite surprised to find highly structured approaches to phenomenological studies on sensitive topics (e.g., "being left out," "insomnia," "being criminally victimized," "life's meaning," "voluntarily changing one's career during midlife," "longing," "adults being abused as children"; Moustakas, 1994, p. 153). But the data analysis procedure, I think, guides a researcher in that direction and presents an overall structure for analysis and ultimately the organization of the report.

Consider the overall organization of a report as suggested by Moustakas (1994). He recommends specific chapters in "creating a research manuscript":

- Chapter 1: Introduction and statement of topic and outline. Topics include an autobiographical statement about experiences of the author leading to the topic, incidents that lead to a puzzlement or curiosity about the topic, the social implications and relevance of the topic, new knowledge and contribution to the profession to emerge from studying the topic, knowledge to be gained by the researcher, the research question, and the terms of the study.

- Chapter 2: Review of the relevant literature. Topics include a review of databases searched, an introduction to the literature, a procedure for selecting studies, the conduct of these studies and themes that emerged in them, a summary of core findings and statements as to how the present research differs from prior research (in question, model, methodology, and data collected).
- Chapter 3: Conceptual framework of the model. Topics include the theory to be used as well as the concepts and processes related to the research design (Chapters 3 and 4 might be combined).
- Chapter 4: Methodology. Topics include the methods and procedures in preparing to conduct the study, in collecting data, and in organizing, analyzing, and synthesizing the data.
- Chapter 5: Presentation of data. Topics include verbatim examples of data collection, data analysis, a synthesis of data, horizonalization, meaning units, clustered themes, textural and structural descriptions, and a synthesis of meanings and essences of the experience.
- Chapter 6: Summary, implications, and outcomes. Sections include a summary of the study, statements about how the findings differ from those in the literature review, recommendations for future studies, the identification of limitations, a discussion about implications, and the inclusion of a creative closure that speaks to the essence of the study and its inspiration for the researcher.

A second model, not as specific, is found in Polkinghorne (1989) where he discusses the "research report." In this model, the researcher describes the procedures to collect data and the steps to move from the raw data to a more general description of the experience. Also, the investigator includes a review of previous research, the theory pertaining to the topic, and implications for psychological theory and application. I especially like Polkinghorne's comment about the impact of such a report:

> Produce a research report that gives an accurate, clear, and articulate description of an experience. The reader of the report should come away with the feeling that "I understand better what it is like for someone to experience that." (Polkinghorne, 1989, p. 46)

## Embedded Rhetorical Structure

Turning to embedded rhetorical structures, the literature provides the best evidence. A writer presents the "essence" of the experience for participants in a study through sketching a short paragraph about it in the narrative or by enclosing this paragraph in a figure. This latter approach is used effectively in a study of the caring experiences of nurses who teach (Grigsby & Megel, 1995). Another structural device is to "educate" the reader through a discussion about phenomenology and its philosophical assumptions. Harper (1981) uses this approach and describes several of Husserl's major tenets as well as the advantages of studying the meaning of "leisure" in a phenomenology.

Finally, I personally like Moustakas's (1994) suggestion: "Write a brief creative close that speaks to the essence of the study and its inspiration to you in terms of the value of the knowledge and future directions of your professional-personal life" (p. 184). Despite the phenomenologist's inclination to bracket himself or herself out of the narrative, Moustakas introduces the reflexivity that psychological phenomenologists can bring to a study, such as casting their initial problem statement within an autobiographical context.

Anderson and Spencer's (2002) phenomenology of how persons living with AIDS image their disease represents many of these overall and embedded writing structures. The overall article has a structured organization, with an introduction, a review of the literature, methods, and results. It follows Colaizzi's (1978) phenomenological methods by reporting a table of significant statements and a table of meaning themes. Anderson and Spencer end with an in-depth, exhaustive description of the phenomenon. They describe this exhaustive description:

> Results were integrated into an essential scheme of AIDS. The lived experience of AIDS was initially frightening, with a dread of body wasting and personal loss. Cognitive representations of AIDS included inescapable death, bodily destruction, fighting a battle, and having a chronic disease. Coping methods included searching for the "right drug," caring for oneself, accepting the diagnosis, wiping AIDS out of their thoughts, turning to God, and using vigilance. With time, most people adjusted to living with AIDS. Feelings ranged from "devastating," "sad," and "angry" to being at "peace" and "not worrying." (Anderson and Spencer, 2002, p. 1349)

Anderson and Spencer began the phenomenology with a quote from a 53-year-old man with AIDS, but did not mention themselves in a reflexive way. They also did not discuss the philosophical tenets behind phenomenology.

## Grounded Theory Structure

From reviewing grounded theory studies in journal article form, qualitative researchers can deduce a general form (and variations) for composing the narrative. The problem with journal articles is that the authors present truncated versions of the studies to fit within the parameters of the journals. Thus, a reader emerges from a review of a particular study without a full sense of the entire project.

## Overall Rhetorical Structure

Most importantly, authors need to present the theory in any grounded theory narrative. As May (1986) comments, "In strict terms, the findings are the theory itself, i.e., a set of concepts and propositions which link them" (p. 148). May continues to describe the research procedures in grounded theory:

- The research questions are broad, and they will change several times during data collection and analysis.
- The literature review "neither provides key concepts nor suggests hypotheses" (May, 1986, p. 149). Instead, the literature review in grounded theory shows gaps or bias in existing knowledge, thus providing a rationale for this type of qualitative study.
- The methodology evolves during the course of the study, so writing it early in a study poses difficulties. However, the researcher begins somewhere, and she or he describes preliminary ideas about the sample, the setting, and the data collection procedures.
- The findings section presents the theoretical scheme. The writer includes references from the literature to show outside support for the theoretical model. Also, segments of actual data in the form of vignettes and quotes provide useful explanatory material. This material helps the reader form a judgment about how well the theory is grounded in the data.
- The final discussion section discusses the relationship of the theory to other existing knowledge and the implications of the theory for future research and practice.

Strauss and Corbin (1990) also provide broad writing parameters for their grounded theory studies. They suggest the following:

- Develop a clear analytic story. This is to be provided in the selective coding phase of the study.
- Write on a conceptual level, with description kept secondary to concepts and the analytic story. This means that one finds little description of the phenomenon being studied and more analytic theory at an abstract level.
- Specify the relationship among categories. This is the theorizing part of grounded theory found in axial coding when the researcher tells the story and advances propositions.
- Specify the variations and the relevant conditions, consequences, and so forth for the relationships among categories. In a good theory, one finds variation and different conditions under which the theory holds. This means that the multiple perspectives or variations in each component of axial coding are developed fully. For example, the consequences in the theory are multiple and detailed.

## Embedded Rhetorical Structure

In grounded theory studies, the researcher varies the narrative report based on the extent of data analysis. Chenitz and Swanson (1986), for example, present six grounded theory studies that vary in the types of analysis reported in the narrative. In a preface to these examples, they mention that the analysis (and narrative) might address one or more of the following: description; the generation of categories through open coding; linking categories around a core category in axial coding, thus developing a substantive, low-level theory; and/or a substantive theory linked to a formal theory.

I have seen grounded theory studies that include one or more of these analyses. For example, in a study of gays and their "coming out" process, Kus (1986) used only open coding in the analysis and identified four stages in the process of coming out: identification, in which a gay person undergoes a radical identity transformation; cognitive changes, in which the individual changes negative views about gays into positive ideas; acceptance, a stage in which the individual accepts being gay as a positive life force; and action, the process of the individual's engaging in behavior that results from accepting being gay, such as self-disclosure, expanding the circle of friends to include gays, becoming politically involved in gay causes, and volunteering for gay groups. Set in contrast to this focus on the process, Brown and I (Creswell & Brown, 1992) followed the coding steps in Strauss and Corbin (1990). We examined the faculty development practices of chairpersons who enhance the research productivity of their faculties. We began with open coding, moved to axial coding complete with a logic diagram, and stated a series of explicit propositions in directional (as opposed to the null) form.

Another embedded narrative feature is to examine the form for stating propositions or theoretical relationships in grounded theory studies. Sometimes, these are presented in "discursive" form, or describing the theory in narrative form. Strauss and Corbin (1990) present such a model in their theory of "protective governing" (p. 134) in the health care setting. Another example is seen in Conrad's (1978) formal propositions about academic change in the academy.

A final embedded structure is the presentation of the "logic diagram," the "mini-framework," or the "integrative" diagram, where the researcher presents the actual theory in the form of a visual model. The elements of this structure are identified by the researcher in the axial coding phase, and the "story" in axial coding is a narrative version of it. How is this visual model presented? A good example of this diagram is found in the Morrow and Smith (1995) study of women who have survived childhood sexual abuse. Their diagram shows a theoretical model that contains the axial coding categories of causal conditions, the central phenomenon, the context, intervening conditions, strategies, and consequences. It is presented with directional arrows indicating the flow of causality from left to right, from causal conditions to consequences. Arrows also show that the context and intervening conditions directly impact the strategies. Presented near the end of the study, this visual form represents the culminating theory for the study.

## Ethnographic Structure

Ethnographers write extensively about narrative construction, from how the nature of the text shapes the subject matter to the "literary" conventions and devices used by authors (Atkinson & Hammersley, 1994). The general shapes of ethnographies and embedded structures are well detailed in the literature.

### Overall Rhetorical Structure

For example, Van Maanen (1988) provides the alternative forms of an ethnography. Many ethnographies are written as realist tales, reports that provide direct, matter-of-fact portraits of studied cultures without much information about how the ethnographers produced the portraits. In this type of tale, a writer uses an impersonal point of view, conveying a "scientific" and "objective" perspective. A confessional tale takes the opposite approach, and the researcher focuses more on his or her fieldwork experiences than on the culture. The final type, the impressionistic tale, is a personalized account of "fieldwork case in dramatic form" (Van Maanen, 1988, p. 7). It has elements of both realist and confessional writing and, in my mind, presents a compelling and persuasive story. In both the confessional and impressionistic tales, the first-person point of view is used, conveying a personal style of writing. Van Maanen (1988) states that other, less frequently written tales also exist—critical tales focusing on large social, political, symbolic, or economic issues; formalist tales that build, test, generalize, and exhibit theory; literary tales in which the ethnographers write like journalists, borrowing fiction-writing techniques from novelists; and jointly told tales in which the production of the studies is jointly authored by the fieldworkers and the informants, opening up shared and discursive narratives.

On a slightly different note, but yet related to the larger rhetorical structure, Wolcott (1994b) provides three components of a good qualitative inquiry that are a centerpiece of good ethnographic writing as well as steps in data analysis. First, an ethnographer writes a "description" of the culture that answers the question "What is going on here?" (p. 12). Wolcott offers useful techniques for writing this description: chronological order, the researcher or narrator order, a progressive focusing, a critical or key event, plots and characters, groups in interaction, an analytical framework, and a story told through several perspectives. Second, after describing the culture using one of these approaches, the researcher "analyzes" the data. Analysis includes highlighting findings, displaying findings, reporting fieldwork procedures, identifying patterned regularities in the data, comparing the case with a known case, evaluating the information, contextualizing the information within a broader analytic framework, critiquing the research process, and proposing a redesign of the study. Of all these analytic techniques, the identification of "patterns" or themes is central to much ethnographic writing. Third, interpretation should be involved in the rhetorical structure. This means that the researcher can extend the analysis, make inferences from the information, do as directed or as suggested by gatekeepers, turn to theory, refocus the interpretation itself, connect with personal experience, analyze or interpret the interpretive process, or explore alternative formats. Of these interpretive strategies, I personally like the approach of interpreting the findings both within the context of the researcher's experiences and within the larger body of scholarly research on the topic.

A more detailed, structured outline for an ethnography is found in Emerson, Fretz, and Shaw (1995). They discuss developing an ethnographic study as a "thematic narrative," a story "analytically thematized, but often in relatively loose ways . . . constructed out of a series of thematically organized units of fieldnote excerpts and analytic commentary" (p. 170). This thematic narrative builds inductively from a main idea or thesis that incorporates several specific analytic themes and is elaborated throughout the study. It is structured as follows:

- First is an introduction that engages the reader's attention and focuses the study, then the researcher proceeds to link his or her interpretation to wider issues of scholarly interest in the discipline.
- After this, the researcher introduces the setting and the methods for learning about it. In this section, too, the ethnographer relates details about entry into and participation in the setting as well as advantages and constraints of the ethnographer's research role.
- The researcher presents analytic claims next, and Emerson and colleagues (1995) indicate the utility of "excerpt commentary" units, whereby an author incorporates an analytic point, provides orientation information about the point, presents the excerpt or direct quote, and then advances analytic commentary about the quote as it relates to the analytic point.
- In the conclusion, the researcher reflects and elaborates on the thesis advanced at the beginning. This interpretation may extend or modify the thesis in light of

the materials examined, relate the thesis to general theory or a current issue, or offer a metacommentary on the thesis, methods, or assumptions of the study.

## Embedded Rhetorical Structure

Ethnographers use embedded rhetorical devices such as figures of speech or "tropes" (Hammersley & Atkinson, 1995). Metaphors, for example, provide visual and spatial images or dramaturgical characterizations of social actions as theater. Another trope is the synecdoche, in which ethnographers present examples, illustrations, cases, and/or vignettes that form a part but stand for the whole. Ethnographers present storytelling tropes examining cause and sequence that follow grand narratives to smaller parables. A final trope is irony, in which researchers bring to light contrasts of competing frames of reference and rationality.

More specific rhetorical devices depict scenes in an ethnography (Emerson et al., 1995). Writers can incorporate details or "write lushly" (Goffman, 1989, p. 131) or "thickly," description that creates verisimilitude and produces for readers the feeling that they experience, or perhaps could experience, the events described (Denzin, 1989b). Denzin (1989b) talks about the importance of using "thick description" in writing qualitative research. By this, he means that the narrative "presents detail, context, emotion, and the webs of social relationships . . . [and] evokes emotionality and self-feelings. . . . The voices, feelings, actions, and meanings of interacting individuals are heard" (p. 83). As an example, Denzin (1989b) first refers to an illustration of "thick" description from Sudnow (1978), and then provides his own version as if it were "thin" description.

- Thick description: "Sitting at the piano and moving into the production of a chord, the chord as a whole was prepared for as the hand moved toward the keyboard, and the terrain was seen as a field relative to the task. . . . There was chord A and chord B, separated from one another. . . . A's production entailed a tightly compressed hand, and B's . . . an open and extended spread. . . . The beginner gets from A to B disjointly" (Sudnow, 1978, pp. 9–10 )
- Thin description: "I had trouble learning the piano keyboard" (Denzin, 1989b, p. 85).

Also, ethnographers present dialogue, and the dialogue becomes especially vivid when written in the dialect and natural language of the culture (see, e.g., the articles on Black English vernacular or "code switching" in Nelson, 1990). Writers also rely on characterization in which human beings are shown talking, acting, and relating to others. Longer scenes take the form of sketches, a "slice of life" (Emerson et al., 1995, p. 85), or larger episodes and tales.

Ethnographic writers tell "a good story" (Richardson, 1990). Thus, one of the forms of "evocative" experimental qualitative writing for Richardson (1990) is the fictional representation form in which writers draw on the literary devices such as flashback, flashforward, alternative points of view, deep characterization, tone shifts, synecdoche, dialogue, interior monologue, and sometimes the omniscient narrator.

Haenfler's (2004) ethnographic study of the core values of the straight edge movement illustrate many of these writing conventions. It falls somewhere between a realist tale, with its review of the literature and extensive method discussion, and a critical tale, with its orientation toward examining closely subculture resistance and the reflexivity of the author as he discusses his involvement as a participant observer. It does follow Wolcott's (1994b) orientation of description with a detailed discussion about the core values of the sXe group, followed by analysis through themes, and ending with a conclusion that discusses an analytic framework for understanding the group. It tells a good, persuasive story, with colorful elements (e.g., T-shirt slogans), "thick" description, and extensive quotes. It does not include some of the literary tropes, such as dialogue, interior monologue, and the tone is one of an omniscient narrator as typically found in the realist tales of Van Maanen (1988).

# Case Study Structure

Turning to case studies, I am reminded by Merriam (1988) that "there is no standard format for reporting case study research" (p. 193). Unquestionably, some case studies generate theory, some are simply descriptions of cases, and others are more analytical in nature and display cross-case or intersite comparisons. The overall intent of the case study undoubtedly shapes the larger structure of the written narrative. Still, I find it useful to conceptualize a general form, and I turn to key texts on case studies to receive guidance.

## Overall Rhetorical Structure

One can open and close the case study narrative with vignettes to draw the reader into the case. This approach is suggested by Stake (1995), who provides a complete outline for the flow of ideas in a case study. These ideas are staged as follows:

- The writer opens with a vignette so that the reader can develop a vicarious experience to get a feel for the time and place of the study.
- Next, the researcher identifies the issue, the purpose, and the method of the study so that the reader learns about how the study came to be, the background of the writer, and the issues surrounding the case.

- This is followed by an extensive description of the case and its context—a body of relatively uncontested data—a description the reader might make if he or she had been there.
- Issues are presented next, a few key issues, so that the reader can understand the complexity of the case. This complexity builds through references to other research or the writer's understanding of other cases.
- Next, several of the issues are probed further. At this point, too, the writer brings in both confirming and disconfirming evidence.
- Assertions are presented, a summary of what the writer understands about the case and whether the initial naturalistic generalizations, conclusions arrived at through personal experience or offered as vicarious experiences for the reader, have been changed conceptually or challenged.
- Finally, the writer ends with a closing vignette, an experiential note, reminding the reader that this report is one person's encounter with a complex case.

I like this general outline because it provides description of the case; presents themes, assertions, or interpretations of the researcher; and begins and ends with realistic scenarios.

A similar model is found in Lincoln and Guba's (1985) substantive case report. They describe a need for the explication of the problem, a thorough description of the context or setting, a description of the transactions or processes observed in that context, saliences at the site (elements studied in depth), and outcomes of the inquiry ("lessons learned").

At a more general level yet, I find Yin's (2003) $2 \times 2$ table of types of case studies helpful. Case studies can be either single-case or multiple-case design and either holistic (single unit of analysis) or embedded (multiple units of analysis) design. He comments further that a single case is best when a need exists to study a critical case, an extreme or unique case, or a revelatory case. Whether the case is single or multiple, the researcher decides to study the entire case, a holistic design, or multiple subunits within the case (the embedded design). Although the holistic design may be more abstract, it captures the entire case better than the embedded design does. However, the embedded design starts with an examination of subunits and allows for the detailed perspective should the questions begin to shift and change during fieldwork.

## Embedded Rhetorical Structure

What specific narrative devices, embedded structures, do case study writers use to "mark" their studies? One might approach the description of the context and setting for the case from a broader picture to a narrower one. For example, in the gunman case (Asmussen & Creswell, 1995), we describe the actual campus incident first in terms of the city in which the situation developed, followed by the campus and, more narrow yet, the actual classroom on campus. This funneling approach narrows the setting from that of a calm city environment to a potentially volatile campus classroom and seems to launch the study into a chronology of events that occur.

Researchers also need to be cognizant of the amount of description in their case studies versus the amount of analysis and interpretation or assertions. In comparing description and analysis, Merriam (1988) suggests that the proper balance might be 60%/40% or 70%/30% in favor of description. In the gunman case, Asmussen and I balanced the elements in equal thirds (33%-33%-33%)—a concrete description of the setting and the actual events (and those that occurred within 2 weeks after the incident); the five themes; and our interpretation, the lessons learned, reported in the discussion section. In our case study, the description of the case and its context did not loom as large as in other case studies. But these matters are up to writers to decide, and it is conceivable that a case study might contain mainly descriptive material, especially if the bounded system, the case, is quite large and complex.

Our gunman study (Asmussen & Creswell, 1995) also represents a single-case study (Yin, 2003), with a single narrative about the case, its themes, and its interpretation. In another study, the case presentation might be that of multiple cases, with each case discussed separately, or multiple case studies with no separate discussions of each case but an overall cross-case analysis (Yin, 2003). Another Yin (2003) narrative format is to pose a series of questions and answers based on the case study database.

Within any of these formats, one might consider structures for building ideas. For example, in our gunman study (Asmussen & Creswell, 1995), we descriptively present the chronology of the events during the incident and immediately after it. The chronological approach seems to work best when events unfold and follow a process; case studies often are bounded by time and cover events over time (Yin, 2003). In addition to this approach, one might build a theory composed of identifying variables (or themes) that are interrelated; use a "suspense" structure with an "answer" to the outcome of the case presented first, followed by the development of an explanation for this outcome; or use an "unsequenced" structure consisting of events, processes, or activities not necessarily presented in the order in which they unfolded in the case (Yin, 2003).

## A Comparison of Narrative Structures

Looking back over Table 9.1, we see many diverse structures for writing the qualitative report. What major differences exist in the structures depending on one's choice of approach?

First, I am struck by the diversity of discussions about narrative structures. I found little crossover or sharing of structures among the five approaches, although, in practice, this undoubtedly occurs. The narrative tropes and the literary devices, discussed by ethnographers and narrative researchers, have applicability regardless of approach. Second, the writing structures are highly related to data analysis procedures. A phenomenological study and a grounded theory study follow closely the data analysis steps. In short, I am reminded once again that it is difficult to separate the activities of data collection, analysis, and report writing in a qualitative study. Third, the emphasis given to writing the narrative, especially the embedded narrative structures, varies among the approaches. Ethnographers lead the group in their extensive discussions about narrative and text construction. Phenomenologists and grounded theory writers spend little time on this topic. Fourth, the overall narrative structure is clearly specified in some approaches (e.g., a grounded theory study, a phenomenological study, and perhaps a case study), whereas it is flexible and evolving in others (e.g., a narrative, an ethnography). Perhaps this conclusion reflects the more structured approach versus the less structured approach, overall, among the five approaches of inquiry.

## Summary

In this chapter, I discussed writing the qualitative report. I began by discussing several rhetorical issues the writer must address. These issues include writing reflexively and with representation, the audience for the writing, the encoding for that audience, and the use of quotes. Then I turned to each of the five approaches of inquiry and presented overall rhetorical structures for organizing the entire study as well as specific embedded structures, writing devices, and techniques that the researcher incorporates into the study. A table of these structures shows the diversity of perspectives about structure that reflects different data analysis procedures and discipline affiliations. I concluded with observations about the differences in writing structures among the five approaches, differences reflected in the variability of approaches, the relationships between data analysis and report writing, the emphasis in the literature of each approach on narrative construction, and the amount of structure in the overall architecture of a study within each approach.

### Additional Readings

A good, thoughtful book on writing qualitative research is Wolcott's popular 2001 book. For examining the issues of reflexivity and representation,

I highly recommend Gilgun (2005), Richardson and St. Pierre (2005), Weis and Fine (2000) and van Manen (2006). For specific applications in the five approaches, see Clandinin and Connelly (2000), Czarniawska (2004), and Denzin (1989b) for narrative research; Moustakas (1994) for phenomenology; Strauss and Corbin (1990, 1998) for grounded theory; Clifford and Marcus (1986), Wolcott (1994b), and Van Maanen (1988) for ethnography; and Stake (1995) and Yin (2003) for case study research.

Clandinin, D. J., & Connelly, F. M. (2000). *Narrative inquiry: Experience and story in qualitative research*. San Francisco: Jossey-Bass.

Clifford, J., & Marcus, G. E. (Eds.). (1986). *Writing culture: The poetics and politics of ethnography*. Berkeley: University of California Press.

Czarniawka, B. (2004). *Narratives in social science research*. Thousand Oaks, CA: Sage.

Denzin, N. K. (1989b). *Interpretive interactionism*. Newbury Park, CA: Sage.

Gilgun, J. F. (2005). "Grab" and good science: Writing up the results of qualitative research. *Qualitative Health Research, 15*, 256–262.

Moustakas, C. (1994). *Phenomenological research methods*. Thousand Oaks, CA: Sage.

Richardson, L., & St. Pierre, E. A. (2005). Writing: A method of inquiry. In N. K. Denzin & Y. S. Lincoln (Eds.), *The Sage handbook of qualitative research* (3rd ed., pp. 959–978). Thousand Oaks, CA: Sage.

Stake, R. (1995). *The art of case study research*. Thousand Oaks, CA: Sage.

Strauss, A., & Corbin, J. (1990). *Basics of qualitative research: Grounded theory procedures and techniques*. Newbury Park, CA: Sage.

Strauss, A., & Corbin, J. (1998). *Basics of qualitative research: Grounded theory procedures and techniques* (2nd ed.). Thousand Oaks, CA: Sage.

Van Maanen, J. (1988). *Tales of the field: On writing ethnography*. Chicago: University of Chicago Press.

van Manen, M. (2006). Writing qualitatively, or the demands of writing. *Qualitative Health Research, 16*, 713–722.

Weis, L., & Fine, M. (2000). *Speed bumps: A student-friendly guide to qualitative research*. New York: Teachers College Press.

Wolcott, H. F. (1994b). *Transforming qualitative data: Description, analysis, and interpretation*. Thousand Oaks, CA: Sage.

Wolcott, H. F. (2001). *Writing up qualitative research* (2nd ed.). Thousand Oaks, CA: Sage.

Yin, R. K. (2003). *Case study research: Design and method* (2nd ed.). Thousand Oaks, CA: Sage.

**Exercises**

1. Show that you understand the overall and embedded rhetorical structures for writing within your approach of inquiry by drafting a complete narrative for

your project. You might model your narrative after a journal article format using your approach.

2. Develop a plan for the narrative structure for a study within your approach of inquiry. To do this, design a matrix with two columns and seven rows. In the first column, list several writing criteria: the overall writing approach, the strategies to display reflexivity and representation, the intended audience for the study, the encoding to be used in the narrative, the approach to using quotes, the general outline of the flow of the ideas in the manuscript, and the embedded rhetorical devices. In the second column, add information about how these criteria will be addressed in your project.

**Exercises**

# 10

# Standards of Validation and Evaluation

Qualitative researchers strive for "understanding," that deep structure of knowledge that comes from visiting personally with participants, spending extensive time in the field, and probing to obtain detailed meanings. During or after a study, qualitative researchers ask, "Did we get it right?" (Stake, 1995, p. 107) or "Did we publish a 'wrong' or inaccurate account?" (Thomas, 1993, p. 39). Is it possible to even have a "right" answer? To answer these questions, researchers need to look to themselves, to the participants, and to the readers. There are multi- or polyvocal discourses at work here that provide insight into the validation and evaluation of a qualitative narrative.

In this chapter, I address two interrelated questions: Is the account valid, and by whose standards? How do we evaluate the quality of qualitative research? Answers to these questions will take us into the many perspectives on validation to emerge within the qualitative community and the multiple standards for evaluation discussed by authors with procedural, interpretive, emancipatory, and postmodern perspectives.

## Questions for Discussion

- What are some qualitative perspectives on validation?
- What are some alternative procedures useful in establishing validation?

- How is reliability used in qualitative research?
- What are some alternative stances on evaluating the quality of qualitative research?
- How do these stances differ by types of approaches to qualitative inquiry?

# Validation and Reliability in Qualitative Research

## Perspectives on Validation

Many perspectives exist regarding the importance of validation in qualitative research, the definition of it, terms to describe it, and procedures for establishing it. In Table 10.1, I illustrate several of the perspectives available on validation in the qualitative literature. These perspectives are viewing qualitative validation in terms of quantitative equivalents, using qualitative terms that are distinct from quantitative terms, employing postmodern and interpretive perspectives, considering validation as unimportant, combining or synthesizing many perspectives, or visualizing it metaphorically as a crystal.

Writers have searched for and found qualitative equivalents that parallel traditional quantitative approaches to validation. LeCompte and Goetz (1982) took this approach when they compared the issues of validation and reliability to their counterparts in experimental design and survey research. They contended that qualitative research has garnered much criticism in the scientific ranks for its failure to "adhere to canons of reliability and validation" (p. 31) in the traditional sense. They applied threats to internal validation in experimental research to ethnographic research (e.g., history and maturation, observer effects, selection and regression, mortality, spurious conclusions). They further identified threats to external validation as "effects that obstruct or reduce a study's comparability or translatability" (p. 51).

Some writers argue that authors who continue to use positivist terminology facilitate the acceptance of qualitative research in a quantitative world. Ely and colleagues (Ely, Anzul, Friedman, Garner, & Steinmetz, 1991) asserted that using quantitative terms tends to be a defensive measure that muddies the waters and that "the language of positivistic research is not congruent with or adequate to qualitative work" (p. 95). Lincoln and Guba (1985) have used alternative terms that, they contended, adhered more to naturalistic research. To establish the "trustworthiness" of a study, Lincoln and Guba (1985) used unique terms, such as "credibility," "authenticity," "transferability," "dependability," and "confirmability," as "the naturalist's equivalents" for "internal validation," "external validation," "reliability," and "objectivity" (p. 300). To operationalize these new terms, they propose techniques such as prolonged engagement in the field and the triangulation

**Table 10.1**    Perspectives and Terms Used in Qualitative Validation

| Study | Perspective | Terms |
|---|---|---|
| LeCompte & Goetz (1982) | Use of parallel, qualitative equivalents to their quantitative counterparts in experimental and survey research | Internal validity<br>External validity<br>Reliability<br>Objectivity |
| Lincoln & Guba (1985) | Use of alternative terms that apply more to naturalistic axioms | Credibility<br>Transferability<br>Dependability<br>Confirmability |
| Eisner (1991) | Use of alternative terms that provide reasonable standards for judging the credibility of qualitative research | Structural corroboration<br>Consensual validation<br>Referential adequacy<br>Ironic validity |
| Lather (1993) | Use of reconceptualized validity in four types | Paralogic validity<br>Rhizomatic validity<br>Situated/embedded voluptuous validity |
| Wolcott (1994b) | Use of terms other than "validity," because it neither guides nor informs qualitative research | Understanding better than validity |
| Angen (2000) | Use of validation within the context of interpretive inquiry | Two types: ethical and substantive |
| Whittemore, Chase, & Mandle (2001) | Use of synthesized perspectives of validity, organized into primary criteria and secondary criteria | Primary criteria: credibility, authenticity, criticality, and integrity<br><br>Secondary criteria: Explicitness, vividness, creativity, thoroughness, congruence, and sensitivity |
| Richardson & St. Pierre (2005) | Use of a metaphorical, reconceptualized form of validity as a crystal | Crystals: Grow, change, alter, reflect externalities, refract within themselves |

of data of sources, methods, and investigators to establish credibility. To make sure that the findings are transferable between the researcher and those being studied, thick description is necessary. Rather than reliability, one seeks dependability that the results will be subject to change and instability. The naturalistic researcher looks for confirmability rather than objectivity in establishing the value of the data. Both dependability and confirmability are established through an auditing of the research process.

Rather than using the term "validation," Eisner (1991) discussed the credibility of qualitative research. He constructed standards such as structural corroboration, consensual validation, and referential adequacy. In structural corroboration, the researcher relates multiple types of data to support or contradict the interpretation. As Eisner (1991) stated, "We seek a confluence of evidence that breeds credibility, that allows us to feel confident about our observations, interpretations, and conclusions" (p. 110). He further illustrated this point with an analogy drawn from detective work: The researcher compiles bits and pieces of evidence to formulate a "compelling whole." At this stage, the researcher looks for recurring behaviors or actions and considers disconfirming evidence and contrary interpretations. Moreover, Eisner (1991) recommended that to demonstrate credibility, the weight of evidence should become persuasive. Consensual validation sought the opinion of others, and Eisner referred to "an agreement among competent others that the description, interpretation, and evaluation and thematics of an educational situation are right" (p. 112). Referential adequacy suggested the importance of criticism, and Eisner described the goal of criticism as illuminating the subject matter and bringing about more complex and sensitive human perception and understanding.

Validation also has been reconceptualized by qualitative researchers with a postmodern sensibility. Lather (1991) commented that current "paradigmatic uncertainty in the human sciences is leading to the re-conceptualizing of validation" and called for "new techniques and concepts for obtaining and defining trustworthy data which avoids the pitfalls of orthodox notions of validation" (p. 66). For Lather, the character of a social science report changes from a closed narrative with a tight argument structure to a more open narrative with holes and questions and an admission of situatedness and partiality. In *Getting Smart*, Lather (1991) advanced a "reconceptualization of validation." She identified four types of validation, including triangulation (multiple data sources, methods, and theoretical schemes), construct validation (recognizing the constructs that exist rather than imposing theories/constructs on informants or the context), face validation (as "a 'click of recognition' and a 'yes, of course,' instead of 'yes, but' experience" (Kidder, 1982, p. 56), and catalytic validation (which energizes participants toward knowing reality to transform it).

In a later article, Lather's (1993) terms became more unique and closely related to feminist research in "four frames of validation." The first, "ironic" validation, is where the researcher presents truth as a problem. The second, "paralogic" validation, is concerned with undecidables, limits, paradoxes, and complexities, a movement away from theorizing things and toward providing direct exposure to other voices in an almost unmediated way. The third, "rhizomatic" validation, pertains to questioning proliferations, crossings, and overlaps without underlying structures or deeply rooted connections. The researcher also questions taxonomies, constructs, and interconnected networks whereby the reader jumps from one assemblage to another and consequently moves from judgment to understanding. The fourth type is situated, embodied, or "voluptuous" validation, which means that the researcher sets out to understand more than one can know and to write toward what one does not understand.

Other writers, such as Wolcott (1990a), have little use for validation. He suggested that "validation neither guides nor informs" his work (p. 136). He did not dismiss validation, but rather placed it in a broader perspective. Wolcott's goal was to identify "critical elements" and write "plausible interpretations from them" (p. 146). He ultimately tried to understand rather than convince, and he voiced the view that validation distracted from his work of understanding what was really going on. Wolcott claimed that the term "validation" did not capture the essence of what he sought, adding that perhaps someone would coin a term appropriate for the naturalistic paradigm. But for now, he said, the term "understanding" seemed to encapsulate the idea as well as any other.

Validation has also been cast within an interpretive approach to qualitative research marked by a focus on the importance of the researcher, a lack of truth in validation, a form of validation based on negotiation and dialogue with participants, and interpretations that are temporal, located, and always open to reinterpretation (Angen, 2000). Angen (2000) suggested that within interpretative research, validation is "a judgment of the trustworthiness or goodness of a piece of research" (p. 387). She espouses an ongoing open dialogue on the topic of what makes interpretive research worthy of our trust. Considerations of validation are not definitive as the final word on the topic, nor should every study be required to address them. Further, she advances two types of validation: ethical validation and substantive validation. Ethical validation means that all research agendas must questions their underlying moral assumptions, their political and ethical implications, and the equitable treatment of diverse voices. It also requires research to provide some practical answers to questions. Our research should also have a "generative promise" (Angen, 2000, p. 389) and raise new possibilities, open up new questions, and stimulate new dialogue. Our research must have transformative value leading to action and change. Our research should also provide nondogmatic answers to the questions we pose.

Substantive validation means understanding one's own understandings of the topic, understandings derived from other sources, and the documentation of this process in the written study. Self-reflection contributes to the validation of the work. The researcher, as a sociohistorical interpreter, interacts with the subject matter to co-create the interpretations derived. Understandings derived from previous research give substance to the inquiry. Interpretive research also is a chain of interpretations that must be documented for others to judge the trustworthiness of the meanings arrived at the end. Written accounts must resonate with their intended audiences, and must be compelling, powerful, and convincing.

A synthesis of validation perspectives comes from Whittemore, Chase, and Mandle (2001), who have analyzed 13 writings about validation, and extracted from these studies key validation criteria. They organized these criteria into primary and secondary criteria. They found four primary criteria: credibility (Are the results an accurate interpretation of the participants' meaning?); authenticity (Are different voices heard?); criticality (Is there a critical appraisal of all aspects of the research?); and integrity (Are the investigators self-critical?). Secondary criteria related to explicitness, vividness, creativity, thoroughness, congruence, and sensitivity. In summary, with these criteria, it seems like the validation standard has moved toward the interpretive lens of qualitative research, with an emphasis on researcher reflexivity and on researcher challenges that include raising questions about the ideas developed during a research study.

Finally, a recent postmodern perspective draws on the metaphorical image of a crystal. Richardson (in Richardson & St. Pierre, 2005) describes this image:

> I propose that the central imaginary for "validation" for postmodern texts is not the triangle—a rigid, fixed, two dimensional object. Rather the central imaginary is the crystal, which combines symmetry and substance with an infinite variety of shapes, substances, transmutations, multidimensionalities, and angles of approach. Crystals grow, change, and are altered, but they are not amorphous. Crystals are prisms that reflect externalities and refract within themselves, creating different colors, patterns, and arrays casting off in different directions. What we see depends on our angle of response—not triangulation but rather crystallization. (p. 963)

Given these many perspectives, I will summarize my own stance:

- I consider "validation" in qualitative research to be an attempt to assess the "accuracy" of the findings, as best described by the researcher and the participants. This view also suggests that any report of research is a representation by the author.

- I also view validation as a distinct strength of qualitative research in that the account made through extensive time spent in the field, the detailed thick description, and the closeness of the researcher to participants in the study all add to the value or accuracy of a study.

- I use the term "validation" to emphasize a process (see Angen, 2000), rather than "verification" (which has quantitative overtones) or historical words such as "trustworthiness" and "authenticity" (recognizing that many qualitative writers do return to words such as "authenticity" and "credibility," suggesting the "staying power" of Lincoln and Guba's 1985 standards; see Whittemore, Chase, & Mandle, 2001). I acknowledge that there are many types of qualitative validation and that authors need to choose the types and terms in which they are comfortable. I recommend that writers reference their validation terms and strategies.

- The subject of validation does arise in several of the approaches to qualitative research (e.g., Stake, 1995; Strauss & Corbin, 1998), but I do not think that distinct validation approaches exist for the five approaches to qualitative research. At best, there might be less emphasis on validation in narrative research and more emphasis on it in grounded theory, case study, and ethnography, especially when the authors talking about these approaches want to employ systematic procedures. I would recommend using validation strategies regardless of type of qualitative approach.

- My framework for thinking about validation in qualitative research is to suggest that researchers employ accepted strategies to document the "accuracy" of their studies. These I call "validation strategies."

## Validation Strategies

It is not enough to gain perspectives and terms; ultimately, these ideas are translated into practice as strategies or techniques. Whittemore, Chase, and Mandle (2001) have organized the techniques into 29 forms that apply to design consideration, data generating, analytic, and presentation. My colleague and I (Creswell & Miller, 2000) have chosen to focus on eight strategies that are frequently used by qualitative researchers. These are not presented in any specific order of importance.

- Prolonged engagement and persistent observation in the field include building trust with participants, learning the culture, and checking for misinformation that stems from distortions introduced by the researcher or informants (Ely et al., 1991; Erlandson, Harris, Skipper, & Allen, 1993; Glesne & Peshkin, 1992; Lincoln & Guba, 1985; Merriam, 1988). In the field, the researcher makes decisions about what is salient to the study, relevant to the purpose of the study, and of interest for focus. Fetterman

(1998) contends that "working with people day in and day out, for long periods of time, is what gives ethnographic research its validation and vitality" (p. 46).

- In triangulation, researchers make use of multiple and different sources, methods, investigators, and theories to provide corroborating evidence (Ely et al., 1991; Erlandson et al., 1993; Glesne & Peshkin, 1992; Lincoln & Guba, 1985; Merriam, 1988; Miles & Huberman, 1994; Patton, 1980, 1990). Typically, this process involves corroborating evidence from different sources to shed light on a theme or perspective.

- Peer review or debriefing provides an external check of the research process (Ely et al., 1991; Erlandson et al., 1993; Glesne & Peshkin, 1992; Lincoln & Guba, 1985; Merriam, 1988), much in the same spirit as interrater reliability in quantitative research. Lincoln and Guba (1985) define the role of the peer debriefer as a "devil's advocate," an individual who keeps the researcher honest; asks hard questions about methods, meanings, and interpretations; and provides the researcher with the opportunity for catharsis by sympathetically listening to the researcher's feelings. This reviewer may be a peer, and both the peer and the researcher keep written accounts of the sessions, called "peer debriefing sessions" (Lincoln & Guba, 1985).

- In negative case analysis, the researcher refines working hypotheses as the inquiry advances (Ely et al., 1991; Lincoln & Guba, 1985; Miles & Huberman, 1994; Patton, 1980, 1990) in light of negative or disconfirming evidence. The researcher revises initial hypotheses until all cases fit, completing this process late in data analysis and eliminating all outliers and exceptions.

- Clarifying researcher bias from the outset of the study is important so that the reader understands the researcher's position and any biases or assumptions that impact the inquiry (Merriam, 1988). In this clarification, the researcher comments on past experiences, biases, prejudices, and orientations that have likely shaped the interpretation and approach to the study.

- In member checking, the researcher solicits participants' views of the credibility of the findings and interpretations (Ely et al., 1991; Erlandson et al., 1993; Glesne & Peshkin, 1992; Lincoln & Guba, 1985; Merriam, 1988; Miles & Huberman, 1994). This technique is considered by Lincoln and Guba (1985) to be "the most critical technique for establishing credibility" (p. 314). This approach, *writ large* in most qualitative studies, involves taking data, analyses, interpretations, and conclusions back to the participants so that they can judge the accuracy and credibility of the account. According to Stake (1995), participants should "play a major role directing as well as acting in case study" research. They should be asked to examine rough drafts of the researcher's work and to provide alternative language, "critical observations or interpretations" (p. 115). For this validation strategy, I convene a focus group composed of participants in my study and ask them to reflect on the accuracy of the account. I do not take back to participants my transcripts or the raw data, but take them my preliminary analyses consisting of description or themes. I am interested in their views of these written analyses as well as what was missing.

• Rich, thick description allows readers to make decisions regarding transferability (Erlandson et al., 1993; Lincoln & Guba, 1985; Merriam, 1988) because the writer describes in detail the participants or setting under study. With such detailed description, the researcher enables readers to transfer information to other settings and to determine whether the findings can be transferred "because of shared characteristics" (Erlandson et al., 1993, p. 32).

• External audits (Erlandson et al., 1993; Lincoln & Guba, 1985; Merriam, 1988; Miles & Huberman, 1994) allow an external consultant, the auditor, to examine both the process and the product of the account, assessing their accuracy. This auditor should have no connection to the study. In assessing the product, the auditor examines whether or not the findings, interpretations, and conclusions are supported by the data. Lincoln and Guba (1985) compare this, metaphorically, with a fiscal audit, and the procedure provides a sense of interrater reliability to a study.

Examining these eight procedures as a whole, I recommend that qualitative researchers engage in at least two of them in any given study. Unquestionably, procedures such as triangulating among different data sources (assuming that the investigator collects more than one), writing with detailed and thick description, and taking the entire written narrative back to participants in member checking all are reasonably easy procedures to conduct. They also are the most popular and cost-effective procedures. Other procedures, such as peer audits and external audits, are more time consuming in their application and may also involve substantial costs to the researcher.

## Reliability Perspectives

Reliability can be addressed in qualitative research in several ways (Silverman, 2005). Reliability can be enhanced if the researcher obtains detailed fieldnotes by employing a good-quality tape for recording and by transcribing the tape. Also, the tape needs to be transcribed to indicate the trivial, but often crucial, pauses and overlaps. Further coding can be done "blind" with the coding staff and the analysts conducting their research without knowledge of the expectations and questions of the project directors, and by use of computer programs to assist in recording and analyzing the data. Silverman also supports intercoder agreement.

Our focus on reliability here will be on intercoder agreement based on the use of multiple coders to analyze transcript data. In qualitative research, "reliability" often refers to the stability of responses to multiple coders of data sets. I find this practice especially used in qualitative health science research and within the form of qualitative research in which inquirers want an external check on the highly interpretive coding process. What seems to be largely missing in the literature (with the exception of Miles and Huberman, 1994, and Armstrong, Gosling, Weinman, & Marteau, 1997) is a discussion about the procedures of actually conducting intercoder agreements checks. One of the key issues is determining what exactly the codings are agreeing on, whether they seek agreement on codes names, the coded passages, or the same passages coded the same way. We also need to decide on whether to seek agreement based on codes, themes, or both codes and themes (see Armstrong et al., 1997).

Undoubtedly, there is flexibility in the process, and researchers need to fashion an approach consistent with the resources and time to engage in coding. At the VA HealthCare System, Ann Arbor, Michigan, I had an opportunity to help design an intercoder agreement process using data related to the HIPPA privacy act (Damschroder, personal communication, March, 2006). In a project at the VA Ann Arbor Health Care System, we used the following steps in our intercoder agreement process:

- We sought to develop a codebook of codes that would be stable and represent the coding analysis of four independent coders. We all used NVivo as a software program to help in this coding.
- To achieve this goal, we read through several transcripts independently and coded each manuscript.
- After coding, say, three to four transcripts, we then met and examined the codes, their names, and the text segments that we coded. We began to develop a preliminary qualitative codebook of the major codes. This codebook contained a definition of each code, and the text segments that we assigned to each code. In this initial codebook, we had "parent" codes and "children" codes. In our initial codebook, we were more interested in the major codes we were finding in the database than in an exhaustive list. We felt that we could add to the codes as the analyses proceeded.
- We then each independently coded three additional transcripts, say, transcripts 5, 6, and 7. Now we were ready to actually compare our codes. We felt that it was more important to have agreement on the text segments we were assigning to codes than to have the same, exact passages coded. Intercoder

agreement to us meant that we agreed that when we assigned a code word to a passage, that we all assigned this same code word to the passage. It did not mean that we all coded the same passages—an ideal that I believe would be hard to achieve because some people code short passages and others longer passages. Nor did it mean that we all bracketed the same lines to include in our code word, another ideal difficult to achieve.

- So we took a realistic stance, and we looked at the passages that we all four coded and asked ourselves whether we had all assigned the same code word to the passage, based on our tentative definitions in the codebook. The decision would be either a "yes" or "no" decision, and we could calculate the percentage of agreement among all four of us on this passage that we all coded. We sought to establish an 80% agreement of coding on these passages (Miles and Huberman, 1994, recommend an 80% agreement). Other researchers might actually calculate a kappa reliability statistic on the agreement, but we felt that a percentage would suffice to report on our published study.

- After we collapsed codes into broader themes, we could conduct the same process with themes, to see if the passages we all coded as themes were consistent in the use of the same theme.

- After the process continued through several more transcripts, we then revised the codebook, and conducted anew an assessment of passages that we all coded and determined if we used the same or different codes or the same or different themes. With each phase in the intercoder agreement process, we achieved a higher percentage of agreed upon codes and themes for text segments.

# Evaluation Criteria

## Qualitative Perspectives

In reviewing validation in the qualitative research literature, I am struck by how validation is sometimes used in discussing the quality of a study (e.g., Angen, 2000). Although validation is certainly an aspect of evaluating the quality of a study, other criteria are useful as well. In reviewing the criteria, I find that here, too, the standards vary within the qualitative community (see my contrast of three approaches to qualitative evaluation, Creswell, 2005). I will first review three general standards and then turn to specific criteria within each of our five approaches to qualitative research.

A methodological perspective comes from Howe and Eisenhardt (1990), who suggest that only broad, abstract standards are possible for qualitative (and quantitative) research. Moreover, to determine, for example, whether a study is a good ethnography cannot be answered apart from whether the study contributes to our understanding of important questions. Howe and Eisenhardt elaborate further, suggesting that five standards be applied to all research. First, they assess a study in terms of whether the research questions drive the data collection and analysis rather than the reverse being the case. Second, they examine the extent to which the data collection and analysis techniques are competently applied in a technical sense. Third, they ask whether the researcher's assumptions are made explicit, such as the researcher's own subjectivity. Fourth, they wonder whether the study has overall warrant, such as whether it is robust, uses respected theoretical explanations, and discusses disconfirmed theoretical explanations. Fifth, the study must have "value" both in informing and improving practice (the "So what?" question) and in protecting the confidentiality, privacy, and truth telling of participants (the ethical question).

A postmodern, interpretive framework forms a second perspective, from Lincoln (1995), who thinks about the quality issue in terms of emerging criteria. She tracks her own thinking (and that of her colleague, Guba) from early approaches of developing parallel methodological criteria (Lincoln & Guba, 1985) to establishing the criteria of "fairness" (a balance of stakeholder views), sharing knowledge, and fostering social action (Guba & Lincoln, 1989) to her current stance. The new emerging approach to quality is based on three new commitments: to emergent relations with respondents, to a set of stances, and to a vision of research that enables and promotes justice. Based on these commitments, Lincoln (1995) then proceeds to identify eight standards:

- The standard set in the inquiry community, such as by guidelines for publication. These guidelines admit that within diverse approaches to research, inquiry communities have developed their own traditions of rigor, communication, and ways of working toward consensus. These guidelines, she also maintains, serve to exclude and legitimate research knowledge and social science researchers.
- The standard of positionality guides interpretive or qualitative research. Drawing on those concerned about standpoint epistemology, this means that the "text" should display honesty or authenticity about its own stance and about the position of the author.
- Another standard is under the rubric of community. This standard acknowledges that all research takes place in, is addressed to, and serves the purposes of the community in which it was carried out. Such communities might be feminist thought, Black scholarship, Native American studies, or ecological studies.
- Interpretive or qualitative research must give voice to participants so that their voice is not silenced, disengaged, or marginalized. Moreover, this standard requires that alternative or multiple voices be heard in a text.
- Critical subjectivity as a standard means that the researcher needs to have heightened self-awareness in the research process and create personal and social transformation. This "high-quality awareness" enables the researcher to understand his or her psychological and emotional states before, during, and after the research experience.

- High-quality interpretive or qualitative research involves a reciprocity between the researcher and those being researched. This standard requires that intense sharing, trust, and mutuality exist.
- The researcher should respect the sacredness of relationships in the research-to-action continuum. This standard means that the researcher respects the collaborative and egalitarian aspects of research and "make[s] spaces for the lifeways of others" (Lincoln, 1995, p. 284).
- Sharing of the privileges acknowledges that in good qualitative research, researchers share their rewards with persons whose lives they portray. This sharing may be in the form of royalties from books or the sharing of rights to publication.

A final perspective utilizes interpretive standards of conducting qualitative research. Richardson (in Richardson & St. Pierre, 2005) identifies four criteria she uses when she reviews papers or monographs submitted for social science publication:

- Substantive contribution. Does this piece contribute to our understanding of social life? Demonstrate a deeply grounded social scientific perspective? Seem "true?"
- Aesthetic merit. Does this piece succeed aesthetically? Does the use of creative analytical practices open up the text and invite interpretive responses? Is the text artistically shaped, satisfying, complex, and not boring?
- Reflexivity. How has the author's subjectivity been both a producer and a product of this text? Is there self-awareness and self-exposure? Does the author hold himself or herself accountable to the standards of knowing and telling of the people he or she has studied?
- Impact. Does this piece affect me emotionally or intellectually? Generate new questions or move me to write? Try new research practices or move me to action? (p. 964)

As an applied research methodologist, I prefer the methodological standards of evaluation, but I can also support the postmodern and interpretive perspectives. What seems to be missing in all of the approaches discussed thus far is their connection to the five approaches of qualitative inquiry. What standards of evaluation, beyond those already mentioned, would signal a high quality narrative study, a phenomenology, a grounded theory study, an ethnography, and a case study?

## Narrative Research

Denzin (1989a) is primarily interested in the problem of "how to locate and interpret the subject in biographical materials" (p. 26). He advances several guidelines for writing an interpretive biography:

- The lived experiences of interacting individuals are the proper subject matter of sociology.
- The meanings of these experiences are best given by the persons who experience them; thus, a preoccupation with method, validation, reliability, generalizability, and theoretical relevance of the biographical method must be set aside in favor of a concern for meaning and interpretation.
- Students of the biographical method must learn how to use the strategies and techniques of literary interpretation and criticism (i.e., bring their method in line with the concern about reading and writing of social texts, where texts are seen as "narrative fictions"; Denzin, 1989a, p. 26).
- When an individual writes a biography, he or she writes himself or herself into the life of the subject about whom the individual is writing; likewise, the reader reads through her or his perspective.

Thus, within a humanistic, interpretive stance, Denzin (1989b) identifies "criteria of interpretation" as a standard for judging the quality of a biography. These criteria are based on respecting the researcher's perspective as well as on thick description. Denzin (1989b) advocates for the ability of the researcher to illuminate the phenomenon in a thickly contextualized manner (i.e., thick description of developed context) so as to reveal the historical, processual, and interactional features of the experience. Also, the researcher's interpretation must engulf what is learned about the phenomenon and incorporate prior understandings while always remaining incomplete and unfinished.

This focus on interpretation and thick description is in contrast to criteria established within the more traditional approach to biographical writing. For example, Plummer (1983) asserts that three sets of questions related to sampling, the sources, and the validation of the account should guide a researcher to a good life history study:

- Is the individual representative? Edel (1984) asks a similar question: How has the biographer distinguished between the reliable and unreliable witnesses?
- What are the sources of bias (about the informant, the researcher, and the informant-researcher interaction)? Or, as Edel (1984) questions, how has the researcher avoided making himself or herself simply the voice of the subject?
- Is the account valid when subjects are asked to read it, when it is compared to official records, and when it is compared to accounts from other informants?

In a narrative study, I would look for the following aspects of a "good" study. The author:

- Focuses on a single individual (or two or three individuals)
- Collects stories about a significant issue related to this individual's life

- Develops a chronology that connects different phases or aspects of a story
- Tells a story that restories the story of the participant in the study
- Tells a persuasive story told in a literary way
- Possibly reports themes that build from the story to tell a broader analysis
- Reflexively brings himself or herself into the study

## Phenomenological Research

What criteria should be used to judge the quality of a phenomenological study? From the many readings about phenomenology, one can infer criteria from the discussions about steps (Giorgi, 1985) or the "core facets" of transcendental phenomenology (Moustakas, 1994, p. 58). I have found direct discussions of the criteria to be missing, but perhaps Polkinghorne (1989) comes the closest in my readings when he discusses whether the findings are "valid" (p. 57). To him, validation refers to the notion that an idea is well grounded and well supported. He asks, "Does the general structural description provide an accurate portrait of the common features and structural connections that are manifest in the examples collected?" (Polkinghorne, 1989, p. 57). He then proceeds to identify five questions that researchers might ask themselves:

1. Did the interviewer influence the contents of the participants' descriptions in such a way that the descriptions do not truly reflect the participants' actual experience?

2. Is the transcription accurate, and does it convey the meaning of the oral presentation in the interview?

3. In the analysis of the transcriptions, were there conclusions other than those offered by the researcher that could have been derived? Has the researcher identified these alternatives?

4. Is it possible to go from the general structural description to the transcriptions and to account for the specific contents and connections in the original examples of the experience?

5. Is the structural description situation specific, or does it hold in general for the experience in other situations? (Polkinghorne, 1989).

My own standards that I would use to assess the quality of a phenomenology would be:

- Does the author convey an understanding of the philosophical tenets of phenomenology?
- Does the author have a clear "phenomenon" to study that is articulated in a concise way?

- Does the author use procedures of data analysis in phenomenology, such as the procedures recommended by Moustakas (1994)?
- Does the author convey the overall essence of the experience of the participants? Does this essence include a description of the experience and the context in which it occurred?
- Is the author reflexive throughout the study?

## Grounded Theory Research

Strauss and Corbin (1990) identify the criteria by which one judges the quality of a grounded theory study. They advance seven criteria related to the general research process:

Criterion #1: How was the original sample selected? What grounds?

Criterion #2: What major categories emerged?

Criterion #3: What were some of the events, incidents, actions, and so on (as indicators) that pointed to some of these major categories?

Criterion #4: On the basis of what categories did theoretical sampling proceed? Guide data collection? Was it representative of the categories?

Criterion #5: What were some of the hypotheses pertaining to conceptual relations (that is, among categories), and on what grounds were they formulated and tested?

Criterion #6: Were there instances when hypotheses did not hold up against what was actually seen? How were these discrepancies accounted for? How did they affect the hypotheses?

Criterion #7: How and why was the core category selected (sudden, gradual, difficult, easy)? On what grounds? (p. 253)

They also advance six criteria related to the empirical grounding of a study:

Criterion #1: Are concepts generated?

Criterion #2: Are the concepts systematically related?

Criterion #3: Are there many conceptual linkages, and are the categories well developed? With density?

Criterion #4: Is much variation built into the theory?

Criterion #5: Are the broader conditions built into its explanation?

Criterion #6: Has process (change or movement) been taken into account? (Strauss & Corbin, 1990, pp. 254–256)

These criteria, related to the process of research and the grounding of the study in the data, represent benchmarks for assessing the quality of a study that the author can mention in his or her research. For example, in a grounded theory dissertation, Landis (1993) not only presented these standards but also assessed for her readers the extent to which her study met the criteria. When I evaluate a grounded theory study, I, too, am looking for the general process and a relationship among the concepts. Specifically, I look for:

- The study of a process, action, or interaction as the key element in the theory
- A coding process that works from the data to a larger theoretical model
- The presentation of the theoretical model in a figure or diagram
- A story line or proposition that connects categories in the theoretical model and that presents further questions to be answered
- A reflexivity or self-disclosure by the researcher about his or her stance in the study

## Ethnographic Research

The ethnographers Spindler and Spindler (1987) emphasize that the most important requirement for an ethnographic approach is to explain behavior from the "native's point of view" (p. 20) and to be systematic in recording this information using note taking, tape recorders, and cameras. This requires that the ethnographer be present in the situation and engage in constant interaction between observation and interviews. These points are reinforced in Spindler and Spindler's nine criteria for a "good ethnography":

Criterion I. Observations are contextualized.

Criterion II. Hypotheses emerge in situ as the study goes on.

Criterion III. Observation is prolonged and repetitive.

Criterion IV. Through interviews, observations, and other eliciting procedures, the native view of reality is obtained.

Criterion V. Ethnographers elicit knowledge from informant-participants in a systematic fashion.

Criterion VI. Instruments, codes, schedules, questionnaires, agenda for interviews, and so forth are generated in situ as a result of inquiry.

Criterion VII. A transcultural, comparative perspective is frequently an unstated assumption.

Criterion VIII. The ethnographer makes explicit what is implicit and tacit to informants.

Criterion IX. The ethnographic interviewer must not predetermine responses by the kinds of questions asked. (Spindler & Spindler, 1987, p. 18)

This list, grounded in fieldwork, leads to a strong ethnography. Moreover, as Lofland (1974) contends, the study is located in wide conceptual frameworks; presents the novel but not necessarily new; provides evidence for the framework(s); is endowed with concrete, eventful interactional events, incidents, occurrences, episodes, anecdotes, scenes, and happenings without being "hyper-eventful"; and shows an interplay between the concrete and analytical and the empirical and theoretical.

My criteria for a good ethnography would include:

- The clear identification of a culture-sharing group
- The specification of a cultural themes that will be examined in light of this culture-sharing group
- A detailed description of the cultural group
- Themes that derive from an understanding of the cultural group
- The identification of issues that arose "in the field" that reflect on the relationship between the researcher and the participants, the interpretive nature of reporting, and sensitivity and reciprocity in the co-creating of the account
- An explanation overall of how the culture-sharing group works
- A self-disclosure and reflexivity by the researcher about her or his position in the research

## Case Study Research

Stake (1995) provides a rather extensive "critique checklist" (p. 131) for a case study report and shares 20 criteria for assessing a good case study report:

1. Is the report easy to read?

2. Does it fit together, each sentence contributing to the whole?

3. Does the report have a conceptual structure (i.e., themes or issues)?

4. Are its issues developed in a serious and scholarly way?

5. Is the case adequately defined?

6. Is there a sense of story to the presentation?

7. Is the reader provided some vicarious experience?

8. Have quotations been used effectively?

9. Are headings, figures, artifacts, appendixes, and indexes used effectively?

10. Was it edited well, then again with a last-minute polish?

11. Has the writer made sound assertions, neither over- nor under-interpreting?

12. Has adequate attention been paid to various contexts?

13. Were sufficient raw data presented?

14. Were data sources well chosen and in sufficient number?

15. Do observations and interpretations appear to have been triangulated?

16. Is the role and point of view of the researcher nicely apparent?

17. Is the nature of the intended audience apparent?

18. Is empathy shown for all sides?

19. Are personal intentions examined?

20. Does it appear that individuals were put at risk? (Stake, 1995, p. 131)

My own criteria for evaluating a "good" case study would include the following:

- Is there a clear identification of the "case" or "cases" in the study?
- Is the "case" (or are the "cases") used to understand a research issue or used because the "case" has (or "cases" have) intrinsic merit?
- Is there a clear description of the "case"?
- Are themes identified for the "case"?
- Are assertions or generalizations made from the "case" analysis?
- Is the researcher reflexive or self-disclosing about his or her position in the study?

# Comparing the Evaluation Standards of the Five Approaches

The standards discussed for each approach differ slightly depending on the procedures of the approaches. Certainly less is mentioned about narrative research and its standards of quality and more is available about the other approaches. From within the major books used for each approach, I have attempted to extract the evaluation standards recommended for their approach to research. To these I have added my own standards that I use in my qualitative classes when I evaluate a project or study presented within each of the five approaches.

## Summary

In this chapter, I discuss validation, reliability, and standards of quality in qualitative research. Validation approaches vary considerably, such as strategies that emphasize using qualitative terms comparable to quantitative terms, the use of distinct terms, perspectives from postmodern and interpretive lenses, syntheses of different perspectives, or descriptions based on metaphorical images. Reliability is used in several ways, one of the most popular being the use of intercoder agreements when multiple coders analyze and then compare their code segments to establish the reliability of the data analysis process. A detailed procedure for establishing intercoder agreement is described in this chapter. Also, diverse standards exist for establishing the quality of qualitative research, and these criteria are based on procedural perspectives, postmodern perspectives, and interpretive perspectives. Within each of the five approaches to inquiry, specific standards also exist; these were reviewed in this chapter. Finally, I advanced the criteria that I use to assess the quality of studies presented to me in classes in each of the five approaches.

## Additional Readings

Key reading on the issue of validation in qualitative research can be found in:

Angen, M. J. (2000). Evaluating interpretive inquiry: Reviewing the validity debate and opening the dialogue. *Qualitative Health Research, 10,* 378–395.

Lincoln, Y. S. (1995). Emerging criteria for quality in qualitative and interpretive research. *Qualitative Inquiry, 1,* 275–289.

Silverman, D. (1993). *Interpreting qualitative data: Methods for analyzing talk, text, and interaction.* London: Sage.

Whittemore, R., Chase, S. K., & Mandle, C. L. (2001). Validity in qualitative research. *Qualitative Health Research, 11,* 522–537.

For understanding further the issue of reliability in qualitative research, look at:

Armstrong, D., Gosling, A., Weinman, J., & Marteau, T. (1997). The place of inter-rater reliability in qualitative resesearch: An empirical study. *Sociology, 31,* 597-606.

Miles, M. B., & Huberman, A.M. (1994). *Qualitative data analysis: A sourcebook of new methods* (2nd ed.). Thousand Oaks, CA: Sage.

Silverman, D. (2005). *Doing qualitative research: A practical handbook* (2nd ed.). London: Sage.

For evaluation standards, look at:

Lincoln, Y. S. (1995). Emerging criteria for quality in qualitative and interpretive research. *Qualitative Inquiry, 1,* 275–289.

Richardson, L., & St. Pierre, E. A. (2005). Writing: A method of inquiry. In N. K. Denzin & Y. S. Lincoln (Eds.), *The Sage handbook of qualitative research* (3rd ed., pp. 959–978). Thousand Oaks, CA: Sage.

Also, look at specific standards used in the methods books in each of the five approaches. In narrative research:

Clandinin, D. J., & Connelly, F. M. (2000). *Narrative inquiry: Experience and story in qualitative research.* San Francisco: Jossey-Bass.

Czarniawka, B. (2004). *Narratives in social science research.* Thousand Oaks, CA: Sage.

Denzin, N. K. (1989a). *Interpretive biography.* Newbury Park, CA: Sage.

In phenomenology:

Moustakas, C. (1994). *Phenomenological research methods.* Thousand Oaks, CA: Sage.

van Manen, M. (1990). *Researching lived experience: Human science for an action sensitive pedagogy.* Albany: State University of New York Press.

In grounded theory:

Charmaz, K. (2006). *Constructing grounded theory.* London: Sage.

Strauss, A., & Corbin, J. (1990). *Basics of qualitative research: Grounded theory procedures and techniques.* Newbury Park, CA: Sage.

In ethnography:

LeCompte, M. D., & Schensul, J. J. (1999). *Designing and conducting ethnographic research.* Walnut Creek, CA: AltaMira.

Madison, D. S. (2005). *Critical ethnography: Methods, ethics, and performance.* Thousand Oaks, CA: Sage.

Wolcott, H. F. (1999). *Ethnography: A way of seeing.* Walnut Creek, CA: AltaMira.

In case study:

Stake, R. (1995). *The art of case study research.* Thousand Oaks, CA: Sage.

Yin, R. K. (2003). *Case study research: Design and methods* (2nd ed.). Thousand Oaks, CA: Sage.

## Exercises

1. Identify one of the procedures for validation mentioned in this chapter and use it in your study. Also, indicate whether your study changed as a result of its use or remained the same.

2. For the approach you used or are planning to use, identify the criteria for assessing the quality of the study and present an argument for each criterion as to how the study meets or will meet each standard.

# 11

# "Turning the Story" and Conclusion

I n this book, I suggest that researchers be cognizant of the procedures of qualitative research and of the differences in approaches of qualitative inquiry. This is not to suggest a preoccupation with method or methodology; indeed, I see two parallel tracks in a study: the substantive content of the study and the methodology. With increased interest in qualitative research, it is important that studies being conducted go forward with rigor and attention to the procedures developed within approaches of inquiry.

The approaches are many, and their procedures for research are well documented within books and articles. A few writers classify the approaches, and some authors mention their favorites. Unquestionably, qualitative research cannot be characterized as of one type, attested to by the multivocal discourse surrounding qualitative research today. Adding to this discourse are perspectives about philosophical, theoretical, and ideological stances. To capture the essence of a good qualitative study, I visualize such a study as comprised of three interconnected circles. As shown in Figure 11.1, these circles include the approach of inquiry, research design procedures, and philosophical and theoretical frameworks and assumptions. The interplay of these three factors contributes to a complex, rigorous study.

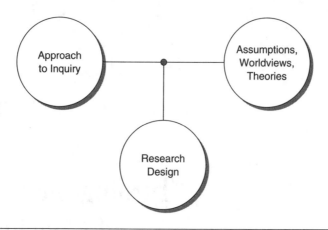

**Figure 11.1**    Visual Diagram of the Three Components of Qualitative Research

## Turning the Story

In this chapter, I again sharpen the distinctions among the approaches of inquiry, but I depart from my side-by-side approach used in prior chapters. I focus the lens in a new direction and "turn the story" of the gunman case (Asmussen & Creswell, 1995) into a narrative study, a phenomenology, a grounded theory, and an ethnography. Before continuing on with this chapter, the reader is advised to reexamine the gunman case study as presented in Appendix F and reviewed in Chapter 5.

Turning the story through different approaches of inquiry raises the issue of whether one should match a particular problem to a approach to inquiry. Much emphasis is placed on this relationship in social and human science research. I agree this needs to be done. But for the purposes of this book, my way around this issue is to pose a *general* problem—"How did the campus react?"—and then construct scenarios for specific problems. For instance, the specific problem of studying a single individual's reaction to the gun incident is different from the specific problem of how several students as a culture-sharing group reacted, but both scenarios are reactions to the general issue of campus reaction to the incident. The general problem that I address is that we know little about how campuses respond to violence and even less about how different constituent groups on campus respond to a potentially violent incident. Knowing this information would help us devise better plans for reacting to this type of problem as well as add to the literature on violence in educational settings. This was the central problem in the gunman case study (Asmussen & Creswell, 1995), and I briefly review the major dimensions of this study.

# A Case Study

This qualitative case study (Asmussen & Creswell, 1995) presented a campus reaction to a gunman incident in which a student attempted to fire a gun at his classmates. Asmussen and I titled this study "Campus Response to a Student Gunman," and we composed this case study with the "substantive case report" format of Lincoln and Guba (1985) and Stake (1995) in mind. These formats called for an explication of the problem, a thorough description of the context or setting and the processes observed, a discussion of important themes, and, finally, "lessons to be learned" (Lincoln & Guba, 1985, p. 362). After introducing the case study with the problem of violence on college campuses, we provided a detailed description of the setting and a chronology of events immediately following the incident and events during the following 2 weeks. Then we turned to important themes to emerge in this analysis—themes of denial, fear, safety, retriggering, and campus planning. In a process of layering of themes, we combined these more specific themes into two overarching themes: an organizational theme and a psychological or social-psychological theme. We gathered data through interviews with participants, observations, documents, and audiovisual materials. From the case emerges a proposed plan for campuses, and the case ends with an implied lesson for the specific Midwestern campus and a specific set of questions this campus or other campuses might use to design a plan for handling future campus terrorist incidents.

Turning to specific research questions in this case, we asked the following. What happened? Who was involved in response to the incident? What themes of response emerged during an 8-month period? What theoretical constructs helped us understand the campus response and what constructs developed that were unique to this case? We entered the field 2 days after the incident and did not use any a priori theoretical lens to guide our questions or the results. The narrative first described the incident, analyzed it through levels of abstraction, and provided some interpretation by relating the context to larger theoretical frameworks. We validated our case analysis by using multiple data sources for the themes and by checking the final account with select participants or member checking.

# A Narrative Study

How might I have approached this same general problem as an interpretive biographical study with a narrative approach? Rather than identifying responses from multiple campus constituents, I would have focused on one individual such as the instructor of the class involved in the incident. I would have tentatively titled the project, "Confrontation of Brothers: An Interpretive Biography of an African American Professor." This instructor, like the gunman, was African American, and his response to such an incident might be situated within racial and cultural contexts. Hence, as an interpretive biographer, I might have asked the following research question: What are the life experiences of the African American instructor of the class, and how do these experiences form and shape his reaction to the incident? This biographical approach would have relied on studying a single individual and situating this individual within his historic background. I would have examined life events or "epiphanies" culled from stories he told me. My approach would have been to restory the stories into an account of his experiences of the gunman that followed a chronology of events. I might have relied on the Clandinin and Connelly (2000) three-dimensional space model to organize the story into the personal, social, and interactional components. Alternatively, the story might have had a plot to tie it together, such as the theoretical perspective. This plot might have spoken to the issues of race, discrimination, and marginality and how these issues played out both within the African American culture and between Black culture and other cultures. These perspectives may have shaped how the instructor viewed the student gunman in the class. I also might have composed this report by discussing my own situated beliefs followed by those of the instructor and the changes he brought about as a result of his experiences. For instance, did he continue teaching? Did he talk with the class about his feelings? Did he see the situation as a confrontation within his racial group? For validation, my narrative story about this instructor would have contained a detailed description of the context to reveal the historical and interactional features of the experience (Denzin, 1989b). I also would have acknowledged that any interpretation of the instructor's reaction would be incomplete, unfinished, and a rendering from my own perspective as a non-African-American.

# A Phenomenology

Rather than study a single individual as in a biography, I would have studied several individual students and examined a psychological concept in the tradition of psychological phenomenology (Moustakas, 1994). My working title might have been: "The Meaning of Fear for Students Caught in a Near Tragedy on Campus." My assumption would have been that this concept of fear was expressed by students during the incident, immediately after it, and several weeks later. I might have posed the following questions. What fear did the students experience, and how did they experience it? What meanings did they ascribe to this experience? As a phenomenologist, I assume that human experience makes sense to those who live it and that human experience can be consciously expressed (Dukes, 1984). Thus, I would bring to the study a phenomenon to explore (fear) and a philosophical orientation to use (I want to study the meaning of the students' experiences). I would have engaged in extensive interviews with up to 10 students, and I would have analyzed the interviews using the steps described by Moustakas (1994). I would have begun with description of my own fears and experiences (epoche) with it as a means to position myself, recognizing that I could not completely remove myself and my interpretation from the situation. Then, after reading through all of the students' statements, I would have located significant statements or quotes about their meanings of fear. These significant statements would then be clustered into broader themes. My final step would have been to write a long paragraph providing a narrative description of what they experienced (textural description) and how they experienced it to (structural description) and combine these two descriptions into a longer description that describes the "essence" of their experiences. This would be the endpoint for the discussion.

## A Grounded Theory Study

If a theory needed to be developed (or modified) to explain the campus reaction to this incident, then I would have used a grounded theory approach. For example, I might have developed a theory around a process—the "surreal" experiences of several students immediately following the incident, experiences resulting in actions and reactions by students. The draft title of my study might have been "A Grounded Theory Explanation of the Surreal Experiences for Students in a Campus Gunman Incident." I might have introduced the study with a specific quote about the surreal experiences:

> In the debriefing by counselors, one female student commented, "I thought the gunman would shoot out a little flag that would say 'bang.'" For her, the event was like a dream.

My research questions might have been: What theory explains the phenomenon of the "surreal" experiences of the students immediately following the incident? What were these experiences? What caused them? What strategies did they use to cope with them? What were the consequences of their strategies? What specific interaction issues and larger conditions influenced their strategies? Consistent with grounded theory, I would not bring into the data collection and analysis a specific theoretical orientation other than to see how the students interact and respond to the incident. Instead, my intent would be to develop or generate a theory. In the results section of this study I would have first identified the open coding categories that I found. Then, I would have described how I narrowed the study to a central category (e.g., the dream element of the process), and made that category the major feature of a theory of the process. This theory would have been presented as a visual model, and in the model I would have included causal conditions that influenced the central category, intervening and context factors surrounding it, and specific strategies and consequences (axial coding) as a result of it occurring. I would have advanced theoretical propositions or hypotheses that explained the dream element of the surreal experiences of the students (selective coding). I would have validated my account by judging the thoroughness of the research process and whether the findings are empirically grounded, two factors mentioned by Corbin and Strauss (1990).

## An Ethnography

In grounded theory, my focus was on generating a theory grounded in the data. In ethnography, I would turn the focus away from theory development to a description and understanding of the workings of the campus community as a culture-sharing group. To keep the study manageable, I might have begun by looking at how the incident, although unpredictable, triggered quite predictable responses among members of the campus community. These community members might have responded according to their roles, and thus I could have looked at some recognized campus microcultures. Students constituted one such microculture, and they, in turn, comprised a number of further microcultures or subcultures. Because the students in this class were together for 16 weeks during the semester, they had enough time to develop some shared patterns of behavior and could have been seen as a culture-sharing group. Alternatively, I might have studied the entire campus community composed of a constellation of groups each reacting differently.

Assuming that the entire campus comprised the culture-sharing group, the title of the study might have been "Getting Back to Normal: An Ethnography of a Campus Response to a Gunman Incident." Notice how this title immediately invites a contrary perspective into the study. I would have asked the following questions: How did this incident produce predictable role performance within affected groups? Using the entire campus as a cultural system or culture-sharing group, in what roles did the individuals and groups participate? One possibility would be that they wanted to get the campus back to normal after the incident by engaging in predictable patterns of behavior. Although no one anticipated the exact moment or nature of the incident itself, its occurrence set in motion rather predictable role performances throughout the campus community. Administrators did not close the campus and start warning, "The sky is falling." Campus police did not offer counseling sessions, although the Counseling Center did. However, the Counseling Center served the student population, not others (who were marginalized), such as the police or groundskeepers, who also felt unsafe on the campus. In short, predictable performances by campus constituencies followed in the wake of this incident.

Indeed, campus administrators routinely held a news conference following the incident. Also, predictably, police carried out their investigation, and students ultimately and reluctantly contacted their parents. The campus slowly returned to normal—an attempt to return to day-to-day business, to steady state, or to homeostasis, as the systems thinkers say. In these predictable role behaviors, one saw culture at work.

As I entered the field, I would seek to build rapport with the community participants, to not further marginalize them or disturb the environment more than necessary through my presence. It was a sensitive time on campus with people who had nerves on edge. I would have explored the cultural themes of the "organization of diversity" and "maintenance" activities of individuals and groups within the culture-sharing campus. Wallace (1970) defines the "organization of diversity" as "the actual diversity of habits, of motives, of personalities, of customs that do, in fact, coexist within the boundaries of any culturally organized society" (p. 23). My data collection would have consisted of observations over time of predictable activities, behaviors, and roles in which people engaged that help the campus return to normal. This data collection would depend heavily on interviews and observations of the classroom where the incident occurred and newspaper accounts. My ultimate narrative of the culture-sharing campus would be consistent with Wolcott's (1994b) three parts: a detailed description of the campus, an analysis of the cultural themes of "organizational diversity" and maintenance (possibly with taxonomies or comparisons; Spradley, 1979, 1980), and interpretation. My interpretation would be counched not in terms of a dispassionate, objective report of the facts, but rather positioned within my own experiences of not feeling safe in a soup kitchen for the homeless (Miller & Creswell, 1998) and my own personal life experiences of having grown up in a "safe" small Midwestern city in Illinois. For an ending to the study, I might have used the "canoe into the sunset" approach (Wolcott, personal communication, November 15, 1996) or the more methodologically oriented ending of checking my account with participants. Here is the "canoe into the sunset" approach:

The newsworthiness of the event will be long past before the ethnographic study is ready, but the event itself is of rather little consequence if the ethnographer's focus is on campus culture. Still, without such an event, the ethnographer working in his or her own society (and perhaps own campus as well) might have a difficult time "seeing" people performing in predictable everyday ways simply because that is the way in which we expect them to act. The ethnographer working "at home" has to find ways in which to make the familiar seem strange. An upsetting event can make ordinary role behavior easier to discern as people respond in predictable ways to unpredictable circumstances. Those predictable patterns are the stuff of culture.

Here is the more methodological ending:

Some of my "facts" or hypotheses may need (and be amenable to) checking or testing if I have carried my analysis in that direction. If I have tried to be more interpretive, then perhaps I can "try out" the account on some of the people described, and the cautions and exceptions they express can be included in my final account to suggest that things are even more complex than the way I have presented them.

## Conclusion

How have I answered my "compelling" question raised at the outset: How does the approach to inquiry shape the design of a study? First, one of the most pronounced ways is in the focus of the study. As discussed in Chapter 4, a theory differs from the exploration of a phenomenon or concept, from an in-depth case, and from the creation of an individual or group portrait. Please examine again Table 4.1 that establishes differences among the five approaches, especially in terms of foci.

However, this is not as clear-cut as it appears. A single case study of an individual can be approached either as a biography or as a case study. A cultural system may be explored as an ethnography, whereas a smaller "bounded" system, such as an event, a program, or an activity, may be studied as a case study. Both are systems, and the problem arises when one undertakes a microethnography, which might be approached either as a case study or as an ethnography. However, when one seeks to study cultural behavior, language, or artifacts, then the study of a system might be undertaken as an ethnography.

Second, an interpretive orientation flows throughout qualitative research. We cannot step aside and be "objective" about what we see and write. Our words flow from our own personal experiences, culture, history, and backgrounds. When we go to the field to collect data, we need to approach the task with care for the participants and sites and to be reflexive about our role and how it shapes what we see, hear, and write. Ultimately, our writing is an interpretation by us of events, people, and activities, and it is only our interpretation. We must recognize that participants in the field, readers, and other individuals reading our accounts will have their own interpretations. Within this perspective, our writing can only be seen as a discourse, one with tentative conclusions, and one that will be constantly changing and evolving. Qualitative research truly has an interpretation element that flows throughout the process of research.

Third, the approach to inquiry shapes the language of the research design procedures in a study, especially the terms used in the introduction to a study, the data collection, and the analysis phases of design. I incorporated these terms into Chapter 6 as I discussed the wording of purpose statements and research questions for different approaches to qualitative research. My theme continued on in Chapter 9 as I talked about encoding the text within an approach to research. The glossary in Appendix A also reinforces this theme as it presents a useful list of terms within each tradition that researchers might incorporate into the language of their studies.

Fourth, the approach to research includes the participants who are studied, as discussed in Chapter 7. A study may consist of one or two individuals (i.e., narrative study), groups of people (i.e., phenomenology, grounded theory), or a entire culture (i.e., ethnography). A case study might fit into all three of these categories as one explores a single individual, an event, or a large social setting. Also in Chapter 7, I highlighted how the approaches vary in the extent of data collection, from the use of mainly single sources of information (i.e., narrative interviews, grounded theory interviews, phenomenological interviews) to those that involve multiple sources of information (i.e., ethnographies consisting of observations, interviews, and documents; case studies incorporating interviews, observations, documents, archival material, and video). Although these forms of data collection are not fixed, I see a general pattern that differentiates the approaches.

Fifth, the distinctions among the approaches are most pronounced in the data analysis phase, as discussed in Chapter 8. Data analysis ranges from unstructured to structured approaches. Among the less structured approaches, I include ethnographies (with the exception of Spradley, 1979, 1980) and narratives (e.g., as suggested by Clandinin & Connelly, 2000, and interpretive forms advanced by Denzin, 1989b). The more structured approaches consist of grounded theory with a systematic procedure and phenomenology (see Colaizzi's 1978 approach and those of Dukes, 1984, and Moustakas, 1994) and case studies (Stake, 1995). These procedures provide direction for the overall structure of the data analysis in the qualitative report. Also, the approach shapes the amount of relative weight given to description in the analysis of the data. In ethnographies, case studies, and biographies, researchers employ substantial description; in phenomenologies, investigators use less description; and in grounded theory, researchers seem not to use it at all, choosing to move directly into analysis of the data.

Sixth, the approach to inquiry shapes the final written product as well as the embedded rhetorical structures used in the narrative. This explains why qualitative studies look so different and are composed so differently, as discussed in Chapter 9. Take, for example, the presence of the researcher. The presence of the researcher is found little in the more "objective" accounts provided in grounded theory. Alternatively, the researcher is center stage in ethnographies and possibly in case studies where "interpretation" plays a major role.

Seventh, the criteria for assessing the quality of a study differ among the approaches, as discussed in Chapter 10. Although some overlap exists in the procedures for validation, the criteria for assessing the worth of a study are available for each tradition.

In summary, when designing a qualitative study, I recommend that the author design the study within one of the approaches of qualitative inquiry. This means that components of the design process (e.g., theoretical framework, research purpose and questions, data collection, data analysis, report writing, verification) will reflect the procedures of the selected approach and they will be composed with the encoding and composing features of that approach. This is not to rigidly suggest that one cannot mix approaches and employ, for example, a grounded theory analysis procedure within a case study design. "Purity" is not my aim. But in this book, I suggested that the reader sort out the approaches first before combining them and see each one as a rigorous procedure in its own right.

I found distinctions as well as overlap among the five approaches, but designing a study attuned to procedures found within one of the approaches suggested in this book will enhance the sophistication of the project and convey a level of methodological expertise for readers of qualitative research.

## Exercises

1. Take the qualitative study you have completed and turn the story into one of the other approaches of qualitative inquiry.

2. In this chapter, I presented the study of campus response to a gunman incident in five ways. Take each scenario and label the parts using the language of each tradition and the terms found in the glossary in Appendix A.

# Appendix A

## An Annotated Glossary of Terms

The definitions in this glossary represent key terms as they are used and defined in this book. Many definitions exist for these terms, but the most workable definitions for me (and I hope for the reader) are those that reflect the content and references presented in this book. I group the terms by approach to inquiry (narrative research, phenomenology, grounded theory, ethnography, case study) and alphabetize them within the approach, and at the end of the glossary I define additional terms that do not conveniently relate to any specific approach.

### Narrative Research

*autobiography* This form of biographical writing is the narrative account of a person's life that he or she has personally written or otherwise recorded (Angrosino, 1989a).

*biographical study* This is the study of a single individual and his or her experiences as told to the researcher or as found in documents and archival materials (Denzin, 1989a). I use the term to connotate the broad genre of narrative writings that includes individual biographies, autobiographies, life histories, and oral histories.

*chronology* This is a common approach for undertaking a narrative form of writing in which the author presents the life in stages or steps according to the age of the individual (Clandinin & Connelly, 2000; Denzin, 1989a).

*epiphanies* These are special events in an individual's life that represent turning points. They vary in their impact from minor epiphanies to major epiphanies, and they may be positive or negative (Denzin, 1989a).

*historical context* This is the context in which the researcher presents the life of the subject. The context may be the subject's family, the subject's society, or the history, social, or political trends of the subject's times (Denzin, 1989a).

*life course stages and experiences* These are stages in an individual's life or key events that become the focus for the biographer (Denzin, 1989a).

*life history* This is a form of biographical writing in which the researcher reports an extensive record of a person's life as told to the researcher (see Geiger, 1986). Thus, the individual being studied is alive and life as lived in the present is influenced by personal, institutional, and social histories (Cole, 1994). The investigator may use different disciplinary perspectives (Smith, 1994), such as the exploration of an individual's life as representative of a culture, as in an anthropological life history.

*narrative research* This is an approach to qualitative research that is both a product and a method. It is a study of stories or narrative or descriptions of a series of events that accounts for human experiences (Pinnegar & Daynes, 2006).

*oral history* In this biographical approach, the researcher gathers personal recollections of events and their causes and effects from an individual or several individuals. This information may be collected through tape recordings or through written works of individuals who have died or are still living. It often is limited to the distinctly "modern" sphere and to accessible people (Plummer, 1983).

*progressive-regressive method* This is an approach to writing a narrative in which the researcher begins with a key event in the subject's life and then works forward and backward from that event (Denzin, 1989a).

*restorying* This is an approach in narrative data analysis in which the researchers retell the stories of individual experiences, and the new story typically has a beginning, middle, and ending (Ollerenshaw & Creswell, 2002).

*single individual* This is the person studied in a narrative research. This person may be an individual with great distinction or an ordinary person. This person's life may be a lesser life, a great life, a thwarted life, a life cut short, or a life miraculous in its unapplauded achievement (Heilbrun, 1988).

*stories* These are aspects that surface during an interview in which the participant describes a situation, usually with a beginning, a middle, and an end, so that the researcher can capture a complete idea and integrate it, intact, into the qualitative narrative (Clandinin & Connelly, 2000; Czarniawska, 2004; Denzin, 1989a).

## Phenomenology

*clusters of meanings* This is the third step in phenomenological data analysis, in which the researcher clusters the statements into themes or meaning units, removing overlapping and repetitive statements (Moustakas, 1994).

*epoche or bracketing* This is the first step in "phenomenological reduction," the process of data analysis in which the researcher sets aside, as far as is humanly possible, all preconceived experiences to best understand the experiences of participants in the study (Moustakas, 1994).

*essential, invariant structure (or essence)* This is the goal of the phenomenologist, to reduce the textural (*what*) and structural (*how*) meanings of experiences to a brief description that typifies the experiences of all of the participants in a study. All individuals experience it; hence, it is invariant, and it is a reduction to the "essentials" of the experiences (Moustakas, 1994).

*hermeneutical phenomenology* A form of phenomenology in which research is oriented toward interpreting the "texts" of life (hermeneutical) and lived experiences (phenomenology) (van Manen, 1990).

*horizonalization* This is the second step in the phenomenological data analysis, in which the researcher lists every significant statement relevant to the topic and gives it equal value (Moustakas, 1994).

*imaginative variation or structural description* Following the textural description, the researcher writes a "structural" description of an experience, addressing *how* the phenomenon was experienced. It involves seeking all possible meanings, seeking divergent perspectives, and varying the frames of reference about the phenomenon or using imaginative variation (Moustakas, 1994).

*intentionality of consciousness* Being conscious of objects always is intentional. Thus, when perceiving a tree, "my intentional experience is a combination of the outward appearance of the tree and the tree as contained in my consciousness based on memory, image, and meaning" (Moustakas, 1994, p. 55).

*lived experiences* This term is used in phenomenological studies to emphasize the importance of individual experiences of people as conscious human beings (Moustakas, 1994).

*phenomenological data analysis* Several approaches to analyzing phenomenological data are represented in the literature. Moustakas (1994) reviews these approaches and then advances his own. I rely on the Moustakas modification that includes the researcher bringing personal experiences into the study, the recording of significant statements and meanings, and the development of descriptions to arrive at the essences of the experiences.

*phenomenological study* This type of study describes the meaning of experiences of a phenomenon (or topic or concept) for several individuals. In this study, the researcher reduces the experiences to a central meaning or the "essence" of the experience (Moustakas, 1994).

*the phenomenon* This is the central concept being examined by the phenomenologist. It is the concept being experienced by subjects in a study, which may include psychological concepts such as grief, anger, or love.

*philosophical perspectives* Specific philosophical perspectives provide the foundation for phenomenological studies. They originated in the 1930s writings of Husserl. These perspectives include the investigator's conducting research with a broader perspective than that of traditional empirical, quantitative science; suspending his or her own preconceptions of experiences; experiencing an object through his or her own senses (i.e., being conscious of an object) as well as seeing it "out there" as real; and reporting the meaning individuals ascribe to an experience in a few statements that capture the "essence" (Stewart & Mickunas, 1990).

*psychological approach* This is the approach taken by psychologists who discuss the inquiry procedures of phenomenology (e.g., Giorgi, 1994; Moustakas, 1994; Polkinghorne, 1989). In their writings, they examine psychological themes for meaning, and they may incorporate their own selves into the studies.

*structural description* From the first three steps in phenomenological data analysis, the researcher writes a description of "how" the phenomenon was experienced by individuals in the study (Moustakas, 1994).

*textural description* From the first three steps in phenomenological data analysis, the researcher writes about *what* was experienced, a description of the meaning individuals have experienced (Moustakas, 1994).

*transcendental phenomenology* According to Moustakas (1994), Husserl espoused transcendental phenomenology, and it later became a guiding concept for Moustakas as well. In this approach, the researcher sets aside prejudgments regarding the phenomenon being investigated. Also, the researcher relies on intuition, imagination, and universal structures to obtain a picture of the experience and uses systematic methods of analysis as advanced by Moustakas (1994).

## Grounded Theory

*axial coding* This step in the coding process follows open coding. The researcher takes the categories of open coding, identifies one as a central phenomenon, and then returns to the database to identify (a) what caused this phenomenon to occur, (b) what strategies or actions actors employed in response to it, (c) what context (specific context) and intervening conditions (broad context) influenced the strategies, and (d) what consequences resulted from these strategies. The overall process is one of relating categories of information to the central phenomenon category (Strauss & Corbin, 1990).

*category* This is a unit of information analyzed in grounded theory research. It is composed of events, happenings, and instances of phenomenon (Strauss & Corbin, 1990) and given a short label. When researchers analyze grounded theory data, their analysis leads, initially, to the formation of a number of categories during the process called "open coding." Then, in "axial coding," the analyst interrelates the categories and forms a visual model.

*causal conditions* In axial coding, these are the categories of conditions I identify in my database that cause or influence the central phenomenon to occur.

*central phenomenon* This is an aspect of axial coding and the formation of the visual theory, model, or paradigm. In open coding, the researcher chooses a central category around which to develop the theory by examining his or her open coding categories and selecting one that holds the most conceptual interest, is most frequently discussed by participants in the study, and is most "saturated" with information. The researcher then places it at the center of his or her grounded theory model and labels it "central phenomenon."

*coding paradigm or logic diagram* In axial coding, the central phenomenon, causal conditions, context, intervening conditions, strategies, and consequences are portrayed in a visual diagram. This diagram is drawn with boxes and arrows indicating the process or flow of activities. It is helpful to view this diagram as more than axial coding; it is the theoretical model developed in a grounded theory study (see Morrow & Smith, 1995).

*conditional matrix* This is a diagram, typically drawn late in a grounded theory study, that presents the conditions and consequences related to the phenomenon under study. It enables the researcher to both distinguish and link levels of conditions and consequences specified in the axial coding model (Strauss & Corbin, 1990). It is a step seldom seen in data analysis in grounded theory studies.

*consequences* In axial coding, these are the outcomes of strategies taken by participants in the study. These outcomes may be positive, negative, or neutral (Strauss & Corbin, 1990).

*constant comparative* This was an early term (Conrad, 1978) in grounded theory research that referred to the researcher identifying incidents, events, and activities and constantly comparing them to an emerging category to develop and saturate the category.

*cconstructivist grounded theory* This is a form of grounded theory squarely in the interpretive tradition of qualitative research. As such, it is less structured than traditional approaches to grounded theory. The constructivist approach incorporates the researcher's views; uncovers experiences with embedded, hidden networks, situations, and relationships; and makes visible hierarchies of power, communication, and opportunity (Charmaz, 2006)

*ontext* In axial coding, this is the particular set of conditions within which the strategies occur (Strauss & Corbin, 1990). These are specific in nature and close to the actions and interactions.

*dimensionalized* This is the smallest unit of information analyzed in grounded theory research. The researcher takes the properties and places them on a continuum or dimensionalizes them to see the extreme possibilities for the property. The dimensionalized information appears in the "open coding" analysis (Strauss & Corbin, 1990).

*discriminant sampling* This is a form of sampling that occurs late in a grounded theory project after the researcher has developed a model.

The question becomes, at this point: How would the model hold if I gathered more information from people similar to those I initially interviewed? Thus, to verify the model, the researcher chooses sites, persons, and/or documents that "will maximize opportunities for verifying the story line, relationships between categories, and for filling in poorly developed categories" (Strauss & Corbin, 1990, p. 187).

*generate or discover a theory* Grounded theory research is the process of developing a theory, not testing a theory. Researchers might begin with a tentative theory they want to modify or no theory at all with the intent of "grounding" the study in views of participants. In either case, an inductive model of theory development is at work here, and the process is one of generating or discovering a theory grounded in views from participants in the field.

*grounded theory study* In this type of study, the researcher generates an abstract analytical schema of a phenomenon, a theory that explains some action, interaction, or process. This is accomplished primarily through collecting interview data, making multiple visits to the field (theoretical sampling), attempting to develop and interrelate categories (constant comparison) of information, and writing a substantive or context-specific theory (Strauss & Corbin, 1990).

*in vivo codes* In grounded theory research, the investigator uses the exact words of the interviewee to form the names for these codes or categories. The names are "catchy" and immediately draw the attention of the reader (Strauss & Corbin, 1990, p. 69).

*intervening conditions* In axial coding, these are the broader conditions—broader than the context—within which the strategies occur. They might be social, economic, and political forces, for example, that influence the strategies in response to the central phenomenon (Strauss & Corbin, 1990).

*memoing* This is the process in grounded theory research of the researcher writing down ideas about the evolving theory. The writing could be in the form of preliminary propositions (hypotheses), ideas about emerging categories, or some aspects of the connection of categories as in axial coding. In general, these are written records of analysis that help with the formulation of theory (Strauss & Corbin, 1990).

*open coding* This is the first step in the data analysis process for a grounded theorist. It involves taking data (e.g., interview transcriptions) and segmenting them into categories of information (Strauss & Corbin, 1990). I recommend that researchers try to develop a small number of categories, slowly reducing the number from, say, 30 to 5 or 6 that become major themes in a study.

*properties* These are other units of information analyzed in grounded theory research. Each category in grounded theory research can be subdivided into properties that provide the broad dimensions for the category. Strauss and Corbin (1990) refer to them as "attributes or characteristics pertaining to a category" (p. 61). They appear in "open coding" analysis.

*propositions* These are hypotheses, typically written in a directional form, that relate categories in a study. They are written from the axial coding model or paradigm and might, for example, suggest why a certain cause influences the central phenomenon that, in turn, influences the use of a specific strategy.

*saturate, saturated, or saturation* In the development of categories and data analysis phase of grounded theory research, I want to find as many incidents, events, or activities as possible to provide support for the categories. In this process, I finally come to a point at which the categories are "saturated"; I no longer find new information that adds to my understanding of the category.

*selective coding* This is the final phase of coding the information. The researcher takes the central phenomenon and systematically relates it to other categories, validating the relationships and filling in categories that need further refinement and development (Strauss & Corbin, 1990). I like to develop a "story" that narrates these categories and shows their interrelationship (see Creswell & Brown, 1992).

*strategies* In axial coding, these are the specific actions or interactions that occur as a result of the central phenomenon (Strauss & Corbin, 1990).

*substantive-level theory* This is a low-level theory that is applicable to immediate situations. This theory evolves from the study of a phenomenon situated in "one particular situational context" (Strauss & Corbin, 1990, p. 174). Researchers differentiate this form of theory from theories of greater abstraction and applicability, called midlevel theories, grand theories, or formal theories.

*theoretical sampling* In data collection for grounded theory research, the investigator selects a sample of individuals to study based on their contribution to the development of the theory. Often, this process begins with a homogeneous sample of individuals who are similar, and, as the data collection proceeds and the categories emerge, the researcher turns to a heterogeneous sample to see under what conditions the categories hold true.

## Ethnography

*analysis of the culture-sharing group* In this step in ethnography, the ethnographer develops themes—cultural themes—in the data analysis. It is a process of reviewing all of the data and segmenting them into a small set of common themes, well supported by evidence in the data (Wolcott, 1994b).

*artifacts* This is the focus of attention for the ethnographer as he or she determines what people make and use, such as clothes and tools (cultural artifacts) (Spradley, 1980).

*behaviors* These are the focus of attention for the ethnographer as he or she attempts to understand what people do (cultural behavior) (Spradley, 1980).

*critical ethnography* This type of ethnography examines cultural systems of power, prestige, privilege, and authority in society. Critical ethnographers study marginalized groups from different classes, races, and genders, with an aim of advocating for the needs of these participants (Madison, 2005; Thomas, 1993).

*cultural portrait* One key component of ethnographic research is composing a holistic view of the culture-sharing group or individual. The final product of an ethnography should be this larger portrait, or overview of the cultural scene, presented in all of its complexity (Spradley, 1979).

*culture* This term is an abstraction, something that one cannot study directly. From observing and participating in a culture-sharing group, an ethnographer can see "culture at work" and provide a description and interpretation of it (H. F. Wolcott, personal communication, October 10, 1996). It can be seen in behaviors, language, and artifacts (Spradley, 1980).

*culture-sharing group* This is the unit of analysis for the ethnographer as he or she attempts to understand and interpret the behavior, language, and artifacts of people. The ethnographer typically focuses on an entire group—one that shares learned, acquired behaviors—to make explicit how the group "works." Some ethnographers will focus on part of the social-cultural system for analysis and engage in a microethnography.

*deception* This is a field issue that has become less and less of a problem since the ethical standards were published in 1967 by the American Anthropological Association. It relates to the act of the researcher intentionally deceiving the informants to gain information. This deception may involve masking the identify of the research, withholding important information about the purpose of the study, or gathering information secretively.

*description of the culture-sharing group* One of the first tasks of an ethnographer is to simply record a description of the culture-sharing group and incidents and activities that illustrate the culture (Wolcott, 1994b). For example, a factual account may be rendered, pictures of the setting may be drawn, or events may be chronicled.

*emic* This term refers to the type of information being reported and written into an ethnography when the researcher reports the views of the informants. When the researcher reports his or her own personal views, the term used is "etic" (Fetterman, 1998).

*ethnography* This is the study of an intact cultural or social group (or an individual or individuals within the group) based primarily on observations and a prolonged period of time spent by the researcher in the field. The ethnographer listens and records the voices of informants with the intent of generating a cultural portrait (Thomas, 1993; Wolcott, 1987).

*etic* This term refers to the type of information being reported and written into an ethnography when the researcher reports his or her own personal views. When the researcher reports the views of the informants, the term used is "emic" (Fetterman, 1998).

*fieldwork* In ethnographic data collection, the researcher conducts data gathering in the "field" by going to the site or sites where the culture-sharing group can be studied. Often, this involves a prolonged period of time with varying degrees of immersion in activities, events, rituals, and settings of the cultural group (Sanjek, 1990).

*function* This is a theme or concept about the social-cultural system or group that the ethnographer studies. Function refers to the social relations among members of the group that help regulate behavior. For example, the researcher might document patterns of behavior of fights within and among various inner-city gangs (Fetterman, 1998).

*gatekeeper* This is a data collection term and refers to the individual the researcher must visit before entering a group or cultural site. To gain access, the researcher must receive this individual's approval (Hammersley & Atkinson, 1995).

*holistic* The ethnographer assumes this outlook in research to gain a comprehensive and complete picture of a social group. It might include the group's history, religion, politics, economy, and/or environment. In this way, the researcher places information about the group into a larger perspective or "contextualizes" the study (Fetterman, 1998).

*immersed* The ethnographic researcher becomes immersed in the field through a prolonged stay, often as long as 1 year. Whether the individual loses perspective and "goes native" is a field issue much discussed in the ethnographic literature.

*interpretation of the culture-sharing group* The researcher makes an interpretation of the meaning of the culture-sharing group. This interpretation may be informed by the literature, personal experiences, or theoretical perspectives (Wolcott, 1994b).

*key informants (or participants)* These are individuals with whom the researcher begins in data collection because they are well informed, are accessible, and can provide leads about other information (Gilchrist, 1992).

*language* This is the focus of attention for the ethnographer as he or she discerns what people say (speech messages) (Spradley, 1980).

*participant observation* The ethnographer gathers information in many ways, but the primary approach is to observe the culture-sharing group and become a participant in the cultural setting (Jorgensen, 1989).

*realist ethnography* A traditional approach to ethnography taken by cultural anthropologists, this approach involves the researcher as an "objective" observer, recording the facts and narrating the study with a dispassionate, omniscient stance (Van Maanen, 1988).

*reciprocity* This field issue addresses the need for the participants in the study to receive something in return for their willingness to be observed and provide information. The researcher needs to consider how he or she will reimburse participants for being allowed to study them.

*reflexivity* This means that the writer is conscious of the biases, values, and experiences that he or she brings to a qualitative research study.

Typically, the writer makes this explicit in the text (Hammersley & Atkinson, 1995).

*structure* This is a theme or concept about the social-cultural system or group that the ethnographer attempts to learn. It refers to the social structure or configuration of the group, such as the kinship or political structure of the social-cultural group. This structure might be exemplified, for example, by an organizational chart (Fetterman, 1998).

## Case Study

*analysis of themes* Following description, the researcher analyzes the data for specific themes, aggregating information into large clusters of ideas and providing details that support the themes. Stake (1995) calls this analysis "development of issues" (p. 123).

*assertions* This is the last step in the analysis, where the researcher makes sense of the data and provides an interpretation of the data couched in terms of personal views or in terms of theories or constructs in the literature.

*bounded system* The "case" selected for study has boundaries, often bounded by time and place. It also has interrelated parts that form a whole. Hence, the proper case to be studied is both "bounded" and a "system" (Stake, 1995).

*case* This is the "bounded system" or the "object" of study. It might be an event, a process, a program, or several people (Stake, 1995). If a single individual is to be studied, then I generally refer the researcher to a narrative research.

*case study* In qualitative research, this is the study of a "bounded system," with the focus being either the case or an issue that is illustrated by the case (or cases) (Stake, 1995). A qualitative case study provides an in-depth study of this "system," based on a diverse array of data collection materials, and the researcher situates this system or case within its larger "context" or setting.

*collective case study* This type of case study consists of multiple cases. It might be either intrinsic or instrumental, but its defining feature is that the researcher examines several cases (e.g., multiple case study) (Stake, 1995).

*context of the case* In analyzing and describing a case, the researcher sets the case within its setting. This setting may be broadly conceptualized (e.g., large historical, social, political issues) or narrowly conceptualized (e.g., the immediate family, the physical location, the time period in which the study occurred) (Stake, 1995).

*cross-case analysis* This form of analysis applies to a collective case (Stake, 1995; Yin, 2003) in which the researcher examines more than one case. It involves examining themes across cases to discern themes that are common to all cases. It is an analysis step that typically follows within-case analysis when the researcher studies multiple cases.

*description* This means simply stating the "facts" about the case as recorded by the investigator. This is the first step in analysis of data in a qualitative case study, and Stake (1995) calls it "narrative description" (p. 123).

*direct interpretation* This is an aspect of interpretation in case study research where the researcher looks at a single instance and draws meaning from it without looking for multiple instances of it. It is a process of pulling the data apart and putting them back together in more meaningful ways (Stake, 1995).

*embedded analysis* In this approach to data analysis, the researcher selects one analytic aspect of the case for presentation (Yin, 2003).

*holistic analysis* In this approach to data analysis, the researcher examines the entire case (Yin, 2003) and presents description, themes, and interpretations or assertions related to the whole case.

*instrumental case study* This is a type of case study with the focus on a specific issue rather than on the case itself. The case then becomes a vehicle to better understand the issue (Stake, 1995). I would consider the gunman case study (Asmussen & Creswell, 1995) mentioned in Chapter 5 of this book to be an instrumental case study.

*intrinsic case study* This is a type of case study with the focus of the study on the case because it holds intrinsic or unusual interest (Stake, 1995).

*multi-site* When sites are selected for the "case," they might be located at different geographical locations. This type of study is considered to be "multi-site." Alternatively, the case might be at a single location and considered a "within-site" study.

*multiple sources of information* One aspect that characterizes good case study research is the use of many different sources of information to provide "depth" to the case. Yin (2003), for example, recommends that the researcher use as many as six different types of information in his or her case study.

*naturalistic generalizations* In the interpretation of a case, an investigator undertakes a case study to make the case understandable. This understanding may be what the reader learns from the case or its application to other cases (Stake, 1995).

*patterns* This is an aspect of data analysis in case study research where the researcher establishes patterns and looks for a correspondence between two or more categories to establish a small number of categories (Stake, 1995).

*purposeful sampling* This is a major issue in case study research, and the researcher needs to clearly specify the type of sampling strategy in selecting the case (or cases) and a rationale for it. It applies to both the selection of the case to study and the sampling of information used within the case. I use Miles and Huberman's (1994) list of sampling strategies and apply it in this book to case studies as well as to other approaches of inquiry.

*within-case analysis* This type of analysis may apply to either a single case or multiple collective case studies. In within-case analysis, the researcher analyzes each case for themes. In the study of multiple cases, the researcher may compare the within-case themes across multiple cases in cross-case analysis.

*within-site* When a site is selected for the "case," it might be located at a single geographical location. This is considered a "within-site" study. Alternatively, the case might be different locations and considered to be "multi-site."

## Other Terms

*approaches to inquiry* This is an approach to qualitative research that has a distinguished history in one of the social science disciplines and that has spawned books, journals, and distinct methodologies. These approaches, as I call them, are known in other books as "strategies of inquiry" (Denzin & Lincoln, 1994) or "varieties" (Tesch, 1990).

*axiological* This qualitative assumption holds that all research is value laden and includes the value systems of the inquirer, the theory, the paradigm used, and the social and cultural norms for either the inquirer or the respondents (Creswell, 2003; Guba & Lincoln, 1988). Accordingly, the researcher admits and discusses these values in his or her research.

*critical race theory (CRT)* This is a theoretical lens used in qualitative research that focuses attention on race and how racism is deeply embedded within the framework of American society (Parker & Lynn, 2002).

*critical theory* This is a theoretical lens used in qualitative research in which a researcher examines the study of social institutions and their transformations through interpreting the meanings of social life; the historical problems of domination, alienation, and social struggles; and a critique of society and the envisioning of new possibilities (Fay, 1987; Madison, 2005; Morrow & Brown, 1994).

*encoding* This term means that the writer places certain features in his or her writing to help a reader know what to expect. These features not only help the reader but also aid the writer, who can then draw on the habits of thought, glosses, and specialized knowledge of the reader (Richardson, 1990). Such features might be the overall organization, code words, images, and other "signposts" for the reader. As applied in this book, the features consist of terms and procedures of a tradition that become part of the language of all facets of research design (e.g., purpose statement, research subquestions, methods).

*epistemological* This is another philosophical assumption for the qualitative researcher. It addresses the relationship between the researcher and that being studied as interrelated, not independent. Rather than "distance," as I call it, a "closeness" follows between the researcher and that being researched. This closeness, for example, is manifest through time in the field, collaboration, and the impact that that being researched has on the researcher.

*feminist research approaches* In feminist research methods, the goals are to establish collaborative and nonexploitative relationships, to place the researcher within the study so as to avoid objectification, and to conduct research that is transformative (Olesen, 2005; Stewart, 1994).

*foreshadowing* This term refers to the technique that writers use to portend the development of ideas (Hammersley & Atkinson, 1995). The wording of the problem statement, purpose statement, and research subquestions foreshadow the methods—the data collection and data analysis—used in the study.

*interpretive qualitative research* This is an approach to qualitative research that has become interwoven into the core characteristics of qualitative research. It recognizes the self-reflective nature of qualitative research and emphasizes the role of the researcher as an interpreter of the data and an individual who represents information. It also acknowledges the importance of language and discourse in qualitative research, as well as issues of power, authority, and domination in all facets of the qualitative inquiry (see Denzin and Lincoln, 2005, and Clarke, 2005).

*issue subquestions* These are subquestions in a qualitative study that follow the central underlying question. They are written to address the major concerns and perplexities to be resolved, the "issue" of a study (Stake, 1995). They typically are few in number and are posed as questions.

*methodological* This assumption holds that a qualitative researcher conceptualizes the research process in a certain way. For example, a qualitative inquirer relies on views of participants, and discusses their views within the context in which they occur, to inductively develop ideas in a study from particulars to abstractions (Creswell, 1994).

*ontological* This is a philosophical assumption about the nature of reality. It addresses the question: When is something real? The answer provided is that something is real when it is constructed in the minds of the actors involved in the situation (Guba & Lincoln, 1988). Thus, reality is not "out there," apart from the minds of actors.

*paradigm or worldview* This is the philosophical stance taken by the researcher that provides a basic set of beliefs that guides action (Denzin & Lincoln, 1994). It defines, for its holder, "the nature of the world, the individual's place in it, and the range of possible relationships to that world" (Denzin & Lincoln, 1994, p. 107). Denzin and Lincoln (1994) further call this the "net that contains the researcher's epistemological, ontological, and methodological premises" (p. 13). In this discussion, I extend this "net" to also include the axiological and rhetorical assumptions.

*postmodernism* This ideological perspective is considered a family of theories and perspectives that have something in common (Slife & Williams, 1995). Postmodernists advance a reaction or critique of the 19th-century Enlightenment and early 20th-century emphasis on technology, rationality, reason, universals, science, and the positivist, scientific method (Bloland,

1995; Stringer, 1993). Postmodernists assert that knowledge claims must be set within the conditions of the world today and in the multiple perspectives of class, race, gender, and other group affiliations.

*procedural subquestions* These are subquestions in a qualitative study that follow the central underlying question. They cover the anticipated needs for information, as Stake (1995) notes, and I have extended Stake's idea to include anticipated procedures in the study for data analysis and reporting the study. In this way, the procedural subquestions foreshadow the procedures to be used in the study.

*qualitative research* This is an inquiry process of understanding based on a distinct methodological tradition of inquiry that explores a social or human problem. The researcher builds a complex, holistic picture, analyzes words, reports detailed views of informants, and conducts the study in a natural setting.

*queer theory* This is a theoretical lens that may be used in qualitative research that focuses on gay, lesbian, or homosexual identity and how it is culturally and historically constituted, linked to discourse, and overlaps gender and sexuality (Watson, 2005).

*research design* I use this term to refer to the entire process of research, from conceptualizing a problem to writing the narrative, not simply the methods such as data collection, analysis, and report writing (Bogdan & Taylor, 1975).

*rhetorical* This assumption means that the qualitative investigator uses terms and a narrative unique to the qualitative approach. The narrative is personal and literary (Creswell, 1994). For example, the researcher might use the first-person pronoun "I" instead of the impersonal third-person voice.

*social science theories* These are the theoretical explanations that social scientists use to explain the world (Slife & Williams, 1995). They are based on empirical evidence that has accumulated in social science fields such as sociology, psychology, education, economics, urban studies, and communication. As a set of interrelated concepts, variables, and propositions, they serve to explain, predict, and provide generalizations about phenomena in the world (Kerlinger, 1979). They may have broad applicability (as in grand theories) or narrow applications (as in minor working hypotheses) (Flinders & Mills, 1993).

*verisimilitude* This is a criterion for a good literary study, in which the writing seems "real" and "alive," transporting the reader directly into the world of the study (Richardson, 1994).

# Appendix B

## A Narrative Research Study

## On the Bus With Vonnie Lee

### Explorations in Life History and Metaphor

*Michael V. Angrosino*

*University of South Florida*

*This article discusses the use of life history as a method of ethnographic research among stigmatized, unempowered people. The author describes and analyzes the process of eliciting the life history of a man with mental retardation. To combine life history interviewing with the detailed observation of behavior in a naturalistic setting is typical of the ethnographic tradition; interviews with people from marginalized social groups (particularly those who are considered mentally "disabled") are, however, often decontextualized and conducted in quasi-clinical settings that emphasize the retrospective reconstruction of a life. By treating a person with mental retardation as a contextualized participant in a world outside the clinical setting and by eliciting the life narrative in the course of following*

SOURCE: This article originally appeared in the *Journal of Contemporary Ethnography*, 23, 14–28. Copyright 1994, Sage Publications, Inc.

*that person as he attempts to make sense of life outside the institution, it is pos-*
*sible to clarify the dynamic in the formation of a metaphor of personal identity.*
*This technique might not be appropriate for all persons with mental disability, but*
*when it can be used, it helps to demonstrate the proposition that mental retar-*
*dation is not a monolithic condition whose victims are distinguished by arbitrary*
*gradations of standardized test scores. Rather, it is only one of many factors that*
*figure into a person's strategy for coping with the world.*

## A Life in Process

### Vonnie Lee

Vonnie Lee Hargrett celebrated his 29th birthday while I was writing this article in the summer of 1993 in the Florida city to which his parents had migrated from a rural part of the state. The family was, in Vonnie Lee's own words, "poor White trash—real crackers." His father was mostly absent, supposedly shuttling around Florida, Georgia, and Alabama seeking work; if he ever did work ("Not like I even once believed he did," Vonnie Lee told me), he never sent any money home, and he disappeared for good ("real good," Vonnie Lee smirked) about 8 years ago. His mother is an alcoholic who has, over the course of the years, taken up with countless men, most of whom were physically abusive to everyone in the family. Several of them were apparently encouraged in their sexual abuse of Vonnie Lee's two sisters; at least two of them also sexually abused Vonnie Lee. The children were sent to school on a come-and-go basis as the mother moved from place to place around town with her different boyfriends. All three children developed serious learning deficits, although only Vonnie Lee seems to have been tagged by a counselor as mentally retarded. He was never in one school long enough to benefit from any special education programs, however, and he stopped going to school altogether by the time he was 12 years old.

During his teen years he lived mostly on the streets in the company of an older man, Lucian, who made a living by "loaning" Vonnie Lee to other men on the street. Vonnie Lee often says, "Lucian, he's like the only real father I ever had—whatever he had he shared with me. I'd-a done anything for him. *Anything.*"

Lucian was found one morning beaten to death in an empty lot. Vonnie Lee, who had been with one of Lucian's clients that night, discovered the body upon his return to their campsite. The police found him, sobbing and gesturing wildly over the body, and took him into custody. He was held briefly on suspicion of murder, but there was no hard evidence linking him to the crime and he was never charged. His disorderly behavior, however, was sufficient to have him "Baker Acted" (involuntarily committed for psychiatric observation under the provisions of the Florida Mental Health Act). He spent the next few years in and out of psychiatric facilities, developing the remarkable—and, to any number of clinicians, the thoroughly frustrating—capacity to turn into the most level-headed, socially appropriate, even intelligent young gentleman after just a short time in treatment. He would be released, make his way back to the streets, survive quite well for a time, then "break up" (a term he explicitly and consistently prefers to "break down") and be carted off to jail or the hospital.

Vonnie Lee was finally remanded to Opportunity House (OH), an agency designed for the habilitation of adults with the dual diagnosis of mental retardation and psychiatric disorder; most of them also have criminal records. There he made sustained academic, social, and vocational progress, and in June 1992 he was deemed ready for "supervised independent living."

One of the key steps in preparing OH clients for independent living is to teach them to use the public transportation system. I had been a member of OH's board of directors since 1982 (a position I was asked to fill as a result of my long-term research involvement with the program) and had also been a frequent volunteer classroom tutor. I was, however, never directly involved with the "social skills habilitation" aspect of the program until I was asked to fill in for an ailing staff member who was supposed to show Vonnie Lee the bus route from his new apartment to the warehouse where he was to begin working. I was not entirely pleased with the prospect; our city, despite its substantial size and pretensions to urban greatness, has a notoriously inadequate bus system, and I knew that even the relatively simple trip from Vonnie Lee's apartment to his work site involved several transfers and could mean long, hot waits at unshaded bus stops.

## Vonnie Lee and Me

I first met Vonnie Lee shortly after his arrival at OH in 1990. The teacher asked me to help him with his reading assignment: a paragraph about some children taking a walk with their dog. (The fact that reading materials for adults with limited reading ability are almost always about children or about topics that would typically engage the imaginations of children is a subtle but nonetheless painful insult that merits at least a parenthetical complaint here.) Vonnie Lee did not have any particular difficulty reading all the words, but he was having trouble with comprehension. After reading the paragraph, he was unable to answer questions requiring recall and synthesis of information. He seemed more depressed than angry over his failure, and so I said, "Let's put the book away for a minute. Why don't you tell me about a time you remember when *you* took a walk." My intention was to allow him to refocus on the elements constituting a simple narrative in his own words instead of on the specific details about the unfamiliar Tom, Sally, and Spot. Instead, he just said, with inexpressible and totally unexpected sadness, "Yeah. Take a walk. Story of my life." I little realized the full import of his remark but did make a mental note to see if he would at some later time be amenable to telling me the "story of my life" along with the other OH clients among whom I was conducting life history research.

We eventually got around to taping some conversations that would lead to the production of his autobiography, but I frankly was stumped. The problem was not that his discourse was jumbled; in fact, it proceeded in the most nearly linear, chronological sequence of any of the stories I worked on at OH. The problem was that even after numerous sessions I could form no clear sense of who or what Vonnie Lee thought he was. Was Vonnie Lee perhaps a person whose mental disorder was—despite his surface demeanor of reasonable intelligence and even a sense of humor—so profound that he couldn't be fitted into my emergent analytical scheme?

Like my other OH life history collaborators, Vonnie Lee worked in an anecdotal style of narrative. That is, rather than say, for example, "I was born in this city. I lived with my mother and father. I remember the house we lived in," he would say, "When I was a real little kid. Yeah. Let me tell you about that." And he would go on to relate an encapsulated anecdote that was meant to represent his life as a "real little kid." Then he would go on and say, "So then I got a little older. Yeah. Here's what it was." And he would launch into another encapsulated story. My problem, though, was that in Vonnie Lee's case the stories were almost devoid of characters, except in marginal scene-setting roles, and of plot, even of the most attenuated type. For example:

> So Hank [one of his mother's boyfriends] says, "Let's you and me go see Ronnie [a dealer in stolen auto parts for whom Hank sometimes worked]." So we're on the bus. It starts over there next to the mall, and it cuts across and then it stops on the corner where it's that hospital. It stopped there a good long while, you know. Then it goes on down 22nd Street. Past the Majik Mart. Past that gas station with the big yellow thing out front.

And on and on the story would go, except that it was essentially a description of the bus route. Vonnie Lee seemed to have a photographic memory of every convenience store, gas station, apartment complex, newspaper machine, and frontyard basketball hoop along the way. But in the process, he completely lost the point (or what I assumed was the point). There was no word about his reaction to all these sights, nor was there any mention of what Hank was doing. Indeed, they never got to Ronnie's place; the anecdote ended when they got off the bus in front of a Salvation Army thrift shop, apparently several blocks from their destination. When I asked him what happened when they got there, he shrugged and said, "Oh, nothing." I sensed that he wasn't trying to cover anything up (he had already made it perfectly clear that Hank was a thief), and I felt certain that he wasn't just goofing around. He truly believed that the point of the story was the bus ride, not the destination. Vonnie Lee was cooperative in responding to direct questions aimed, on my part, at identifying key players in his life and the events that linked them together. But on his own initiative, he was inclined only to offer what he seemed to feel were these deeply revelatory bus itineraries.

## Bus Trip

On the day I picked up Vonnie Lee at OH to show him the bus routes, we drove first to his new apartment complex. I parked my car and we walked up to the corner bus stop. Vonnie Lee was visibly excited, more animated and seemingly more happy than I had ever seen him. It was a crushingly hot Florida summer day and thunderstorms threatened, but he seemed so elated that my own spirits were lifted. "I bet you're really excited about having your own place," I ventured (violating the first rule of life history interviewing by putting words into the mouth of an informant). "Nah," he replied, "I like the streets to live on—but they won't let me or else I go back to lockup." Nothing daunted, I went on, "But it must be great to have a real job." "In that old dump? Hell no!" he retorted. So what was he so happy about? It dawned on me that the bus itself was the object of his joy, as I watched him bounce into the vehicle when it finally lumbered to a stop. The symbol of the city's bus line is a large red heart, and Vonnie Lee made a dash for a seat directly under a poster bearing that logo; from time to time during the ride, he would reach up and touch it lovingly.

Vonnie Lee seemed to be very familiar with the route we were taking. "Yeah, I walked it about 13 million times," he said with contempt. But now as we sat on the nearly empty bus, he kept swiveling from one side to the other, calling out local landmarks with great glee. At one point, we passed an elderly lady laboriously dragging several large plastic supermarket bags across the street. "I know her type," he sneered. "Uses up every last damn dime she got and she can't ride the bus back home. Drags her ass around like some goddam retard."

We reached the junction where we needed to transfer. "Oh, *here's* where I do it!" Vonnie Lee shouted ecstatically. "I *love* this street, but I never get a chance to come here no more!" The street in question is one of the city's shabbiest, lined with unpleasant-looking bars, secondhand clothing stores, and unkempt, garagelike structures from which used furniture, carpeting "seconds," rebuilt appliances, and sundry "recyclables" are delivered. The place where Vonnie Lee was to work was on a street like this one but which required a further transfer to reach; he was, I had to admit, quite right in characterizing such a place as an "old dump," but his mood betrayed not the slightest hint of regret.

We waited for a very long time at the transfer stop. Two heavily made-up young women were lolling in front of one of the bars but were making no real attempt to secure business; they seemed stunned by the heat and shook their heads wearily at the spectacle of Vonnie Lee jumping up and down to catch a glimpse of the approaching bus. When it came at last, it was more crowded than the first had been, and Vonnie Lee's face clouded briefly when he saw that the favored seat under the heart was already taken. He resigned himself to a less desirable place but kept turning his head toward the heart as if to reassure himself, even as he resumed his practice of announcing every building on the street. He was less familiar with this street than with the first, and his litany seemed to be serving the purpose of fixing the sights in his own mind as well as of enlightening me.

Vonnie Lee seemed sorry to get off when we reached our stop, but he brightened immediately when he saw the street down which our third and final lap would take us. It was a street very much like the second, although it led off to a part of town he hardly knew at all; the thrill of the new gave him added zest. It began to rain while we waited and waited for the third bus, and we found only modest shelter in the boarded-up doorway of what had once been a storefront church. Vonnie Lee's spirits didn't sag in the least, even when the bus arrived, packed full of damp and irritable riders. He managed to find a standing spot near enough the heart logo and immediately set about his recitation of the sights. Some of the people nearby looked a little annoyed, but no one said anything. The crowd was as thoroughly depressed and defeated as Vonnie Lee was giddy.

It bears mentioning that the city's buses are *very slow*. Not only do they run infrequently, but once they do arrive they appear to obey an unstated mandate to stop at *every* marked stop, whether or not anyone wants to get on or off. As a result, the trip from the apartment to the warehouse, which might have taken at most 20 minutes by car, ended up consuming an hour and a half by bus. The other anomaly in the bus system that I had ample time to observe that day (and I had lived in the city for nearly two decades at that time without ever having ridden a city bus) was that all the riders (including my own wet and bedraggled self) looked like stereotyped versions of either very poor or mentally/physically disabled people. I came to realize that no bus route connected one "nice" part of town with another; all of them took off into and covered most of their distance within "bad" sections. (Since my trip with Vonnie Lee, the transit authority has added several routes connecting upscale residential neighborhoods with the downtown business and government districts, but they are all "express" runs that zoom right through the intervening "bad" spots.) It was clear that the bus system had been designed primarily for domestic workers going to and from the posh homes and business offices and for blue-collar workers traveling from low-rent districts to downscale factory zones. In many big cities, going to work by bus is a perfectly appropriate thing for even the most affluent of business people to do. But in our city, the bus is the very embodiment of stigma—the slow, inconvenient transport of the poor, the powerless, and the socially marginal.

When we reached our stop, literally in front of the plumbing supply warehouse where Vonnie Lee was to work, it began pouring again. We dashed inside where Vonnie Lee's supervisor, Mr. Washington, was very gracious in showing us around and then allowing us to wait out the storm. The warehouse was cramped and dingy, but it seemed to be doing a brisk business ("People are going to need toilets, even during a recession," Mr. Washington noted), and the supervisor and most of the other workmen showed a genuine interest in Vonnie Lee's welfare. (Several other OH clients had been employed there over the years; the owner of the business had a mentally retarded brother who died young, and he looked on his employment program as a way of honoring his memory.) "I love it!" Vonnie Lee shouted as one of the other men took him through the back door to show him a tiny commissary where they could buy soft drinks and snacks.

"It's the bus he loves—coming here on the bus," Mr. Washington said to me when we were alone. I admitted that Vonnie Lee did seem to have had an unaccountably good time on the ride. "Yeah. I've seen it before. Ask him about it, why don't you?"

The return ride was a replay of the first; Vonnie Lee had already memorized all the new landmarks. "Why do you like the bus so much?" I asked at last—the question that had been obvious all day but had seemed too silly and irrational to bring up. And, as he always did when I put a direct question to him, he gave me a straightforward answer. He repeated his answer on tape later on, telling me,

Like I always said, we was dirt poor at home. Mama never had no car or nothing. Most of them guys was even more worthlesser than Daddy. Why that woman has a thing for big losers I'll never know! Now every once in a while one of 'em took me on the bus. And poor old Lucian—he didn't like to get on the bus because he said everybody looked at him funny, but still we did it now and again just to show we *could*. I mean—it's only the lowdownest who can't *never* do it.

Jeez! I'd walk a street and say, "If I was a bigshot, I'd be on the bus right now!" The bad thing—a man just can't ride one end of the street to another like he was some retard with no place to go. A man gotta go *somewheres,* and I never knew how to get *anywheres* like from one to the other one. I nearly peed my pants when they told me I could learn; they never thought I could before now. I got kinda scared when they told me Ralph couldn't take me and you was gonna do it; I thought, "Hey, he don't really work at OH. Maybe he don't know how and he'll screw me up." But then I figured you'd figure it out and then you'd show me.

## The Meaning of the Bus

It all became clear to me. The bus—to the "nice" people the symbol of poverty, the despised underside of the glittery urban lifestyle touted by the city's boosters—was for Vonnie Lee a potent symbol of empowerment. Coming from a family that was *too poor even to take the bus* was a humiliation that had scarred his young life. He spent his years grimly walking, walking, learning the details of the streets and yet yearning for the time when he could be chauffeured high above those streets in the style to which he felt himself entitled. For Vonnie Lee, the payoff for all his hard work in overcoming both his background and his numerous "break up" reversals was neither the apartment nor the job but the fact that he was finally deemed worthy to learn how to ride the bus between the two. So many of the OH clients with whom I spoke longed to see themselves in positions of power, and their dreams of driving fancy cars, although unrealistic, were at least recognizable ambitions. I had completely missed Vonnie Lee's ideal; because he saw escape and empowerment in the bus (something to which *anyone,* even a person with mental retardation, could reasonably aspire), I had ignored the fact that it *was* a dream for him and that it gave shape and meaning to his life.

I finally saw why, when telling his life story without specific prompts, he did so in the form of bus routes. For Vonnie Lee, those rare rides were the stuff of which his dreams were made; they embodied his values, his aspirations, and even his self-image. My other informants' stories led me to conclude that they had developed stable self-images that survived all the vicissitudes of their lives. Vonnie Lee's self-image, on the other hand, was bound up not in who he was but in who he wanted to be: a man on a bus, going somewhere. Since his earlier rides had been dry runs, as it were, they didn't add up to a consistent pattern, and he went along as someone else's adjunct (and the someone else was, at best, only a temporarily significant other); he never felt that they represented sort of defined closure. As a result, he did not feel impelled to "finish" those stories, as the real finish—the point at which he was ready to believe himself to be someone—was in the future.

Mr. Washington had indeed seen a number of his charges who liked to ride the bus because it was so liberatingly different from the heavily supervised minivan that shuttled the OH clients around prior to their graduation. But Vonnie Lee's fixation on the bus went even further than the supervisor could have imagined. When we finally got back to OH, I took a careful look around Vonnie Lee's room as I helped him pack up some of his belongings. Taped to his mirror was an outline drawing of a heart; it had been cleanly scissored out of a coloring book about seasons and holidays that one of his "lower functioning" roommates was using in class. Vonnie Lee had carefully colored the big valentine with a neon red marker. Before that day, I would have assumed that he was, like some of my other informants, pretending to have received at least one passionate proposal of marriage on Valentine's Day. But now I knew immediately that it was not a valentine at all but the closest thing he could find to the bus company logo.

## Discussion

Vonnie Lee's autobiography, and the story of my interaction with him, is part of a long-term research project whose methodology and conceptual framework were described in some detail in earlier writings (Angrosino 1989, 1992; Angrosino and Zagnoli 1992). That project was designed to demonstrate three points: that individual identity is conceptualized and communicated as much through the form as through the content of autobiographical material (Crocker 1977; Hankiss 1981; Howarth 1980; Olney 1972), that autobiographies are best interpreted as extended metaphors of self (Fitzgerald 1993; Norton 1989), and that even persons with conditions that interfere with their ability to construct conventionally coherent narratives nevertheless sustain self-images (Zetlin and Turner 1984) and can communicate those images to others of the same culture by using culturally recognizable metaphorical forms.

In earlier analyses based on this research, I relied essentially on literary theory as applied to autobiography to define "metaphor." In accordance with that view, I was less concerned with specific expressive metaphors ("The house I grew up in was a toilet") than with the way in which an entire life was reconstructed in the narration around a master concept of self. The informants profiled in the other studies had adopted clearly defined social roles (the "blame attributor," the "tactical dependent," the "denier," the "passer") and told their stories in ways that marshaled rhetorical devices ("antithesis," "compensation," "allusion," "anecdote," "oratory," "dialogue") to buttress the presentation of those roles. In so doing, the roles became a dominating metaphor of the stigmatization experienced by the informants.

Although this perspective on metaphor was a useful framework for analyzing the stories of some informants who, in their various ways, perceived a continuity between their early experiences and their current lives, it was inadequate in Vonnie Lee's case. Although reasonably articulate about his past, Vonnie Lee is a person who adamantly refuses to live in the past; his orientation is so thoroughly toward the future that he resists characterizing himself in terms of what he has always been. Far from operating on the assumption that he is a product of his past (even if only in reaction to it), Vonnie Lee sees his life as beginning only when he makes a definitive break with that past. For Vonnie Lee, the past is not even prologue; it is, for all intents and purposes, irrelevant as a predictor of his future.

The dominating metaphor of Vonnie Lee's life, then, emerges not out of retrospective narrative but out of the actions he is currently taking to remake himself into his desired new image. For this reason, it is important not to limit the dialogue of discovery to retrospective interviews conducted in a time and place of their own. Rather, it is crucial to conduct what amounts to a personalized ethnography of this informant—to catch him in the act of self-creation, as it were. He does not use metaphor to symbolize the asserted continuities of his broken life as do the other informants; his metaphorical image is created in the actions that define his trajectory of "becoming."

For an ethnographer who works in an applied field (such as the formulation of policy for and the delivery of services to people with a defined disorder, such as mental retardation), this research demonstrates the benefits of the in-depth autobiographical interview methodology for establishing the human dimensions of mentally disordered persons, who are all too frequently described in terms of deviations from standardized norms. Vonnie Lee's story goes one step further: It demonstrates the desirability of contextualizing the autobiographical interview within the ongoing life experience of the subject rather than treating it as a retrospective review.

Such contextualization is, to be sure, an article of faith among anthropological ethnographers and is widely accepted by other social scientists working in the ethnographic tradition. It is, however, a conclusion that has rarely been applied to studies of "deviant," "stigmatized," or "marginalized" people.

There is a great deal of published material based on the life histories of people with mental retardation, but, as Whittemore, Langness, and Koegel (1986) point out in their critical survey of that literature, those materials are almost entirely lacking in any sense of an insider's perspective. Much of that literature is more focused on the experiences of caregivers, the assumption being that the person with retardation is unable to speak coherently on his or her own behalf. It is true that retarded people in clinical settings are interviewed with an eye to telling their life stories, but such accounts (which are rarely published) presuppose a clinically defined disorder and focus on the psychodynamics of the illness; the whole person is subsumed into the "disorder," and the interview itself is part of the process of correction and therapy.

My work is more closely allied with the tradition pioneered at the University of California, Los Angeles, by Edgerton and his colleagues in the "sociobehavioral" group, who have made it a practice to study the lives of their subjects in their entirety. "Because these lives change in response to various environmental demands, just as they develop in reaction to maturational changes, we emphasize process" (Edgerton 1984, 1–2) rather than retrospection; moreover, they do so by providing detailed descriptions of the communities where the subjects live so as to situate the life histories outside the strictly clinical milieu. Nevertheless, even this approach begins with the acceptance of clinically defined disorder, such that the contextualized life history serves mainly to illuminate the process of "adjustment" to a presumed mainstream norm.

My encounter with Vonnie Lee taught me that his worldview was not a failed approximation of how a "normal" person would cope, nor was it, when taken on its own terms, intrinsically disordered. His fixation on the bus is only "disordered" or an "attempt at adjustment" if we assume that the rest of us are not without our own idées fixes regarding the world and our place in it. Were we to suffer the misfortune of being labeled "retarded," would all of our ideas, attitudes, and practices stand up to scrutiny as being unimpeachably "normal"? Once we start looking for evidence of "disorder," then "disorder" is almost certainly what we find. If anything, Vonnie Lee's logic is more clearly worked out and better integrated than that of more sophisticated people; his "retardation" may, indeed, lie in the way he has purified his obsession down to its basics rather than veiling it in varieties of symbolic discourse as "normal" people do. Interviewing him in a way that arose out of a normal activity did not merely "contextualize" his disorder, it removed the emphasis on disorder altogether. Like a conventional anthropologist conducting participant observation in a community other than his or her own, I only began to make progress when I stopped thinking that there was something "exotic" in Vonnie Lee's approach to the world and started asking him simply what things meant to him. This insight might not come as a surprise to theoreticians who work in the autobiographical genre, or to anthropologically oriented ethnographers in general, but it is certainly a different point of view from that typically seen among professional service providers in the mental health/mental retardation field.

It is certainly true that Vonnie Lee's is only one story. I have been asked by several people who have read drafts of this article, "But is he typical of retarded people in his ability to concentrate and integrate his life experiences?" The honest answer is that I don't know. But in a larger sense, to ask the question is to assume that "mental retardation" is a defined, bounded category fixed within the parameters of clinical, statistical norms. Vonnie Lee is "retarded" in the sense that he has been so labeled and has been dealt with by "the system" as a retarded person for much of his life. And yet he copes with the world around him in a way that, although out of the statistical "norm," is not entirely dysfunctional—as long as we stop trying to see his experiences as illustrations of disorder. Mental retardation is a broad and heterogeneous category; I do not doubt that many persons so diagnosed would have great difficulty in expressing as coherent a worldview as Vonnie Lee's. This kind of methodology would almost certainly not work with all people so diagnosed.

What this fragment of a research project demonstrates is that for at least some people with mental retardation, it is possible to do what anthropological ethnographers have long done: get away from asking retrospective questions that only emphasize the "exoticism" of the subjects and, instead, allow questions to flow naturally out of observations of the subjects in their ordinary round of activities. That method has long been a way to see cultural differences as variations in human responses to certain common problems; here it is a way of seeing that a person like Vonnie Lee might be extreme in some of his responses but is still part of the same continuum of experience. Perhaps in his specialness and individual quirkiness, Vonnie Lee *is* typical after all—not of "mentally retarded persons" but of human beings who learn how to use elements of the common culture to serve their individual purposes.

# References

Angrosino, M. V. 1989. *Documents of interaction: Biography, autobiography, and life history in social science perspective.* Gainesville: University of Florida Press.

———. 1992. Metaphors of stigma: How deinstitutionalized mentally retarded adults see themselves. *Journal of Contemporary Ethnography* 21:171–99.

Angrosino, M. V., and L. J. Zagnoli. 1992. Gender constructs and social identity: Implications for community-based care of retarded adults. In *Gender constructs and social issues,* edited by T. Whitehead and B. Reid, 40–69. Urbana: University of Illinois Press.

Crocker, J. C. 1977. The social functions of rhetorical form. In *The social uses of metaphor,* edited by D. J. Sapir and J. C. Crocker, 33–66. Philadelphia: University of Pennsylvania Press.

Edgerton, R. B. 1984. Introduction. In *Lives in process: Mildly retarded adults in a large city,* edited by R. B. Edgerton. Washington, DC: American Association on Mental Deficiency.

Fitzgerald, T. K. 1993. Limitations of metaphor in the culture-communication dialogue. Paper presented at the annual meeting of the Southern Anthropological Society, Savannah, GA, 25 March.

Hankiss, A. 1981. Ontologies of the self: On the mythological rearranging of one's life history. In *Biography and society: The life history approach in the social sciences,* edited by D. Bertaux, 203–10. Beverly Hills, CA: Sage.

Howarth, W. L. 1980. Some principles of autobiography. In *Autobiography: Essays theoretical and critical,* edited by J. Olney, 86–114. Princeton, NJ: Princeton University Press.

Norton, C. S. 1989. *Life metaphors: Stories of ordinary survival.* Carbondale: Southern Illinois University Press.

Olney, J. 1972. *Metaphors of self: The meaning of autobiography.* Princeton, NJ: Princeton University Press.

Whittemore, R. D., L. L. Langness, and P. Koegel. 1986. The life history approach to mental retardation. In *Culture and retardation,* edited by L. L. Langness and H. G. Levine, 1–18. Dordrecht, Netherlands: D. Reidel.

Zetlin, A. G., and J. L. Turner. 1984. Self-perspectives on being handicapped: Stigma and adjustment. In *Lives in process: Mildly retarded adults in a large city,* edited by R. B. Edgerton, 93–120. Washington, DC: American Association on Mental Deficiency.

# Appendix C

## A Phenomenological Study

### Cognitive Representations of AIDS

*Elizabeth H. Anderson*

**Margaret Hull Spencer**

*Cognitive representations of illness determine behavior. How persons living with AIDS image their disease might be key to understanding medication adherence and other health behaviors. The authors' purpose was to describe AIDS patients' cognitive representations of their illness. A purposive sample of 58 men and women with AIDS were interviewed. Using Colaizzi's (1978) phenomenological method, rigor was established through application of verification, validation, and validity. From 175 significant statements, 11 themes emerged. Cognitive representations included imaging AIDS as death, bodily destruction, and just*

AUTHORS' NOTE: This study was funded in part by a University of Connecticut Intramural Faculty Small Grants and School of Nursing Dean's Fund.

We wish to thank Cheryl Beck, D.N.Sc., and Deborah McDonald, Ph.D., for reading an earlier draft. Special thanks go to Stephanie Lennon, BSN, for her research assistance.

Address reprint requests to Elizabeth H. Anderson, Ph.D., A.P.R.N., Assistant Professor, University of Connecticut School of Nursing, 231 Glenbrook Road, U-2026, Storrs, CT 06269-2026, USA.

SOURCE: The material in this appendix originally appeared in *Qualitative Health Research, 12*(10), 1338–1352. Copyright 2002, Sage Publications, Inc.

265

*a disease. Coping focused on wiping AIDS out of the mind, hoping for the right drug, and caring for oneself. Inquiring about a patient's image of AIDS might help nurses assess coping processes and enhance nurse-patient relationships.*

A 53-year-old man with a history of intravenous drug use, prison, shelters, and methadone maintenance described AIDS as follows:

> My image of the virus was one of total destruction. It might as well have killed me, because it took just about everything out of my life. It was just as bad as being locked up. You have everything taken away from you. The only thing to do is to wait for death. I was afraid and I was mad. Mostly I didn't care about myself anymore. I will start thinking about the disease, and I'll start wondering if these meds are really going to do it for me.

To date, 36 million people worldwide (Centers for Disease Control and Prevention [CDC], 2001b) are infected with Human Immunodeficiency Virus (HIV) that develops into end-stage Acquired Immunodeficiency Syndrome (AIDS). In the United States, 448,060 have died of AIDS-related illnesses, and more than 322,000 persons are living with AIDS, the highest number ever reported (CDC, 2001a).With HIV/AIDS, 95% adherence to antiretroviral (ART) drug regimens is necessary for complete viral suppression and prevention of mutant strains (Bartlett & Gallant, 2001). Adherence to ART regimens can slow the disease process but does not cure HIV or AIDS. Persons with AIDS experience numerous side effects associated with ART drugs, which can lead to missed doses, profound weight loss, and decreased quality of life (Douaihy & Singh, 2001). The incidence of HIV/AIDS is reduced through prevention that is dependent on life-long commitment to the reduction of high-risk drug and sexual behaviors. To achieve maximum individual and public health benefits, it might be helpful to explore patients' lived experience of AIDS within the framework of the self-regulation model of illness.

In the Self-Regulation Model of Illness Representations, patients are active problem solvers whose behavior is a product of their cognitive and emotional responses to a health threat (Leventhal, Leventhal, & Cameron, 2001). In an ongoing process, people transform internal (e.g., symptoms) or external (e.g., laboratory results) stimuli into cognitive representations of threat and/or emotional reactions that they attempt to understand and regulate. The meaning placed on a stimulus (internal or external) will influence the selection and performance of one or more coping procedure (Leventhal, Idler, & Leventhal, 1999). Emotions influence the formation of illness representations and can motivate a person to action or dissuade him or her from it. Appraisal of the consequences of coping efforts is the final step in the model and provides feedback for further information processing.

Although very individual, illness representations are the central cognitive constructs that guide coping and appraisal of outcomes. A patient's theory of illness is based on many factors, including bodily experience, previous illness, and external information. An illness representation has five sets of attributes: (a) identity (i.e., label, symptoms), (b) time line (i.e., onset, duration), (c) perceived cause (i.e., germs, stress, genetics), (d) consequences (i.e., death, disability, social loss), and (e) controllability (i.e., cured, controlled) (Leventhal, Idler, et al., 1999; Leventhal, Leventhal, et al., 2001).

Attributes have both abstract and concrete form. For example, the attribute "identity" can have an abstract disease label (e.g., AIDS) and concrete physical symptoms (e.g., nausea and vomiting). Symptoms are convenient and available cues or suggestions that can shape an illness representation and help a person correctly or incorrectly interpret the experience. Although symptoms are not medically associated with hypertension, patients who believed medications reduced their symptoms reported greater adherence and better blood pressure control (Leventhal, Leventhal, et al., 2001).

Understanding how individuals cognitively represent AIDS and their emotional responses can facilitate adherence to therapeutic regimens, reduce high-risk behaviors, and enhance quality of life. Phenomenology provides the richest and most descriptive data (Streubert & Carpenter, 1999) and thus is the ideal research process for eliciting cognitive representations. Consequently, the purpose of this study was to explore patients' experience and cognitive representations of AIDS within the context of phenomenology.

## Review of the Literature

Vogl et al. (1999), in a study of 504 ambulatory patients with AIDS who were not taking protease inhibitor (PI) drugs, found the most prevalent symptoms were worry, fatigue, sadness, and pain. Both the number of symptoms and the level of symptom distress was associated with psychological distress and poorer quality of life. Persons with a history of intravenous drug use reported more symptoms and greater symptom distress. In contrast, a telephone survey and chart review of 45 men and women with HIV/AIDS suggested that PI therapy was associated with weight gain, improved CD4 counts, decreased HIV RNA viral loads, fewer opportunistic infections, and better quality of life (Echeverria, Jonnalagadda, Hopkins, & Rosenbloom, 1999).

Reporting on pain from patients' perspective, Holzemer, Henry, and Reilly (1998) noted that 249 AIDS patients reported experiencing moderate level of pain, but only 80% had effective pain control. A higher level of pain was associated with lower quality of life. In a phenomenological study focusing on pain, persons with HIV/AIDS viewed pain as not only physical but also an experience of loss, not knowing, and social (Laschinger & Fothergill, 1999).

Turner (2000), in a hermeneutic study of HIV-infected men and women, found that AIDS-related multiple loss was an intense, repetitive process of grief. Two constitutive patterns emerged: Living with Loss and Living beyond Loss. Likewise, Brauhn (1999), in a phenomenological study of 12 men and 5 women, found that although persons with HIV/AIDS experienced their illness as a chronic disease, their illness had a profound and pervasive impact on their identity. Participants planned for their future with cautious optimism but could identify positive aspects about their illness.

McCain and Gramling (1992), in a phenomenological study on coping with HIV disease, reported three processes: Living with Dying, Fighting the Sickness, and Getting Worn Out. Koopman et al. (2000) found that among 147 HIV-positive persons, those with the greatest level of stress in their daily lives had lower incomes, disengaged behaviorally/emotionally in coping with their illness, and approached interpersonal relationships in a less secure or more anxious manner. With somewhat similar results, Farber, Schwartz, Schaper, Moonen, and McDaniel (2000) noted that adaptation to HIV/AIDS was associated with lower psychological distress, higher quality of life, and more positive personal beliefs related to the world, people, and self-worth. Fryback and Reinert (1999), in a qualitative study of women with cancer and men with HIV/AIDS, found spirituality to be an essential component to health and well-being. Respondents who found meaning in their disease reported a better quality of life than before diagnosis.

Dominguez (1996) summarized the essential structure of living with HIV/AIDS for women of Mexican heritage as struggling in despair to endure a fatal, transmittable, and socially stigmatizing illness that threatens a woman's very self and existence. Women were seen as suffering in silence while experiencing shame, blame, and concern for children. In a phenomenological study of five HIV-infected African American women, 12 themes emerged, ranging from violence, shock, and denial to uncertainty and survival (Russell & Smith, 1999). The researchers concluded that women have complex experiences that need to be better understood before effective health care interventions can be designed.

No studies reported AIDS patients' cognitive representations or images of AIDS. Consequently, this study focused on how persons with AIDS cognitively represented and imaged their disease.

# Method

## Sample

A purposive sample of 41 men and 17 women with a diagnosis of AIDS participated in this phenomenological study. Participants were predominately Black (40%), White (29%), and Hispanic (28%). Average age was 42 years ($SD = 8.2$). The majority had less than high school education (52%) and were never married (53%), although many reported being in a relationship. Mean CD4 count was 153.4 ($SD = 162.8$) and mean viral load, 138,113 ($SD = 270,564.9$). Average time from HIV diagnosis to interview was 106.4 months ($SD = 64.2$). Inclusion criteria were (a) diagnosis of AIDS, (b) 18 years of age or older, (c) able to communicate in English, and (d) Mini-Mental Status exam score > 22.

## Research Design

In phenomenology, the researcher transcends or suspends past knowledge and experience to understand a phenomenon at a deeper level (Merleau-Ponty, 1956). It is an attempt to approach a lived experience with a sense of "newness" to elicit rich and descriptive data. Bracketing is a process of setting aside one's beliefs, feelings, and perceptions to be more open or faithful to the phenomenon (Colaizzi, 1978; Streubert & Carpenter, 1999). As a health care provider for and researcher with persons with HIV/AIDS, it was necessary for the interviewer to acknowledge and attempt to bracket those experiences. No participant had been a patient of the interviewer.

Colaizzi (1978) held that the success of phenomenological research questions depends on the extent to which the questions touch lived experiences distinct from theoretical explanations. Exploring a person's image of AIDS taps into a personal experience not previously studied or shared clinically with health care providers.

## Procedure

After approval from the university's Institutional Review Board and a city hospital's Human Subject Review Committee, persons who met inclusion criteria were approached and asked to participate. Interviews were conducted over 18 months at three sites dedicated to persons with HIV/AIDS: a hospital-based clinic, a longterm care facility, and a residence. All interviews were tape-recorded and transcribed verbatim. Participants were involved in multiple life situations and were unavailable for repeated interviews related to personal plans, discharge, returning to life on the street, or progression of the disease. One participant died within 4 weeks of the interview. Interviews lasted between 10 and 40 minutes and proceeded until no new themes emerged. Persons who reported not thinking about AIDS provided the shortest interviews. Consequently, to obtain greater richness of data and variation of images, we interviewed 58 participants (Morse, 2000). The first researcher conducted all 58 interviews.

After obtaining informed consent, each participant was asked to verbally respond to the following: "What is your experience with AIDS? Do you have a mental image of HIV/AIDS, or how would you describe HIV/AIDS? What feelings come to mind? What meaning does it have in your life?" As the richness of cognitive representations emerged, it became apparent that greater depth could be achieved by asking participants to draw their image of AIDS and provide an explanation of their drawing. Eight participants drew their image of AIDS.

Background information was obtained through a paper-and-pencil questionnaire. Most recent CD4 and Viral Load laboratory values were obtained from patient charts. Based on institution policy, participants at the long-term care facility and residence received a $5.00 movie pass. Clinic participants received $20.00.

## Data Analysis

Colaizzi's (1978) phenomenological method was employed in analyzing participants' transcripts. In this method, all written transcripts are read several times to obtain an overall feeling for them. From each transcript, significant phrases or sentences that pertain directly to the lived experience of AIDS are identified. Meanings are then formulated from the significant statements and phrases. The formulated meanings are clustered into themes allowing for the emergence of themes common to all of the participants' transcripts. The results are then integrated into an indepth, exhaustive description of the phenomenon. Once descriptions and themes have been obtained, the researcher in the final step may approach some participants a second time to validate the findings. If new relevant data emerge, they are included in the final description.

Methodological rigor was attained through the application of verification, validation, and validity (Meadows & Morse, 2001). Verification is the first step in achieving validity of a research project. This standard was fulfilled through literature searches, adhering to the phenomenological method, bracketing past experiences, keeping field notes, using an adequate sample, identification of negative cases, and interviewing until saturation of data was achieved (Frankel, 1999; Meadows & Morse, 2001). Validation, a within-project evaluation, was accomplished by multiple methods of data collection (observations, interviews, and drawings), data analysis and coding by the more experienced researcher, member checks by participants and key informants, and audit trails. Validity is the outcome goal of research and is based on trustworthiness and external reviews. Clinical application is suggested through empathy and assessment of coping status (Kearney, 2001).

## Results

From 58 verbatim transcripts, 175 significant statements were extracted. Table 1 includes examples of significant statements with their formulated meanings. Arranging the formulated meanings into clusters resulted in

| Table 1 | Selected Examples of Significant Statements of Persons With AIDS and Related Formulated Meanings |
|---|---|
| *Significant Statement* | *Formulated Meaning* |
| In the beginning, I had a sense that I did have it, so it wasn't an unexpected thing although it did bother me. I know it was a bad thing to let it traumatize so. | AIDS is such a traumatizing reality that people have difficulty verbalizing the word "AIDS." |
| [AIDS] a disease that has no cure. Meaning of dread and doom and you got to fight it the best way you can. You got to fight it with everything you can to keep going. | AIDS is a dangerous disease that requires every fiber of your being to fight so you can live. |
| I see people go from somebody being really healthy to just nothing—to skin and bones and deteriorate. I've lost a lot of friends that way. It's nothing pretty. I used to be a diesel mechanic. I can't even carry groceries up a flight of stairs anymore. | As physical changes are experienced, an image of AIDS wasting dominates thoughts. |
| First image—death. Right away fear and death. That's because I didn't know any better. Now it's destruction. Pac-man eating all your immune cells up and you have nothing to fight with. | Overwhelming image of AIDS is one of death and destruction, with no hope of winning. |

| Table 2 | Example of Two Theme Clusters With Their Associated Formulated Meanings |
|---|---|

**Dreaded bodily destruction**

Physical changes include dry mouth, weight loss, mental changes

Expects tiredness, loss of vision, marks all over the body

Holocaust victims

Confined to bed with sores all over

Extreme weight loss

Horrible way to die

Changes from being really healthy to skin and bones

Bodily deterioration

**Devouring life**

Whole perspective on life changed

Never had a chance to have a family

Life has stopped

No longer able to work

Will never have normal relations with women

Uncertain what's going to happen from day to day

Worked hard and lost everything

11 themes. Table 2 contains two examples of theme clusters that emerged from their associated meanings.

*Theme 1: Inescapable death.* Focusing on negative consequences of their disease was the pervading image for many persons with AIDS. Responding quickly and spontaneously, AIDS was described as "death, just death," "leprosy," "a nightmare," "a curse," "black cloud," and "an evil force getting back at you." The sense of not being able to escape was evident in descriptions of AIDS as "The blob. It's a big Jell-O thing that comes and swallows you up" and "It's like I'm in a hole and I can't get out." Another stated, "AIDS, it's a killer and it will get you at any God-given time."

A sense of defeat was evident in a Hispanic man's explanation that with AIDS you are a "goner." He stated, "With HIV you still have a chance to fight. Once that word 'AIDS' starts coming up in your records, you bought a ticket [to death]."

A 29-year-old woman, diagnosed with HIVand AIDS 9 months before the interview, drew a picture of a grave with delicate red and yellow flowers and wrote on the tomb stone "RIP Devoted Sister and Daughter." Over the grave, she drew a black cloud with the sun peeking around the edge, which she described as symbolizing her family's sadness at her death.

*Theme 2: Dreaded bodily destruction.* In this cluster, respondents focused on physical changes associated with their illness. AIDS was envisioned as people who were skin and bones, extremely weak, in pain, losing their minds, and lying in bed waiting for the end. Descriptions were physically consistent but drawn froma variety of experiences, such as seeing a family member or friend die from AIDS or from pictures of holocaust victims. It is an ending that is feared and a thought that causes deep pain. Body image became a marker for level of wellness or approach of death.

One woman described her image of AIDS as a skeleton crying. An extremely tall, thin man awaiting a laryngectomy on the eve of his 44th birthday described his image of AIDS by saying, "Look at me." Another recalled Tom Hanks in the movie *Philadelphia* (Saxon & Demme, 1993): "The guy in the hospital and how he aged and how thin he got. You start worrying about . . . you don't want to end like that. I don't like the image I see when I see AIDS." A 53-year-old man with a 10-year history of HIV/AIDS drew his image of AIDS as a devil with multiple ragged horns, bloodshot eyes, and a mouth with numerous sharp, pointed teeth. He described the mouth as "teeth with blood dripping down and sucking you dry." Another man drew AIDS as an angry purple animal with red teeth. He stated the color purple symbolized a "bruise" and the red teeth "destruction." The extensive physical and emotional devastation of AIDS was evident in the drawing by a 36-year-old Black woman, who pictured herself lying on a bed surrounded by her husband and children. She wrote, "Pain from head to toe, no hair, 75 pounds, can't move, can't eat, lonely and scared. Family loving you and you can't love them back."

*Theme 3: Devouring life.* Persons grieved for their past lives. A 41-year-old man described AIDS as, "It's not like I can walk around the corner or go to the park with friends because it has devoured your life." Another man noted, "My life has stopped." A 48-year-old woman stated, "I feel like I have no life. It has changed my whole perspective."

With the diagnosis of AIDS, dreams of marrying, having children, or working were no longer perceived as possible. The impact on each one's life was measured differently from loss of ability to work to loss of children, family, possessions, and sense of oneself. The thought of leaving children, family, and friends was extremely difficult but considered a reality. A woman with four children aged 8 to 12 years stated,

It's not a disease that you would want to have because it's really bad. I know I get upset sometimes because I have it. You know you are going to die and I have kids. I really don't want to leave them. I want to see them grow up and everything. I know that's not going to happen.

Consistently, participants felt a deep rupture in life as illustrated in the following statement: "It just took my whole life and turned it upside down. I can't do a lot of the things I used to. I lost a house because of it. Everything I worked for I lost." A 44-year-old Hispanic mother of two boys reported with sadness,

It has affected my life. I have lost my children by not being able to take care of them. It has changed my freedom and relationships. Being sick all the time and I couldn't take care of my little one, so he was taken away from me.

A Black woman described the far-reaching effect AIDS had on her life as follows:

Everything is different about me now. The way I look, the way I talk, the way I walk, the way I feel on a daily basis. I miss my life before, I really do. I miss it a lot. I don't think about it because it makes me sad.

*Theme 4: Hoping for the right drug.* In this theme, people focused on pharmacological treatment/cure for AIDS. Hope was evident as participants expressed anticipation that a medication recently started would help them or a cure would be found in their lifetime. One person described it as "You start becoming anxious and you're hoping that you get some kind of good news today about a new pill or something that's going to help you with the disease." Another, diagnosed within the last 3 years, wondered, "With all the new meds and everything, they say you can live a normal life and a long life. Time will tell, I guess."

Some participants had been told that there were no drugs available for them. A 31-year-old woman, diagnosed for 16 years, reported, "They haven't been able to find a medicine that won't keep me from being sick, so I'm not taking any HIV meds." Others spoke of waiting to see how their bodies responded to newly prescribed ART medications. Hispanic man articulated his search:

I try not to let it bother me because my viral load and everything is real low. The meds are not working for me. We [health care provider and patient] are still trying to find the right one. As long as I'm still living, that's what I'm happy about.

The hope of finding a cure was on the minds of many. A 53-year-old man diagnosed for 10 years noted, "I'm just happy to be here now and hope to be here when they find something." Another stated, "Just hope [for a cure] and hold on." In contrast, a 41-year-old man diagnosed for 9 years stated, "There is no cure and I don't see any coming either." A 56-year-old man, living 13 years with HIV/AIDS, expressed a similar view: "I don't think there is a cure, not right around the corner anyhow. Not in my lifetime."

*Theme 5: Caring for oneself.* Persons with AIDS attempted to control the progress of their disease by caring for themselves. This was evident in the following responses: "If I don't take care of myself, I know I can die from it [AIDS]" and "It's a deadly disease if you don't take care of yourself." A Hispanic man explained, "We never know how long we are going to live. I have to take care of myself if I want to live a couple of years." One woman spoke of her fears and efforts to cope:

> I'm scared—losing the weight and losing the mind and whatnot. I'm scared, but I don't let it get me down. I think about it and whatever is going to happen. I can't stop it. I try to take care of myself and go on.

How to take care of oneself was not always articulated. Eating and taking prescribed medications seemed to be a major focus. "When I get up I know that my first priority is to eat and take my medication." This singleness of purpose is further illustrated in the statement, "I can't think of anything else other than keeping myself healthy so that I can live a little longer. Take my medications. Live a little longer."

*Theme 6: Just a disease.* In this cluster of images, people cognitively represented the cause of AIDS as "an unseen virus," "like any infection," "a common cold," and "a little mini bug the size of a mite." Minimizing the external cause, one participant viewed AIDS as an "inconvenience" and another as having been dealt a "bad card."

Some normalized AIDS by imaging it as a chronic disease. Like people with cancer or diabetes, persons with AIDS felt the need to get on with their lives and not focus on their illness. The supposition was that if medications were taken and treatments followed they could control their illness the same as persons do with cancer or diabetes. The physical or psychological consequences that occur with other chronic diseases were not mentioned. The following two excerpts illustrate the disease image:

> It's just a disease. Since I go to support groups and everything, they tell me to look at it as if it were cancer or diabetes and just do what you have to do. Take your medicine, leave the drugs alone, and you will acquire a long life.

And

> [AIDS is] a controllable disease, not a curse. I'm going to control it for the rest of my life. I feel lucky. There is nothing wrong with me. I'm insisting on seeing it that way. It may not be right, but it keeps me going good.

Sometimes, the explanations for AIDS were scientifically incorrect but presented a means for coping. One man described AIDS: "It's just a disease. It's a form of cancer and that's been going on for years and they just come up with the diagnosis."

*Theme 7: Holding a wildcat.* In this theme, people focused on hypervigilance during battle. While under permanent siege, every fiber of their being was used to fight "a life-altering disease." A 48-year-old man diagnosed 6 months before the interview stated, "I have to pay attention to it. It's serious enough to put me out of work." Another man, diagnosed for 6 years, was firm in his resolve: "I'm a fighter and I'm never going to give up until they come up with a cure for this." These images were essentially positive as can be seen in the following description of AIDS in which a scratch by a wildcat is not "super serious."

> To me HIV is sort of like you've got a wildcat by the head staring you in the face, snapping and snarling. As long as you are attentive, you can keep it at bay. If you lose your grip or don't maintain the attentiveness, it will reach out and scratch you. Which in most cases is not a super serious thing, but it's something of a concern that it will put you in the hospital or something like that. You got to follow the rules quite regimentally and don't let go. If you let go, it will run you over.

Vigilance was used not only to control one's own disease progression but also to protect others. A woman diagnosed for 3 years noted,

> Just being conscious of it because when you got kids and when you got family that you live with, you have to be extremely cautious. You got to realize it at all times. It has to just be stuck in your mind that you have it and don't want to share it. Even attending to one of your children's cuts.

*Theme 8: Magic of not thinking.* Some made a strong effort to forget their disease and, at times, their need for treatment. A few reported no image of AIDS. Thinking about AIDS caused anger, anxiety, sadness, and depression. Not thinking about AIDS seemed to magically erase the reality, and it provided a means for controlling emotions and the disease. A 41-year-old man who has lived with his disease 10 years described AIDS:

It's a sickness, but in my mind I don't think that I got it. Because if you think about having HIV, it comes down more on you. It's more like a mind game. To try and stay alive is that you don't even think about it. It's not in the mind.

The extent to which some participants tried not to think about AIDS can be seen in the following descriptions in which the word *AIDS* was not spoken and only referred to as "it." A 44-year-old Hispanic woman stated, "It's a painful thing. It's a sad thing. It's an angry thing. I don't think much of it. I try to keep it out of my mind." Another woman asserted, "It's a terrible experience. It's very bad, I can't even explain it. I never think about it. I try not to think about it. I just don't think about it. That's it, just cross it out of my mind."

*Theme 9: Accepting AIDS.* In this theme, cognitive representations centered on a general acceptance of the diagnosis of AIDS. Accepting the fact of having AIDS was seen as vital to coping well. People with AIDS readily assessed their coping efforts.

A Hispanic woman noted, "I'm not in denial any more." A 39-year-old Hispanic man who has had the disease for 8 years stated, "Like it or not you have to deal with this disease." Another noted, "You have to live with it and deal with it and that's what I'm trying to do." A 56-year-old man who has had the disease for 13 years summarized his coping:

Either you adjust or you don't adjust. What are you going to do? That's life. It's up to you. I'm happy. I eat well and I take care of myself. I go out. I don't let this put me in a box. Sometimes you don't like it, but you have to accept it because you really can't change it.

Individuals diagnosed more recently struggled to accept their disease. A Black man diagnosed for 2 years vacillated in his acceptance: "I hate that word. I'm still trying to accept it, I think. Yes, I am trying to accept it." However, he stated that he avoids conversation about HIV/AIDS and is not as open with his family. Another man diagnosed 3 years prior noted,

I still don't believe that it's happen to me and it's taken all this time to get a grip on it or to deal with it. I still haven't got a grip on it, but I'm trying. It's finally sinking in that I do have it and I'm starting to feel lousy about it.

Neither of these last participants mentioned the word "HIV" or "AIDS."

*Theme 10: Turning to a higher power.* In this theme, cognitive representations of AIDS were associated with "God," "prayer," "church," and "spirituality." Some saw AIDS as a motivation to change their lives and reach for

God. An Hispanic man living with HIV/AIDS for 6 years stated, "If I didn't have AIDS, I'd probably still be out there drinking, drugging, and hurting people. I turned my life around. I gave myself over to the Lord and Jesus Christ." Another noted, "It [AIDS] worries me. What I do is a lot of praying. It really makes me reach for God."

Others saw religion as a means to help them cope with AIDS. One person expressed it as "I know I can make it from the grace of God. My Jesus Christ is my Savior and that's what's keeping me going every day." One man reported how his spirituality not only helped him cope but also made him a better person:

> At one point I just wanted to give up. If it wasn't for knowing the love of Jesus I couldn't have the strength to keep going. I feel today that I'm a better person spiritually. Maybe not healthwise, but more understanding of this disease.

In contrast, a man diagnosed in jail attributed AIDS to a punishment from God: "Sometimes God punishes you. It's like I told my wife. I should have cleaned up my act."

*Theme 11: Recouping with time.* Although the initial fear and shock was overwhelming, time became a healer such that images, feelings, and processes of coping changed. A sense of imminent doom hurled some into constant preoccupation with their illness, despondency, and increased addiction. Living with HIV/AIDS facilitated change. One woman noted, "When I first found out, I wanted to kill myself and just get it over with. But now it's different. I want to live and just live out the rest of my life." Another described her transition as "At first I thought I was going to be all messed up, all dried up and looking weird and stuff like that, but I don't think of those things anymore. I just keep living life."

As time passed, negative behaviors were replaced with knowledge about their illness, efforts at medication adherence, and a journey of personal growth facilitated by people who believed in them. One man reported that his initial image changed from being in bed with tubes coming out of his nose and Kaposi sarcoma over his body to living a normal life except for not being able to work.

Change was evident in one man's image of AIDS as a time line. He drew a wide vertical line beginning at the top with the first phase, diagnosis, colored red because "it means things are not good, like a red light on a machine." The next phase was shaded blue and labeled "medication, education, and acceptance" to reflect the sky that he could see from his inpatient bed. The final stage was colored bright yellow and labeled "hope."

A 40-year-old Hispanic man drew a chronicle of his life with five addictive substances beginning with alcohol to the injection of heroin. He then sketched four views of himself showing the end stage of his disease: a standing skeleton without face, hair, clothes, or shoes; a sad-faced person without hair lying in a hospital bed; and a grave with flowers. The final picture drawn was of a drug-free person with a well-developed body, smiling face, hair, shoes, shirt, and shorts, symbolizing his readiness for a vacation in Florida. In contrast, a 53-year-old man reported that in 14 years he had no change in his image of AIDS as a "black cloud."

Results were integrated into an essential schema of AIDS. The lived experience of AIDS was initially frightening, with a dread of body wasting and personal loss.

Cognitive representations of AIDS included inescapable death, bodily destruction, fighting a battle, and having a chronic disease. Coping methods included searching for the "right drug," caring for oneself, accepting the diagnosis, wiping AIDS out of their thoughts, turning to God, and using vigilance. With time, most people adjusted to living with AIDS. Feelings ranged from "devastating," "sad," and "angry" to being at "peace" and "not worrying."

## Discussion

In this study, persons with AIDS focused on the end stage of wasting, weakness, and mental incapacity as a painful, dreaded, inevitable outcome. An initial response was to ignore the disease, but symptoms pressed in on their reality and forced a seeking of health care. Hope was manifested in waiting for a particular drug to work and holding on until a cure is found. Many participants saw a connection between caring for themselves and the length of their lives.

Some participants focused on the final outcome of death, whereas others spoke of the emotional and social consequences of AIDS in their lives. Efforts were made to regulate mood and disease by increased attentiveness, controlling thoughts, accepting their illness, and turning to spirituality. Some coped by thinking of AIDS as a chronic illness like cancer or diabetes.

As noted earlier, McCain and Gramling (1992) identified three methods of coping with HIV, namely, Living with Dying, Fighting the Sickness, and Getting Worn Out. Images of Dying and Fighting were strong in Themes 1 (Inescapable Death) and 7 (Holding a Wildcat). Participants in this study were well aware of whether they were coping. Many spoke about accepting or dealing with AIDS, whereas others could not stand the word, tried to wipe it out of their minds, or referred to AIDS as "it."

Consistent with Fryback and Reinert's study (1999), Theme 10, Turning to a Higher Power, emerged as a means of coping as participants faced their mortality. Like Turner's (2000) sample, participants in the current study experienced many changes/losses in their lives and reflected on death and dying. Similar to Turner's theme of Lessons Learned, some participants saw AIDS as a turning point in their lives.

Aligned with Brauhn's (1999) study, chronic disease emerged as an image. In contrast to Brauhn's sample, these participants used the nomenclature of chronic illness to minimize the negative aspects of AIDS. It can be posited that the lack of cautious optimism in planning their future was not present in this study because the entire sample had AIDS.

## Theoretical Elements

As Diefenbach and Leventhal (1996) noted, cognitive representations were highly individual and not always in accord with medical facts. Consistent with research in other illnesses, persons with AIDS had cognitive representations reflecting attributes of consequences, causes, disease time line, and controllability (Leventhal, Leventhal, et al., 2001). In particular, we identified three themes that centered on anticipated or experienced consequences associated with AIDS. Inescapable Death and Dreaded Bodily Destruction involved negative physical consequences that are understandable at end stage in a disease with no known cure. The theme Devouring Life focused on the far-reaching emotional, social, and economic consequences experienced by participants. The Just a Disease theme reflected cognitive representations of the cause of AIDS and Recouping with Time had elements of a disease time line from diagnosis to burial.

Six themes (Hoping for the Right Drug, Caring for Oneself, Holding a Wildcat, Magic of Not Thinking, Accepting AIDS, and Turning to a Higher Power) were similar to the controllability attribute of illness representations. Previous research centered on controlling a disease or condition through an intervention by the individual or an expert, such as taking a medication or having surgery (Leventhal, Leventhal, et al., 2001). This finding was substantiated in the themes Hoping for the Right Drug and Caring for Oneself. Unique to this study, persons with AIDS attempted to control not only their emotions but also their disease through vigilance, avoidance, acceptance, and spirituality coping methods. This is particularly evident in the statement that "To try and stay alive is that you don't even think about it." This study extends previous research on illness representations to persons with AIDS and contributes to the theory of Self-Regulation by suggesting that in AIDS coping methods function like the attribute controllability. Of note is that eight participants drew and described their dominant image of AIDS. These drawings provide a unique revelation of participants' concerns, fears, and beliefs. Having participants draw images of AIDS provides a new method of assessing a person's dominant illness representation.

## Implications for Nursing

Inquiring about a patient's image of AIDS might be an efficient, cost-effective method for nurses to assess a patient's illness representation and coping processes as well as enhance nurse-patient relationships. Patients who respond that AIDS is "death" or "they wipe it out of their minds" might need more psychological support.

Many respondents used their image of AIDS as a starting point to share their illness experiences. As persons with AIDS face their mortality, reminiscing with someone who treasures their stories can be a priceless gift. Asking patients about their image of AIDS might touch feelings not previously shared and facilitate patients' self-discovery and acceptance of their illness.

## Future Research

Cognitive representations have been identified with AIDS. From this research, it can be posited that how a person images AIDS might influence medication adherence, high-risk behavior, and quality of life. If persons with AIDS believed that there is no hope for them, would they adhere to a difficult medication regimen or one with noxious side effects? Would a person who experienced emotional and social consequences of AIDS be more likely to protect others from contracting the disease? Would it be reasonable to expect that persons who focus on fighting AIDS or caring for themselves would be more likely to adhere to medication regimens? Do persons who turn to a higher power, accept their diagnosis, or minimize the disease have a better quality of life? Further research combining images of AIDS and objective measures of medication adherence, risk behaviors, and quality of life is needed to determine if there is an association between specific illness representations and adherence, risk behaviors, and/or quality of life.

## References

Bartlett, J. G, & Gallant, J. E. (2001). *Medical management of HIV infection, 2001-2002*. Baltimore: Johns Hopkins University, Division of Infectious Diseases.

Brauhn, N. E. H. (1999). *Phenomenology of having HIV/AIDS at a time when medical advances are improving prognosis*. Unpublished doctoral dissertation, University of Iowa.

Centers for Disease Control and Prevention. (2001a). HIV and AIDS—United States, 1981-2000. *MMWR 2001, 50*(21), 430–434.

Centers for Disease Control and Prevention. (2001b). The global HIV and AIDS epidemic, 2001. *MMWR 2001, 50*(21), 434–439.

Colaizzi, P. F. (1978). Psychological research as the phenomenologist views it. In R.Valle & M. King (Eds.), *Existential phenomenological alternatives in psychology* (pp. 48–71). New York: Oxford University Press.

Diefenbach, M. A., & Leventhal, H. (1996). The common-sense model of illness representation: Theoretical and practical considerations. *Journal of Social Distress and the Homeless, 5*(1), 11–38.

Dominguez, L. M. (1996). *The lived experience of women of Mexican heritage with HIV/AIDS.* Unpublished doctoral dissertation, University of Arizona.

Douaihy, A., & Singh, N. (2001). Factors affecting quality of life in patients with HIV infection. *AIDS Reader, 11*(9), 450–454, 460–461.

Echeverria, P. S., Jonnalagadda, S. S., Hopkins, B. L., & Rosenbloom, C. A. (1999). Perception of quality of life of persons with HIV/AIDS and maintenance of nutritional parameters while on protease inhibitors. *AIDS Patient Care and STDs, 13*(7), 427–433.

Farber, E.W., Schwartz, J. A., Schaper, P. E., Moonen, D. J., & McDaniel, J. S. (2000). Resilience factors associated with adaptation to HIV disease. *Psychosomatics, 41*(2), 140–146.

Frankel, R. M. (1999). Standards of qualitative research. In B. F. Crabtree & W. L. Miller (Eds.), *Doing qualitative research* (2nd ed., pp. 333–346). Thousand Oaks, CA: Sage.

Fryback, P. B., & Reinert, B. R. (1999). Spirituality and people with potentially fatal diagnoses. *Nursing Forum, 34*(1), 13–22.

Holzemer, W. L., Henry, S. B., & Reilly, C. A. (1998). Assessing and managing pain in AIDS care: The patient perspective. *Journal of the Association of Nurses in AIDS Care, 9*(1), 22–30.

Kearney, M. H. (2001). Focus on research methods: Levels and applications of qualitative research evidence. *Research in Nursing & Health, 24,* 145–153.

Koopman, C., Gore, F. C., Marouf, F., Butler, L. D., Field, N., Gill, M., Chen, X., Israelski, D., & Spiegel, D. (2000). Relationships of perceived stress to coping, attachment and social support among HIV positive persons. *AIDS Care, 12*(5), 663–672.

Laschinger, S. J., & Fothergill-Bourbonnais, F. (1999). The experience of pain in persons with HIV/AIDS. *Journal of the Association of Nurses in AIDS Care, 10*(5), 59–67.

Leventhal, H., Idler, E. L., & Leventhal, E. A. (1999). The impact of chronic illness on the self system. In R. J. Contrada & R. D. Ashmore (Eds.), *Self, social identity, and physical health* (pp. 185-208). New York: Oxford University Press.

Leventhal, H., Leventhal, E. A., & Cameron, L. (2001). Representations, procedures, and affect in illness self-regulation: A perceptual-cognitive model. In A. Baum, T. A. Revenson, & J. E. Singer (Eds.), *Handbook of health psychology* (pp. 19–47). Mahwah, NJ: Lawrence Erlbaum.

McCain, N. L., & Gramling, L. F. (1992). Living with dying: Coping with HIV disease. *Issues in Mental Health Nursing, 13*(3), 271–284.

Meadows, L. M., & Morse, J. M. (2001). Constructing evidence within the qualitative project. In J. M. Morse, J. M. Swansen, & A. Kuzel (Eds.), *Nature of qualitative evidence* (pp. 187–200). Thousand Oaks, CA: Sage.

Merleau-Ponty, M. (1956). What is phenomenology? *Cross Currents, 6,* 59–70.

Morse, J. M. (2000). Determining sample size. *Qualitative Health Research, 10*(1), 3–5.

Russell, J. M., & Smith, K. V. (1999). A holistic life view of human immunodeficiency virus-infected African American women. *Journal of Holistic Nursing, 17*(4), 331–345.

Saxon, E. (Producer), & Demme, J. (Producer/Director). (1993). *Philadelphia* [motion picture]. Burbank, CA: Columbia Tristar Home Video.

Streubert, H. J., & Carpenter, D. R. (1999). *Qualitative research in nursing: Advancing the humanistic imperative* (2nd ed.). New York: Lippincott.

Turner, J. A. (2000). *The experience of multiple AIDS-related loss in persons with HIV disease: A Heideggerian hermeneutic analysis.* Unpublished doctoral dissertation, Georgia State University.

Vogl, D., Rosenfeld, B., Breitbart, W., Thaler, H., Passik, S., McDonald, M., et al. (1999). Symptom prevalence, characteristics, and distress in AIDS outpatients. *Journal of Pain and Symptom Management, 18*(4), 253–262.

# Appendix D

## A Grounded Theory Study

### Constructions of Survival and Coping by Women Who Have Survived Childhood Sexual Abuse

**Susan L. Morrow**

*University of Utah*

**Mary Lee Smith**

*Arizona State University*

*This qualitative study investigated personal constructs of survival and coping by 11 women who have survived childhood sexual abuse. In-depth interviews, a 10-week focus group, documentary evidence, and follow-up participant checks and collaborative*

AUTHORS' NOTE: We thank Arlene Metha, Gail Hackett, Carole Edelsky, B. J. Moore, Lucille Pope, Helga Kansy, and the research collaborators for their valuable input related to the structure and process of this research. Susan L. Morrow conducted the research for this article while at Arizona State University, and the design and analysis were the collaborative activities of both Susan L. Morrow and Mary Lee Smith. Correspondence concerning this article should be addressed to Susan L. Morrow, Department of Educational Psychology, 327 Milton Bennion Hall, University of Utah, Salt Lake City, Utah 84112.

SOURCE: The material in this appendix is reprinted from Morrow, S. L., & Smith, M. L. (1995). Constructions of survival and coping by women who have survived childhood sexual abuse. *Journal of Counseling Psychology, 42*, 24–33. Copyright 1995, American Psychological Association. Used by permission.

*analysis were used. Over 160 individual strategies were coded and analyzed, and a theoretical model was developed describing (a) causal conditions that underlie the development of survival and coping strategies, (b) phenomena that arose from those causal conditions, (c) context that influenced strategy development, (d) intervening conditions that influenced strategy development, (e) actual survival and coping strategies, and (f) consequences of those strategies. Subcategories of each component of the theoretical model were identified and are illustrated by narrative data. Implications for counseling psychology research and practice are addressed.*

The sexual abuse of children appears to exist at epidemic levels, with estimates that 20%–45% of women and 10%–18% of men in the United States and Canada have been sexually abused as children; experts agree that these figures are underestimates (Geffner, 1992; Wyatt & Newcomb, 1990). Approximately one third of students seeking counseling in one university counseling center reported having been sexually abused as children (Stinson & Hendrick, 1992). Because of the breadth and severity of psychological and physical symptoms consequent to childhood sexual abuse, the confusion surrounding treatment methods, and the large number of "normal" individuals seeking counseling who display severe psychological symptoms (Courtois, 1988; Geffner, 1992; Lundberg-Love, Marmion, Ford, Geffner, & Peacock, 1992; Russell, 1986), a theoretical framework is needed to better understand the consequences of childhood sexual abuse.

Two primary modes of understanding and responding to consequences of childhood sexual abuse are symptom and construct approaches (Briere, 1989). Researchers and practitioners alike have adopted a symptom-oriented approach to childhood sexual abuse. It is characteristic of both academic and lay literatures to portray consequences of sexual abuse in lengthy lists of symptoms (Courtois, 1988; Russell, 1986). Briere (1989), however, encouraged a broader perspective, advocating the identification of overarching constructs and core effects—as opposed to symptoms—of sexual victimization.

Mahoney (1991) explicated core ordering processes—tacit, deep-structural processes of valence, reality, identity, and power—that underlie personal meanings or constructions of reality. He emphasized the importance of understanding tacit theories of self and world that guide the development of patterns of affect, thinking, and behavior. A construct-oriented approach to the study of survival and coping offers the possibility of developing a conceptual framework that will bring order into the chaos of symptomatology that currently characterizes the field, as well as relating those symptoms to core ordering processes.

A number of authors (Johnson & Kenkel, 1991; Long & Jackson, 1993; Roth & Cohen, 1986) have related coping theories (Horowitz, 1979; Lazarus & Folkman, 1984) to sexual-abuse trauma. However, traditional coping theories have tended to problematize emotion-focused and avoidant coping styles commonly used by women and abuse survivors (Banyard & Graham-Bermann, 1993). Strickland (1978) stressed the importance of practitioners accurately assessing [an] individual's life situations in determining the efficacy of certain coping strategies. Banyard and Graham-Bermann (1993) emphasized the need to examine power as a mediator in the coping process. The child who is a victim of sexual abuse is inherently powerless; therefore, particular attention must be paid to a reexamination of coping strategies with this population.

The purpose of the present research was to understand the lived experiences of women who had been sexually abused as children and to generate a theoretical model for the ways in which they survived and coped with their abuse. As Hoshmand (1989) noted, qualitative research strategies are particularly appropriate to address meanings and perspectives of participants. In addition, she suggested that naturalistic methods offer the researcher access to deep-structural processes.

Considerable attention has been given to the truthfulness of claims of childhood sexual abuse, particularly when alleged victims have forgotten or repressed all or part of their abuse experiences. Loftus (1993) outlined the difficulties inherent in determining the veridicality of retrieved memories, urging caution on the part of psychologists working in the area of sexual abuse and calling for ongoing research into the nature of true repressed memories. While acknowledging the importance of Loftus's concerns, a constructivist approach orients toward "assessing the viability (utility) as opposed to the validity (truth) of an individual's unique worldview" (Neimeyer & Neimeyer, 1993, p. 2). In accordance with this view, each volunteer's self-identification as an abuse survivor was the criterion for inclusion in the present investigation and her definition of survival and coping the starting point for the investigation. We accepted the stories of participants at face value as their phenomenological realities.

The primary method of investigating those realities was grounded theory (Glaser & Strauss, 1967), a qualitative research method designed to aid in the systematic collection and analysis of data and the construction of a theoretical model. The data analysis was based on transcriptions of semistructured, in-depth interviews; videotapes of a 10-week group that focused on what survival and coping meant to the research participants; documentary evidence, including participants' journals and other relevant writings; and Susan L. Morrow's field notes and journals.

# Method

Qualitative research methods are particularly suited to uncovering meanings people assign to their experiences (Hoshmand, 1989; Polkinghorne, 1991). Chosen to clarify participants' understandings of their abuse experiences, the methods used involved (a) developing codes, categories, and themes inductively rather than imposing predetermined classifications on the data (Glaser, 1978), (b) generating working hypotheses or assertions (Erickson, 1986) from the data, and (c) analyzing narratives of participants' experiences of abuse, survival, and coping.

## Participants

Research participants were 11 women, with ages ranging from 25 to 72, who had been sexually abused as children. One woman was African American, 1 was West Indian, and the remainder were Caucasian. Three were lesbians, 1 was bisexual, and 7 were heterosexual. Three women were physically disabled. Participants' educational levels ranged from completion of the Graduate Equivalency Degree to having a master's degree. Abuse experiences varied from a single incident of molestation by a family friend to 18 years of ongoing sadistic abuse by multiple perpetrators. Age of initial abuse ranged from infancy to 12 years of age; abuse continued as late as age 19. All participants had been in counseling or recovery processes lasting from one 12-step meeting to years of psychotherapy.

## Procedure

*Entry into the field.* Research participants were recruited in a large southwestern metropolitan area through therapists known for expertise in their work with the survivors of sexual abuse. Each therapist was sent a letter describing the study in detail; a similar letter was enclosed to give to clients who might benefit from or be interested in participating in the study. Interested clients, in turn, called Susan L. Morrow, the investigator. Of the 12 respondents, 11 became research participants. The 12th declined to participate for personal reasons.

When prospective participants contacted Morrow, the purpose and scope of the study were reviewed and an appointment was made for an initial interview. Informed consent was discussed in detail at the beginning of the interview, with an emphasis on confidentiality and the potential emotional consequences of participation. After a participant signed the consent, audio- or videotaping commenced. Each participant chose her own pseudonym for the research and was promised the opportunity to review quotes and other information about her before publication.

*Data sources.* Each of the 11 survivors of sexual abuse participated in a 60- to 90-min in-depth, open-ended interview, during which two questions were asked: "Tell me, as much as you are comfortable sharing with me right now, what happened to you when you were sexually abused," and "What are the primary ways in which you survived?" Morrow's responses included active listening, empathic reflection, and minimal encouragers.

After the initial interviews, 7 of the 11 interviewees became focus-group participants. Four were excluded from the group: 2 who were interviewed after the group had started and 2 who had other commitments. The group provided an interactive environment (Morgan, 1988) that focused on survival and coping. In the initial meeting, participants brainstormed about the words *victim*, *survivor*, and *coping*. Subsequent group sessions built on the first, with participants exploring emerging categories from the data analysis and their own research questions, which had been invited by Morrow. Morrow took a participant-observer role, moving from less active involvement in the beginning to a more fully participatory role toward the end (Adler & Adler, 1987).

A central feature of the analysis was Morrow's self-reflectivity (Peshkin, 1988; Strauss, 1987). Morrow's own subjective experiences were logged, examined for tacit biases and assumptions, and subsequently analyzed.

Documentary evidence completed the data set. These data consisted of participants' journals, kept both in conjunction with and independent of the project, artistic productions, and personal writings from earlier periods of participants' lives.

*Data collection, analysis, and writing.* A central concern for rigor in qualitative research is evidentiary adequacy—that is, sufficient time in the field and extensiveness of the body of evidence used as data (Erickson, 1986). The data consisted of over 220 hours of audio- and videotapes, which documented more than 165 hours of interviews, 24 hours of group sessions, and 25 hours of follow-up interactions with participants over a period of more than 16 months. All of the audiotapes and a portion of the videotapes were transcribed verbatim by Morrow. In addition, there were over 16 hours of audio-taped field notes and reflections. The data corpus consisted of over 2,000 pages of transcriptions, field notes, and documents shared by participants.

The analytic process was based on immersion in the data and repeated sortings, codings, and comparisons that characterize the grounded theory approach. Analysis began with open coding, which is the examination of minute sections of text made up of individual words, phrases, and sentences. Strauss and Corbin (1990) described open coding as that which "fractures the data and allows one to identify some categories, their properties and dimensional locations" (p. 97). The language of the participants guided the development of code and category labels, which were identified with short descriptors, known as *in vivo codes*, for survival and coping strategies. These codes and categories were systematically compared and contrasted, yielding increasingly complex and inclusive categories.

Morrow also wrote analytic and self-reflective memos to document and enrich the analytic process, to make implicit thoughts explicit, and to expand the data corpus. Analytic memos consisted of questions, musings, and speculations about the data and emerging theory. Self-reflective memos documented Morrow's personal reactions to participants' narratives. Both types of memos were included in the data corpus for analysis. Analytic memos were compiled and an analytic journal was kept for cross-referencing codes and emerging categories. Large poster boards with movable tags were used to facilitate the arranging and rearranging of codes within categories.

Open coding was followed by axial coding, which puts data "back together in new ways by *making connections between a category and its subcategories*" (italics in original, Strauss & Corbin, 1990, p. 97). From this process, categories emerged and were assigned in vivo category labels. Finally, selective coding ensued. Selective coding was the integrative process of "selecting the core category, systematically relating it to other categories, validating those relationships [by searching for confirming and disconfirming examples], and filling in categories that need[ed] further refinement and development" (Strauss & Corbin, 1990, p. 116).

Codes and categories were sorted, compared, and contrasted until saturated—that is, until analysis produced no new codes or categories and when all of the data were accounted for in the core categories of the grounded theory paradigm model. Criteria for core status were (a) a category's centrality in relation to other categories, (b) frequency of a category's occurrence in the data, (c) its inclusiveness and the ease with which it related to other categories, (d) clarity of its implications for a more general theory, (e) its movement toward theoretical power as details of the category were worked out, and (f) its allowance for maximum variation in terms of dimensions, properties, conditions, consequences, and strategies (Strauss, 1987).

In keeping with Fine's (1992) recommendations that researchers move beyond the stances of ventriloquists or mere vehicles for the voices of those being researched, we sought to engage the participants as critical members of the research team. Consequently, after completion of the group, the 7 group members were invited to become coanalysts of data from the focus group. Four elected to do so. Not choosing to extend their original commitment,

2 terminated their participation at that point; a 3rd declined because of physical problems. The 4 coanalysts (termed *participant-coresearchers*) continued to meet with Morrow for more than a year. They acted as the primary source of participant verification, analyzing videotapes of the group sessions in which they had participated, suggesting categories, and revising the emerging theory and model. Participant-coresearchers used their natural intuitive analytic skills as well as grounded theory principles and procedures that had been taught to them by Morrow to collaborate in the data analysis.

Morrow met weekly with an interdisciplinary qualitative research collective throughout the data gathering, analysis, and writing of the research account. The group provided peer examination of the analysis and writing, as recommended by LeCompte and Goetz (1982), thereby enhancing researcher and theoretical sensitivity, overcoming selective inattention, and enhancing receptiveness to the setting (Glaser, 1978; Lincoln & Guba, 1985).

Accountability was achieved through ongoing consultations with participants and colleagues and by maintaining an audit trail that outlined the research process and the evolution of codes, categories, and theory (Miles & Huberman, 1984). The audit trail consisted of chronological narrative entries of research activities, including pre-entry conceptualizations, entry into the field, interviews, group activities, transcription, initial coding efforts, analytic activities, and the evolution of the survival and coping theoretical model. The audit trail also included a complete list of the 166 in vivo codes that formed the basis for the analysis.

Because of the human cognitive bias toward confirmation (Mahoney, 1991), an active search for disconfirming evidence was essential to achieving rigor (Erickson, 1986). Data were combed to disconfirm various assertions made as a result of the analysis. Discrepant case analysis, also advised by Erickson (1986), was conducted, and participants were consulted to determine reasons for discrepancies.

# Results

The grounded theory model for surviving and coping with childhood sexual abuse, evolving from Strauss and Corbin's (1990) framework and developed from the present investigation, is present in Figure 1.

## Causal Conditions of Phenomena Related to Sexual Abuse

Two types of causal conditions emerged from the data, which ultimately led to certain phenomenological experiences related to sexual abuse. These causal conditions were (a) cultural norms and (b) forms of sexual abuse.

Cultural norms of dominance and submission, violence, maltreatment of women, denial of abuse, and powerlessness of children formed the bedrock on which sexual abuse was perpetrated. Paula's (all names used are pseudonyms) experiences reflected a number of these norms: Her father enforced his dominance by physically and sexually abusing Paula's mother and calling Paula and her mother "cunt," "whore," and "fat pig." He was an avid reader of pornography and regularly invited Paula into the bathroom, where he showed her pictures from his magazines. He took photographs of her in the bathtub or sunbathing by the pool. She stated that most of his abuse of her was ". . . real, real physical. [He] beat the shit out of us." His sexual abuse of her was "covert." Audre commented the following after disclosing that her sexual abuser had beaten her "only" once: "You know, he never whipped me like that again. Never again. And he never had to. . . . Whenever I would resist him at any point, he'd just look at me." Dominance, violence, and the powerlessness of children converged in Audre's life to set the stage for her abuse, as did the denial of abuse or the potential for abuse by significant people in her life and in the lives of other victims. After being sexually abused by an elderly neighbor, Liz brought home a picture he had taken to show her parents. Liz reported, "My mother got right down in my face and said, 'He didn't do anything to you, did he?'" Frightened, Liz replied, "No, he didn't do anything to me."

The second causal condition consisted of the various forms of sexual abuse that had been perpetrated. Abuses ranged from innuendos and violations of privacy to rape and vaginal penetration with loaded guns. These forms of abuse were classified through the data analysis into five categories: (a) nonphysical sexual abuses, (b) physical molestation, (c) being forced to perform sexual acts, (d) penetration, and (e) sexual torture. Nonphysical sexual abuses, perpetrated on all of the victims, consisted of perpetrators engaging in sexual talk, photographing the child in sexual poses or nude, exposing the genitals to the child, engaging in sexual teasing and jokes, performing sexual activities in front of the child, and inviting the child to participate in sexual activity. Physical molestation, also experienced by all of the participants, included sexual touching, pinching, poking, tickling, and stroking the child with objects; removing the child's covers or clothes; holding the child in such a way that sexual contact was made; masturbating the child; washing and examining the child's genitals in excess of actions necessary for health and cleanliness; and performing cunnilingus on the child. Of the participants, 7 had been forced to perform sexual acts, such as masturbation, fellatio, or cunnilingus. At least 5 of the victims had been penetrated vaginally, orally, or anally with fingers, hands, penises, guns, knives, or other implements; four others were uncertain about penetration because of

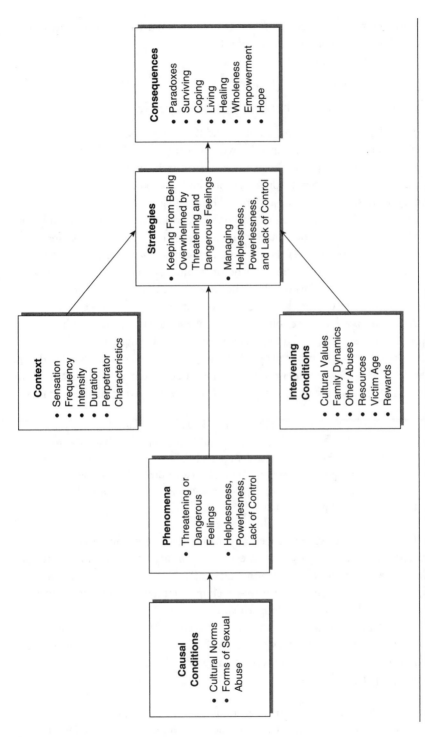

**Figure 1** Theoretical Model for Surviving and Coping With Childhood Sexual Abuse

amnesic episodes. Six remembered being subjected to sexual tortures of a sadistic nature beyond those already described.

## Phenomena Resulting From Cultural Norms and Forms of Sexual Abuse

Causal conditions—cultural norms and the forms of sexual abuse to which victims were subjected—resulted in two core categories of subjective phenomena as reported by participants: (a) being overwhelmed by feelings victims experienced as threatening or dangerous and (b) experiencing helplessness, powerlessness, and lack of control. These categories support and extend Herman's (1992) description of traumatic reactions, in which she found that "the salient characteristic of the traumatic event is its power to inspire helplessness and terror" (p. 34). This research indicates that terror is but one of the overwhelming emotions characteristic of trauma experienced by survivors of sexual abuse. Most, but not all, of the survivors in the study experienced terror; all experienced overwhelming emotions of fear, pain, or rage.

Meghan foreshadowed one of these phenomena the first night of the group, when she said, "To keep from feeling my feelings, I have become a very skilled helper of other people." Throughout the data, others echoed her words. The analytic moment in which this category emerged is illustrated in the following analytic memo written by Morrow (in vivo codes are in italics):

> I'm reaching a higher level of abstraction. Is the overarching category *protection from feelings*? Many categories are subsumed under it: One *talks* to get out the *stories*; the *feelings* are less intense. Fake orgasm (*sex*) because you don't have any physical *feelings*. *Art* was used to deal with *feelings*, express *anger*, *release* the pressure of the *feelings*, use *chemicals* to deal with *feelings* (and a whole complex interaction here). . .

Existing and emergent codes and categories were compared and contrasted with this category; the category was modified to accommodate the data, producing the phenomenon that was labeled *being overwhelmed by threatening or dangerous feelings*—feelings that participants described as subjectively threatening or dangerous.

In addition to being overwhelmed by feelings, participants experienced what was termed *helplessness, powerlessness, and lack of control*. Lauren provided an exemplar of the second category, illustrating the pervasiveness of her perpetrator's power:

> He stands there. A silhouette at first and then his face and body come into view. He is small, but the backlighting intensifies his figure and he seems huge,

like a prison guard. He is not always there but it feels like he might as well be. When he's not there, I search the distance for him and he appears. He seems to be standing there for hours. As if he's saying, you are weak, I am in control.

Not only did Lauren experience powerlessness during her abuse, but her lack of control invaded her dreams and her moments alone.

## Context in Which Survival and Coping Strategies Developed

Strategies for survival and coping were developed in response to being overwhelmed by threatening or dangerous feelings and experiencing helplessness, powerlessness, and lack of control. These strategies were influenced by particular contextual markers related to both the causal conditions—particularly the forms of sexual abuse—and the resultant phenomena. These contextual markers included (a) sensations, (b) frequency, (c) intensity, (d) duration, and (e) perpetrator characteristics.

Sensations experienced by victims during sexual abuse ranged from arousal to pain, varying from mild to severe intensity. The frequency and duration of sexual abuse ranged from a single instance to years of ongoing sexual abuse, which occurred as often as daily or as infrequently as once every summer. Perpetrator characteristics varied from one to multiple perpetrators of both genders, who were always older and larger than their victims and ranged in relationship from blood relatives to strangers. The phenomena—being overwhelmed by threatening or dangerous feelings and experiencing helplessness, powerlessness, and lack of control—also varied as to types of physical and emotional sensations; ranged in intensity, frequency, and duration; and frequently continued for years after the original abuse had ended.

## Intervening Conditions Influencing Survival and Coping Strategies

In addition to context, there were also intervening conditions, which were broad, general conditions that influenced participants' choices of survival and coping strategies. Intervening conditions included (a) cultural values, (b) family attitudes, values, beliefs, and dynamics, (c) other abuses present, (d) age of the victim, (e) rewards that accompanied the abuse, and (f) outside resources. Cultural values that were particularly influential were those of a religious nature related to sex and sexual abuse: "Guilt, I believe, is the driving force in Catholicism. . . . I felt guilt after I was molested. . . . I see the Catholic stuff as running in tandem with the issues of being a sexual-abuse survivor." One woman uncovered a family norm that condoned incest when her uncle bragged, "We were one big fuckin' family. . . . Everybody screwed everybody." Alcohol and alcoholic dynamics were part of almost every family, and it was rare that emotional or physical abuse was not an accompaniment of sexual violation. When perpetrators provided rewards or favors to their victims, victims were more likely to cooperate but expressed more confusion than did those who were not rewarded.

The ages at which participants had been abused ranged from infancy through 19 years of age. The data analysis revealed only one pattern related to the age of the victim when she was abused. In keeping with the literature on dissociation (Kluft, 1985), all of the participants who had developed severe dissociative patterns had been sexually abused in infancy or early childhood.

Only one participant experienced outside intervention in her abuse, although all had since turned to and found emotional support from friends, partners, or therapists. As in Liz's case ("He didn't do anything to you, did he?"), potential helpers were unwilling or unable to see that abuse was happening. However, in one case, a grandmother—who knew of and was powerless to stop the abuse—provided the support that the survivor now believes saved her life and sanity.

## Strategies for Surviving and Coping With Childhood Sexual Abuse

In the presence of the context and intervening conditions described above, two overarching phenomena led to the development of two parallel core strategies for survival and coping: (a) keeping from being overwhelmed by threatening or dangerous feelings and (b) managing helplessness, powerlessness, and lack of control. Because so few resources were available for help, most of the strategies described by participants were internally oriented and emotion focused. The strategies within each core category are illustrated in Figure 2.

*Keeping from being overwhelmed by threatening or dangerous feelings.* Being sexually abused produced confusing and intense emotions in the child victims. Lacking the cognitive skills to process overwhelming feelings of grief, pain, and rage, these children developed strategies to keep from being overwhelmed. These strategies were (a) reducing the intensity of troubling feelings, (b) avoiding or escaping feelings, (c) exchanging the overwhelming feelings for other, less threatening ones, (d) discharging or releasing feelings, (e) not knowing or remembering experiences that generated threatening feelings, and (f) dividing overwhelming feelings into manageable parts.

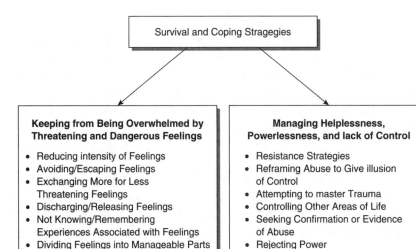

**Figure 2**    Survival and Coping Strategies of Women Who Have Survived
Childhood Sexual Abuse

The first strategy used by participants in this research was reducing the
intensity of the feelings. Participants used various methods to reframe their
abuse so that their resultant feelings were less intense; to dull, numb, or not
experience negative feelings that emerged or threatened to emerge; or to
comfort themselves. By mentally or verbally reframing their abuse, victims
found ways to excuse their perpetrators or to minimize the importance of
the trauma. Lisa reported, "I never, never blamed him. . . . He was just a
boy. . . . He didn't know any better." To modify the intense feelings that
arose, participants dulled and numbed those feelings with substances such as
alcohol, drugs, cigarettes, and food and by sleeping or becoming depressed.
Liz became depressed to tone down the rage she did not allow herself to feel.
Participants kept feelings from emerging in a number of ways. Paula commented," The feelings are in the words"; thus, one strategy for not feeling
was not to talk. Meghan analyzed her experiences instead: "I lived in my
head." As these emotions emerged, participants "stuffed" or consciously
repressed them. Liz said, "I didn't mind how much it bothered me, I learned
to repress the emotions," while Lisa swallowed her feelings with cinnamon
rolls. Participants used a variety of ways to find comfort. Amaya found
comfort outside herself: "The grandmother, she was a very spiritual
woman. . . . She used to rock and sing to us." Others, unable to find comfort from outside, nurtured themselves with animals or dolls. "I used to play
with paper dolls. . . . They were my friends. They could never hurt me."

Participants used a variety of means to meet unmet emotional needs: "I used sex for validation 'cause that makes me pretty and that means you love me." Meghan became "mother hen" from the time she was little, receiving approval, attention, and appreciation from her family. Participants coped spiritually in a number of ways, some finding spiritual solace or relief by praying to or raging against God, while others rejected religious systems that they saw as being supportive of their abuse. Some sought alternative spiritual paths. Kitty believed that God would not give her any more than she could handle.

The second strategy for keeping from being overwhelmed was avoiding or escaping the threatening or dangerous feelings. In many instances, similar substrategies (e.g., drugs or alcohol) facilitated different processes. In a previous example, alcohol was used to dull and numb feelings as one way to reduce the intensity of those feelings. In some of the examples that follow, alcohol was used to escape. Strategies for escaping and leaving took both problem- and emotion-focused directions (Folkman & Lazarus, 1980, 1985) and included attempts to physically avoid or escape abuse, ignore the abuse, escape its reality, or leave mentally or emotionally. In their attempts to physically escape abuse, participants went to their rooms, ran away, moved out, married young, or separated themselves from others: "I isolated forever." When physical escape seemed impossible, some victims thought of dying or actually attempted suicide when they were children to adolescents in an effort to escape their abuse. To prevent either sexual abuse or related physical abuse, participants attempted to distract their perpetrators, tried talking them out of abusing them, or told them to stop. Velvia remembered, "I kept wanting it to be like it was and I kept asking him, 'Let's just read.'. . ." They also reported having developed heightened intuition about danger or having lied to others about their abuse to avoid being punished or further abused. Participants attempted to escape their abuse by hiding, both literally and figuratively. Ananda found refuge in a canyon, while Meghan strove for invisibility by being very, very good. Danu's conflict revealed itself in her poetry: "I didn't want to be/ 'miss smarty pants.'/ I tried to be quieter/ more secret and private./ I knew it would be safer/ if no one noticed me." Lauren and Kitty hid their bodies with oversized clothes. To ignore or escape the reality of their abuse, participants wished, fantasized, denied, avoided, and minimized: "I avoid things . . . the other side of denial. I won't look at it." Lauren "left the story behind," and the abuse gradually became less and less real in her mind until it was forgotten. Sometimes victims simply left mentally or emotionally. Kitty said, "Mind, take me outa there!" and it did. Some experienced tunnel vision, floating, "spacing out," or separating from their bodies or other people. Ananda described "a kind of spiritual leaving this planet."

Another way that the research participants avoided being overwhelmed was to exchange threatening or dangerous feelings for other, less distressing ones. Overwhelming feelings could be exchanged by overriding the feelings with other, more intense feelings; replacing them with less threatening, substitute feelings; or distracting themselves with activities that produced innocuous feelings. Participants overrode dirty feelings by physically scrubbing them away. Some used self-induced physical pain, such as self-mutilation, to override emotional pain. Kitty commented, "Physical pain keeps me from feeling my feelings. That's where my anorexia came from. . . . The physical pain of not eating. I can't feel things when I'm in pain." The women who experienced feelings of pain and grief as dangerous developed an ability to switch immediately to anger or rage, substituting the latter emotions for the pain that threatened to overwhelm them. Others bypassed the more threatening feelings of anger or rape, switching to tears. "I have [anger] for about two seconds, then I cry; it turns into sadness." Participants also distracted themselves from their feelings by turning to activities that produced innocuous or pleasant feelings: "The crunching kind of distracts me from the pain inside"; "I looked at other things."

The fourth strategy for keeping from being overwhelmed was discharging or releasing feelings. Verbal activities included writing in journals or talking to "get the feelings out." The use of humor was especially effective. Mimicking her usual 12-step meeting greeting ("My name is Paula, and I'm an alcoholic"), one participant declared, "Hello, I'm Paula, and I'm sorry!" They also shouted or screamed to release tension. Paula, a highly competitive athlete, used physical strategies that ranged from athletics to self-harm. She cut crosses in her skin and vomited to release her feelings: "I'll go purge and, uh, I'll feel elated, and better, and I also got rid of some of the feelings as a way of letting go." Artistic endeavors also facilitated release: "To this day, if I get those feelings, I can draw, and not necessarily feel better, but less pressure."

Not knowing or remembering experiences associated with threatening or dangerous feelings was the fifth strategy—a complex category involving head memories, head knowledge, clues or evidence, bodily sensations, intuition, and feelings or emotions. Head memories were one of the most haunting and difficult aspects of having been sexually abused. Virtually every participant had experienced some degree of memory loss surrounding her abuse, as illustrated by Velvia's comments: "There are some things that I remember, but only up to a certain point, and I don't know what happened next. . . . [T]he place where it stops sticks in my head. . . ." Some participants depended on head knowledge to know that they had been sexually abused. Audre disclosed, "The only reason I know about [the abuse] is because my abuser called me about a year ago to tell me." Detective work was rampant in survivors' searches for outside evidence or clues of their abuse. Some sought verification from siblings or nonoffending parents. Others depended on feelings about places or photographs to cue them about when their abuse had occurred: "We moved to a big huge house when I was 11. And that's when I think that it started, 'cause I don't remember anything in the old house." Survivors experienced "body memories," or physical sensations, frequently in the absence of head memories or knowledge. Kitty suffered intense pelvic pain whenever she talked about abuse: "Somebody'd be talking about being attacked, and I would experience all this pain in my stomach and in my female part of me." Others experienced nausea, trembling, and abreactions as a result of talking about sexual abuse. Intuition also contributed to a survivor's knowledge that she had been abused. Participants reported that intuition—in the form of a sudden awareness or hunch—was a powerful source of knowing at the moment of insight but that it could quickly fade to disbelief. Feelings or emotions were experienced as the least trustworthy of all evidence, particularly if unaccompanied by other forms of knowing. Despite the intensity of feelings of terror, deep sadness, and shame, women in the study were far more likely to believe they were "crazy" than to trust their feelings or emotions as evidence of sexual abuse: "I'm having all these feelings and all these symptoms . . . but maybe it has to with my mother dropped me on my head or she dressed me funny. . . ."

Dividing overwhelming feelings into manageable parts was a complex process of partitioning emotions into different compartments or separating them from cognitions, sensations, behaviors, or intuitions. Dividing was one of the ways in which memories were lost and knowing was jeopardized. Participants exercised three forms of dividing: "disassociating," dividing up overwhelming emotions, and dividing up cognitive functions. Participants typically used the lay term *disassociate* rather than *dissociate* to explain the process of altering consciousness. Although disassociation was used to escape feelings, it also provided the gateway for dividing. Dividing up overwhelming emotions took place as overwhelming or disparate emotional states were compartmentalized in order to make them more manageable. On one end of a continuum were facades or masks that hid the more vulnerable aspects of self. Participants had also developed different parts of themselves. The more rigid divisions were characterized by some degree of amnesia or distortion of behavior, motor coordination, self-perception, or time characteristic of dissociative disorders (Braun, 1986):

> I'm not sure that I really thought that I did survive . . . going away and seeing myself laying there on the bed—I can see my face, I can see the little girl

laying there with her head kind of turned, her eyes closed, sweat or something, you know. She's—her head's wet—me—I guess it must be me.

In addition to dividing emotional states, participants separated cognitive functions such as actions, emotions, thoughts, bodily sensations, and intuitions, congruent with Braun's (1988) BASK (behavior, affect, sensation, knowledge) model of dissociation. Kitty learned to crawl out of her body: "I could see me screaming, but I couldn't hear." She "was actually frozen and could do nothing. . . . I wondered at the time why couldn't I do something? Why couldn't I move?"

*Managing helplessness, powerlessness, and lack of control.* In addition to developing strategies to keep from being overwhelmed by emotions, participants had developed strategies for managing powerlessness in the face of their abuse. Six categories of survival and coping strategies were used to manage helplessness, powerlessness, and lack of control: (a) creating resistance strategies, (b) reframing abuse to create an illusion of control or power, (c) attempting to master the trauma, (d) attempting to control other areas of life besides the abuse, (e) seeking confirmation or evidence from others, and (f) rejecting power.

One way in which participants managed their lack of power was to resist or rebel. Meghan refused to eat. Kitty spoke of her resistance: "Those fuckers aren't gonna get me. I'm not gonna kill myself. . . . [T]hat's when they win." Some reframed the abuse to create an illusion of control or power. Meghan believed that she could control her abuse: "If somehow I could be good enough and do things right enough, she wouldn't be like that anymore." Survivors attempted to master the trauma, at times recapitulating their abuse: "If I can create pain that I can feel, and I'm in control, it's different. It's totally different." Others turned to helping abused people. Participants frequently tried to control other areas of life besides the abuse. Barbara became ". . . a savior. I ride a white horse, rescue." Meghan stated, "I couldn't manage the abuse, but I could manage the household." All of the participants sought confirmation or evidence from others in order to control their own perceived reality. Only Liz rejected power. "I don't want to be like her. . . . She was very powerful. . . . I'm afraid of power in myself, even."

## Consequences of Strategies for Survival and Coping

The strategies used by participants were not without consequences. In every case, those strategies succeeded in keeping them from being overwhelmed by feelings or aided them in managing helplessness, powerlessness, and lack of control. However, while their strategies for survival and coping were successful, that success was also costly.

Two women saw the creation of alter personalities—their primary survival and coping strategy—as a sane alternative to psychosis, or "going crazy." However, they both paid the price of living fragmented lives.

When asked what being overwhelmed by feelings meant to her, Meghan responded, "Screaming metal . . . pain and anguish that goes on and on and on and never stops." She has continued to spiral back through depression, pain, and anguish that, at times, feel as if they will never end. Paradoxically, her strategies worked to keep overwhelming feelings at bay until she actively began the therapy process. As she has faced the emotions she buried, she has been overwhelmed many times.

Participants had fears, wishes, or dreams of dying, yet all are alive today. But while all still live, they did not feel they survived intact; as Barbara disclosed, "I'm not sure I survived," and as Liz said, "Part of me died."

Another paradox arose during the examination of the consequences of the strategy to manage helplessness, powerlessness, and lack of control. Often, the very strategies adopted by participants to exercise power or control backfired, ultimately taking control of the survivors. One woman, whose childhood refusal to eat resulted in her doctor prescribing crackers and cream cheese for breakfast (the only food she would eat), found in adulthood that she turned repeatedly—and sometimes compulsively—to these same foods.

Many times, participants commented that they were barely surviving—that they were in pain, exhausted, or overwhelmed. However, surviving and coping [were] what participants did best. Liz declared, "My will to survive is strong, stronger than I realized." In a conversation among the participant-coresearchers, Meghan said angrily, "I don't want to be surviving. I want to be living. I want to have some fun. I want to be happy. And that's what's not happening right now." Liz responded, "First you have to survive. You have to survive it. And that's where I'm getting to, is the realization that I'm surviving this stuff again."

Each of the survivors echoed Meghan's feelings. Four had become drug- and alcohol-free in their efforts to move beyond mere survival to healing, wholeness, and empowerment. Paula disclosed, "I'm just startin' to realize that this is worth it. [My drawings are] more elaborate, they're bigger, I'm using more mediums, they're more detailed." Velvia used the word "empowerment" to describe a process that went beyond survival. Amaya wrote,

> Today I got in touch with *mi otro yo* [my other me]. . . .
>
> She is so powerful, so sure of herself, so strong, so real, so alive.

I did not die like I thought I would when I felt her.

Instead, I got in touch with the missing part of my inner power and wholeness.

The pain, grief, and terror that the survivors had experienced and continued to wrestle with are very real, and the healing process is long and arduous. However, throughout the research, participants expressed hope. Despite her terror and pain, Kitty reflected, "I have hope in my life. . . . There's just a little bit of sunlight coming in. There's a little bit of heaven up there that comes inside of my soul and heals."

## Discussion

Although the counseling literature is rich with descriptions of specific outcomes of childhood sexual abuse, this study is distinctive in its systematic examination of the survival and coping strategies from the perspectives of women who were sexually abused as children. A theoretical model of the survival and coping strategies of 11 participants was constructed through qualitative data analysis, which included engaging participants in the analytic process in order to ensure that the model reflected their personal constructs. This model establishes, from a multitude of strategies and symptoms, a coherent, construct-focused framework for understanding the often-confusing constellation of behavior patterns of the survivors of abuse.

Cultural norms set the stage for sexual abuse. As Banyard and Graham-Bermann (1993) emphasized, it is important for researchers and practitioners to examine the social milieu in which particular stressors are experienced. In relation to childhood sexual abuse, an examination of social forces helps to shift the focus of coping from a purely individual analysis to an individual-in-context analysis, thereby normalizing the victim's experience and reducing self-blame.

The powerlessness of girls, which can be attributed to the societal positioning of women and children, to their physical size, and to undependable resources for intervention available to abuse victims, explains the overwhelming predominance of emotion-focused over problem-focused coping strategies used by the participants in this study. In addition, the context of denial and secrecy surrounding sexual abuse in the lives of girls and women may further exacerbate a preference for emotion-focused coping.

The present analysis is congruent with Long and Jackson's (1993) findings that victims of childhood sexual abuse attempted to have an impact on the actual abuse situation by using problem-focused strategies, while they managed their distress through emotion-focused coping. The two core strategies—keeping from being overwhelmed by threatening and dangerous feelings and managing helplessness, powerlessness, and lack of control— parallel Long and Jackson's emotion-focused and problem-focused strategies. Long and Jackson found that few victims attempted problem-focused strategies and speculated that resources may not have been available, either in fact or in the cognitive appraisals of victims. The present research demonstrated that options for problem-focused coping were, in fact, not readily available. In addition, specific cultural and family norms served to convince children of the limited efficacy of problem-focused solutions.

Researchers and practitioners may need to think beyond the categories of emotion- and problem-focused coping strategies (Banyard & Graham-Bermann, 1993). L. Benishek proposed that certain so-called emotion-focused strategies, such as dissociation, may, in fact, be problem focused (personal communication, December 1, 1993). Indeed, according to Banyard and Graham-Bermann (1993), "There are times when emotion-focused strategies may be used as problem-focused solutions to a stressful dilemma" (p. 132). Additional qualitative research in this area may prove fruitful.

Mahoney (1991) described core ordering processes as deep-structure processes that "lie at the core of every person's lifelong efforts to organize and reorganize their experience" (p. 179). Of his four proposed core ordering processes (valence, reality, identity, and power), the present analysis yielded two: valence, which encompasses processes of motivation and emotion, and power, which is characterized by processes of control and ability. These two processes correspond, respectively, to the core strategies found in this research related to participants' keeping from being overwhelmed with feelings and managing helplessness, powerlessness, and lack of control. Because this research was pursued inductively without imposing preexisting categories on the data, the core ordering processes of identity and reality did not emerge. However, it would be appropriate to reanalyze the data with these categories in mind. The process of identity, for example, can be seen in Liz's statement about seeing herself lying on the bed during her abuse. "She . . . I guess it must be me." Although the present research did not address identity or reality, it provided a more detailed understanding of the processes of valence and power, particularly as they were experienced by the survivors of sexual abuse in this investigation.

The emergent theoretical model of survival and coping was, in effect, Morrow's interpretation of 11 participants' constructions of their survival and coping. As is frequently the case in qualitative research, the results of this analysis are unique to the particular investigator, participants, and context of this study. The transferability of this theoretical model for survival and coping takes place as the reader examines these results in the context of specific circumstances of interest.

Feminist researchers have expressed concern about the potential for the exploitation of women and other marginalized groups in academic research and have urged investigators to examine closely what participants receive in exchange for their contributions (Landrine, Klonoff, & Brown-Collins, 1992). Their recommendations have influenced the present investigation in two ways. First, the categories that emerged from this research made sense to and were useful, in a practical sense, to the participants themselves. When the developing model for survival and coping was presented to the participant-coresearchers, one woman took the information home to her husband, with whom she had experienced painful and confusing dynamics surrounding her abuse. Her response endorsed the applicability of this model in practice, not only for spouses or partners, but for families and the therapeutic relationship as well: ". . . [I]t felt like months and months . . . of stuff that just felt so hard . . . trudging through this sludge—it was like the clarity! It was just unbelievable . . . the closeness between us." It appears that presenting this model to clients and significant others has potential, as a psychoeducational tool, to ease the difficult and perilous journey that individuals must travel as they work through abuse trauma and its consequences.

In addition, the collaborative research process itself has implications for research with the survivors of sexual abuse. Participant-coresearchers described their experiences of collaborative meaning-making as "important" and "empowered." Coparticipatory data analysis therefore holds promise as an empowering model for researchers and participants alike.

Finally, from a standpoint of the "psychology of human effectiveness" (Gelso & Fassinger, 1992, p. 293), the resilience and resourcefulness of the participants in this investigation cannot be overstated. What appears at first glance to be a profusion of dysfunctional symptoms becomes, upon closer examination, rational and reasonable coping strategies given the extremity of the stressors to which these women, as children, were subjected. For example, dividing various aspects of the self into alter personalities enabled victims to disperse trauma among various parts of the self, thereby decreasing the potential for being overwhelmed. In addition, multiplicity provided for self-nurturing and furnished a cognitive structure in which valuable functions and personality characteristics were preserved until they could be safely reintegrated. This investigation focused on the strengths of the survivors of sexual abuse and encourages practitioners to view clients who have been sexually abused in light of those strengths, rather than from a perspective that emphasizes pathology (Adams & Betz, 1993; Hill, 1993; Howard, 1992). Given the prevalence of sexual abuse, adaptation to childhood trauma must be considered a part of the process of normal development for a large number of individuals. The present findings may facilitate a reevaluation of that adaptation and offer clients and their therapists a conceptual framework to facilitate healing.

## References

Adams, E. M., & Betz, N. E. (1993). Gender differences in counselors' attitudes toward and attributions about incest. *Journal of Counseling Psychology, 40,* 210–216.

Adler, P. A., & Adler, P. (1987). *Membership roles in field research.* Newbury Park, CA: Sage.

Banyard, V. L., & Graham-Bermann, S. A. (1993). Can women cope? A gender analysis of theories of coping with stress. *Psychology of Women Quarterly, 17,* 303–318.

Braun, B. G. (1986). Issues in the psychotherapy of multiple personality disorder. In B. G. Braun (Ed.), *Treatment of multiple personality disorder* (pp. 3–28). Washington, DC: American Psychiatric Press.

Braun, B. G. (1988). The BASK (behavior, affect, sensation, knowledge) model of dissociation. *Dissociation, 1,* 4–23.

Briere, J. (1989). *Therapy for adults molested as children: Beyond survival.* New York: Springer.

Courtois, C. A. (1988). *Healing the incest wound: Adult survivors in therapy.* New York: Norton.

Erickson, F. (1986). Qualitative methods in research on teaching. In M. C. Wittrock (Ed.), *Handbook of research on teaching* (3rd ed., pp. 119–161). New York: Macmillan.

Fine, M. (1992). *Disruptive voices: The possibilities of feminist research.* Ann Arbor: University of Michigan Press.

Folkman, S., & Lazarus, R. S. (1980). An analysis of coping in a middle-aged community sample. *Journal of Health and Social Behavior, 21,* 219–239.

Folkman, S., & Lazarus, R. S. (1985). If it changes it must be a process: Study of emotion and coping during three stages of a college examination. *Journal of Personality and Social Psychology, 48,* 150–170.

Geffner, R. (1992). Current issues and future directions in child sexual abuse. *Journal of Child Sexual Abuse, 1*(1), 1–13.

Gelso, C. J., & Fassinger, R. E. (1992). Personality, development, and counseling psychology: Depth, ambivalence, and actualization. *Journal of Counseling Psychology, 39,* 275–298.

Glaser, B. G. (1978). *Theoretical sensitivity.* Mill Valley, CA: Sociology Press.

Glaser, B. G., & Strauss, A. L. (1967). *The discovery of grounded theory: Strategies for qualitative research.* Hawthorne, NY: Aldine.

Herman, J. L. (1992). *Trauma and recovery: The aftermath of violence: From domestic abuse to political terror.* New York: Basic Books.

Hill, C. E. (1993). Editorial. *Journal of Counseling Psychology, 40,* 252–256.

Horowitz, M. (1979). Psychological response to serious life events. In V. Hamilton & D. M. Warburton (Eds.), *Human stress and cognition: An information processing approach* (pp. 237–265). Chichester, England: Wiley.

Hoshmand, L.L.S. (1989). Alternate research paradigms: A review and teaching proposal. *The Counseling Psychologist, 17,* 3–79.

Howard, G. S. (1992). Behold our creation! What counseling psychology has become and might yet become. *Journal of Counseling Psychology, 39,* 419–442.

Johnson, B. K., & Kenkel, M. B. (1991). Stress, coping, and adjustment in female adolescent incest victims. *Child Abuse & Neglect, 15,* 293–305.

Kluft, R. P. (Ed.). (1985). *Childhood antecedents of multiple personality.* Washington, DC: American Psychiatric Press.

Landrine, H., Klonoff, E. A., & Brown-Collins, A. (1992). Cultural diversity and methodology in feminist psychology. *Psychology of Women Quarterly, 16,* 145–163.

Lazarus, R. S., & Folkman, S. (1984). *Stress, appraisal, and coping.* New York: Springer.

LeCompte, M. D., & Goetz, J. P. (1982). Problems of reliability and validity in ethnographic research. *Review of Educational Research, 52,* 31–60.

Lincoln, Y. S., & Guba, E. G. (1985). *Naturalistic inquiry.* Beverly Hills, CA: Sage.

Loftus, E. F. (1993). The reality of repressed memories. *American Psychologist, 48,* 518–537.

Long, P. J., & Jackson, J. L. (1993). Childhood coping strategies and the adult adjustment of female sexual abuse victims. *Journal of Child Sexual Abuse, 2*(2), 23–39.

Lundberg-Love, P. K., Marmion, S., Ford, K., Geffner, R., & Peacock, L. (1992). The long-term consequences of childhood incestuous victimization upon adult women's psychological symptomatology. *Journal of Child Sexual Abuse, 1*(1), 81–102.

Mahoney, M. J. (1991). *Human change processes: The scientific foundations of psychotherapy.* New York: Basic Books.

Miles, M. B., & Huberman, A. M. (1984). *Qualitative data analysis: A sourcebook of new methods.* Beverly Hills, CA: Sage.

Morgan, D. L. (1988). *Focus groups as qualitative research.* Newbury Park, CA: Sage.

Neimeyer, G. J., & Neimeyer, R. A. (1993). Defining the boundaries of constructivist assessment. In G. J. Neimeyer (Ed.), *Constructivist assessment: A casebook* (pp. 1–30). Newbury Park, CA: Sage.

Peshkin, A. (1988). In search of subjectivity: One's own. *Educational Researcher, 17,* 17–21.

Polkinghorne, D. E. (1991). Two conflicting calls for methodological reform. *The Counseling Psychologist, 19,* 13–114.

Roth, S., & Cohen, L. J. (1986). Approach, avoidance, and coping with stress. *American Psychologist, 41,* 813–819.

Russell, D. E. H. (1986). *The secret trauma: Incest in the lives of girls and women.* New York: Basic Books.

Stinson, M. H., & Hendrick, S. S. (1992). Reported childhood sexual abuse in university counseling center clients. *Journal of Counseling Psychology, 39,* 370–371.

Strauss, A. L. (1987). *Qualitative analysis for social scientists*. Cambridge, England: Cambridge University Press.

Strauss, A., & Corbin, J. (1990). *Basics of qualitative research: Grounded theory procedures and techniques*. Newbury Park, CA: Sage.

Strickland, B. R. (1978). Internal-external expectancies and health-related behaviors. *Journal of Consulting and Clinical Psychology, 46,* 1192–1211.

Wyatt, G. E., & Newcomb, M. (1990). Internal and external mediators of women's sexual abuse in childhood. *Journal of Consulting and Clinical Psychology, 58,* 758–767.

**Appendix D: Grounded Theory Study**

# Appendix E

## An Ethnography

### Rethinking Subcultural Resistance

*Core Values of the Straight Edge Movement*

### *Ross Haenfler*

*University of Colorado–Boulder*

"By focusing their message at their families, subcultural peers, mainstream youth, and the larger society, sXe created a multilayered resistance that individuals could customize to their own interests."

*This article reconceptualizes subcultural resistance based on an ethnographic examination of the straight edge movement. Using the core values of straight edge, the author's analysis builds on new subcultural theories and suggests a framework for how members construct and understand their subjective experiences of being a part of a subculture. He suggests that adherents hold both individual and collective meanings of resistance and express their resistance via personal and political methods. Furthermore, they consciously enact resistance at the micro, meso, and macro levels, not solely against an ambiguous*

AUTHOR'S NOTE: I would like to thank Patti Adler for her support and guidance. I would also like to thank the reviewers for their helpful comments.

SOURCE: This article originally appeared in the *Journal of Contemporary Ethnography*, 33(4), 406–436. Copyright 2004, Sage Publications, Inc.

*"adult" culture. Resistance can no longer be conceptualized in neo-Marxist terms of changing the political or economic structure, as a rejection only of mainstream culture, or as symbolic stylistic expression. Resistance is contextual and many layered rather than static and uniform.*

*Keywords: resistance; straight edge; subculture; youth; punk*

Resistance has been a core theme among both subcultural participants and the scholars who study them. Early subcultural theorists associated with Birmingham University's Centre for Contemporary Cultural Studies (CCCS) concentrated on the ways youth symbolically resisted mainstream or "hegemonic" society through style, including clothing, demeanor, and vernacular (Hebdige 1979). Subcultures emerged in resistance to dominant culture, reacting against blocked economic opportunities, lack of social mobility, alienation, adult authority, and the "banality of suburban life"(Wooden and Blazak 2001, 20). Theorists found that young working-class white men joined deviant groups to resist conforming to what they saw as an oppressive society (Hebdige 1979; Hall and Jefferson 1976). Scholars have given a great deal of attention to whether these youth subcultures resist or reinforce dominant values and social structure (Hebdige 1979; Willis 1977; Brake 1985; Clarke, Hall, Jefferson, and Roberts 1975). The CCCS emphasized that while subcultural style was a form of resistance to subordination, ultimately resistance merely reinforced class relations (Cohen 1980; Willis 1977). Therefore, any such resistance was illusory; it gave subculture members a feeling of resistance while not significantly changing social or political relations (Clarke et al. 1975). In fact, according to this view, subcultures often inadvertently reinforce rather than subvert mainstream values, recasting dominant relationships in a subversive style (see Young and Craig 1997).

The CCCS has drawn substantial criticism for ignoring participants' subjectivity, failing to empirically study the groups they sought to explain, focusing too much on Marxist/class–based explanations and grand theories, reifying the concept of subculture, and overemphasizing style (Muggleton 2000; Clarke [1981] 1997; Blackman 1995; Widdicombe and Wooffitt 1995). Based on solid ethnographic work, contemporary theorists have acknowledged the fluidity of subcultures and retooled the notion of resistance to include the subjective understandings of participants. Leblanc (1999), studying female punks, found that resistance included both a subjective and objective component. Leblanc redefined resistance broadly as political behavior, including discursive and symbolic acts. Postmodern theorists have further questioned CCCS ideas of resistance, suggesting that many narratives can simultaneously be true, contingent on one's perspective. They encourage us to examine subcultural quests for authenticity from the participants' points of view, paying particular attention to the individualistic, fragmented, and heterogeneous natures of subcultures (Muggleton 2000; Rose 1994; Grossberg 1992). Viewed in this way, subcultural involvement is more a personal quest for individuality, an expression of a "true self," rather than a collective challenge. In fact, most members have an "anti-structural subcultural sensibility" (Muggleton 2000, 151), view organized movements with suspicion, and instead criticize "mainstream society" in individualized ways (Gottschalk 1993, 369).

Each of these critiques demands a broader understanding of resistance that accounts for members' individualistic orientations. Resistance may be "political behavior" broadly defined, but how individuals express and understand their involvement needs further attention. My analysis builds on new subcultural theories and suggests a framework for how members construct and understand their subjective subcultural experiences. I suggest that adherents hold both individual and collective meanings of resistance and express their resistance via personal and political methods. Furthermore, they consciously enact resistance at the micro, meso, and macro levels, emerging at least partly in reaction to other subcultures instead of solely against an ambiguous "adult" culture. Resistance can no longer be conceptualized in neo-Marxist terms of changing the political or economic structure, as a rejection only of mainstream culture, or as symbolic stylistic expression. A conceptualization of resistance must account for individual opposition to domination, "the politicization of the self and daily life" (Taylor and Whittier 1992, 117) in which social actors practice the future they envision (Scott 1985; Melucci 1989, 1996). Resistance is contextual and many layered rather than static and uniform.

As a relatively unstudied movement, straight edge (sXe) provides an opportunity to rethink and expand notions of resistance. The straight edge[1] movement emerged on the East Coast of the United States from the punk subculture of the early 1980s. The movement arose primarily as a response to the punk scene's nihilistic tendencies, including drug and alcohol abuse, casual sex, violence, and self-destructive "live-for-the-moment" attitudes. Its founding members adopted a "clean-living" ideology, abstaining from alcohol, tobacco, illegal drugs, and promiscuous sex. Early sXe youth viewed punk's self-indulgent rebellion as no rebellion at all, suggesting that in many ways punks reinforced mainstream culture's intoxicated lifestyle in a mohawked, leatherjacketed guise.

Straight edge remains inseparable from the hardcore[2] (a punk genre) music scene. Straight edge bands serve as the primary shapers of the group's ideology and collective identity. Hardcore "shows"[3] (small concerts) are an important place for sXers[4] to congregate, share ideas, and build solidarity. Since its beginnings, the movement has expanded around the globe, counting tens of thousands of young people among its members. In the United States, the typical sXer is a white, middle-class male, aged fifteen to twenty-five. Straight edgers clearly distinguish themselves from their peers by marking a large X, the movement's symbol, on each hand before attending punk concerts. While scholars have thoroughly researched other postwar youth subcultures such as hippies, punks, mods, skinheads, and rockers (e.g., Hall and Jefferson 1976; Hebdige 1979; Brake 1985), we know little about sXe, despite its twenty-year history.

The basic tenets of sXe are quite simple: members abstain, completely, from drug, alcohol, and tobacco use and usually reserve sexual activity for caring relationships, rejecting casual sex. These sXe "rules" are absolute; there are no exceptions, and a single lapse means an adherent loses any claim to the sXe identity. Members commit to a lifetime of clean living. They interpret their abstention in a variety of ways centered on resistance, self-realization, and social transformation. Clean living is symbolic of a deeper resistance to mainstream values, and abstinence fosters a broader ideology that shapes sXers' gender relationships, sense of self, involvement in social change, and sense of community.

This article fills a gap in the literature by giving an empirical account of the sXe movement centered on a description of the group's core values. I begin by providing a very brief overview of several previous subcultures, to place sXe in a historical context. I then discuss my involvement in the sXe scene and the methods I employed throughout my research. Next, I examine the group's core values, focusing on how members understand their involvement.[5] Finally, I provide a new framework for analyzing members' experiences that encompasses the multitude of meanings, sites, and methods of resistance.

## Previous Youth Subcultures

Studies of hippies, skinheads, and punks demonstrate both similarities and profound differences between these groups and the sXe movement. Hippies evolved in the mid-1960s from the old beatnik and folknik subcultures (Irwin 1977; Miller 1999). Their lifestyle was a reaction to the stifling homogeneity of the 1950s, emphasizing communalism over conformity and deliberate hedonism over reserve (Miller 1991). "If it feels good, then do it so long as it doesn't hurt anyone else" was the scene's credo. Hippie core values included peace, racial harmony, equality, liberated sexuality, love, and communal living (Miller 1991). They rejected compulsive consumerism, delayed gratification, and material success (Davis 1967). "Dope," however, was one of the group's most visible characteristics (Miller 1991; Irwin 1977). Dope differed from drugs; dope, such as LSD and marijuana, was good, while drugs, such as speed and downers, were bad. For hippies, dope expanded the mind, released inhibitions, boosted creativity, and was part of the revolution. It was the means to discovering a new ethic, heightening awareness, and "understanding and coping with the evils of American culture" (Miller 1991, 34). LSD "gave the mind more power to choose, to evaluate, even, perhaps, to reason" (Earisman 1968, 31). Like dope, sex, in its own way, was revolutionary. "Free love" rejected the responsibilities normally associated with sexual relationships: marriage, commitment, and children (Earisman 1968). By practicing what most at the time would call promiscuous sex, the hippies deliberately threw their irreverence for middle-class values in the face of dominant society (Irwin 1977).

Skinheads received a great deal of attention during the 1990s, as reports of their growing membership in neo-Nazi groups infiltrated both popular media and scholarly work (Bjorgo and Wilte 1993; Moore 1994; Young and Craig 1997). Skinheads emerged in late-1960s Britain as an offshoot of the mod subculture (Cohen 1972; Hebdige 1979). While most of the fashion-conscious mods listened to soul music, frequented discotheques, and dressed in impeccably pressed trousers and jackets, the "hard mods," who eventually became the skinheads, favored ska and reggae, local pubs, and a working-class "uniform" of heavy boots, close-cropped hair, Levi jeans, plain shirts, and braces (suspenders) (Brake 1985). While the mods attempted to emulate the middle-class, hip 1960s style, the skins were ardently working class. Nearly everything about skinheads revolved around their working-class roots. Hard work and independence were among their core values; they abhorred people, such as some hippies, who they believed "live off the system." Skinheads were extremely nationalistic and patriotic, adorning themselves with tattoos, T-shirts, and patches of their country's flag. After a long day at work, they enjoyed drinking beer with their friends at the local pub. Although there were some women skins, males dominated the subculture and often reinforced traditional patriarchal ideals of masculinity.

The original skinheads borrowed heavily from the West Indian culture, adopting their music, mannerisms, and style, including among their number a variety of races. While they were not violently racist at the level of the current neo-Nazi groups, these skins, both black and white, engaged in violence against Pakistani immigrants ("Pakibashing") (Hebdige 1979, 56). Eventually, with reggae's turn to Rastafarianism and black pride, many white skinheads became increasingly racist. At the turn of the century, three main types of skinheads prevailed: neo-Nazis (racist), skinheads against racism (e.g., Skinheads Against Racial Prejudice), and nonpolitical skinheads, who took neither a racist nor an antiracist stand (Young and Craig 1997). Skinheads were quite visible at punk, ska, and Oi! music shows, though the nonpolitical and antiracist skins were more prevalent. Very rarely, a skinhead was also sXe.[6]

In many ways, punk was a reaction to "hippie romanticism" and middle-class culture; punk celebrated decline and chaos (Brake 1985, 78; Fox 1987; O'Hara 1999). In mid-1970s Britain, youth faced a lack of job opportunities or, at best, the prospect of entering a mainstream world they found abhorrent (Henry 1989). They attempted to repulse dominant society by valuing anarchy, hedonism, and life in the moment. Early punks borrowed heavily from the styles of Lou Reed, David Bowie ("Ziggy Stardust"), and other glam-rock and new-wave artists. Adorned with safety pins, bondage gear, heavy bright makeup, torn clothing, flamboyant hairstyles, and spiked leather jackets, punks lived by their motto "No Future," celebrating rather than lamenting the world's decline. They embraced alienation, and their "nihilist aesthetic" included "polymorphous, often willfully perverse sexuality, obsessive individualism, a fragmented sense of self" (Hebdige 1979, 28).

Like the skinheads, punks disdained hippies; the preeminent punk band the Sex Pistols titled one of their live recordings "Kill the Hippies" (Heylin 1998, 117). Unlike the skins, and like the hippies, however, punks chose to reject society, conventional work, and patriotism. Many used dangerous drugs to symbolize "life in the moment" and their self-destructive, nihilistic attitude (Fox 1987). Straight edge emerged relatively early in the punk scene and has shared certain values and styles with punks, hippies, and skins ever since. While some punks today are sXe, the two scenes have become relatively distinct, and the sXe movement has replaced many of the original antisocial punk values with prosocial ideals.

## Method

My first encounter with sXe occurred in 1989 at the age of fifteen through my involvement in a Midwest punk rock scene. As I attended punk shows and socialized with the members, I noticed that many kids scrawled large Xs on their hands with magic marker before they went to a concert. I eventually learned that the X symbolized the clean-living sXe lifestyle and that many punks in our scene had taken on a totally drug- and alcohol-freeway of life. Having tried the alcohol-laden life of most of my peers, I quickly discovered it was not for me. I despised feeling I had to "prove" myself (and my manhood) again and again by drinking excessively. I could not understand why the "coolest," the most highly regarded men were often the ones who most degraded women. Furthermore, given my family's history of alcoholism, I wanted to avoid my relatives' destructive patterns. Finally, the local sXers' involvement in progressive politics and activist organizations connected with my interest in social justice and environmentalism. My association with sXers led me to adopt the sXe ideology as what I viewed, at the time, to be an alternative to peer pressure and a proactive avenue to social change. After a period of careful consideration (like many punks, I was suspicious of "rules"), I made known my commitment to avoid consuming alcohol, drugs, and tobacco, and the group accepted me as one of their own. Since then, I have attended more than 250 hardcore shows, maintained the lifestyle, and associated with many sXers on a fairly regular basis. The data I present result from more than fourteen years of observing the sXe movement in a variety of settings and roles and interviewing members of the scene.

During college, my involvement with sXe waned, and for several years I had little contact with the group. After completing my undergraduate career, I moved to "Clearweather," a metropolitan area in the western United States, to begin graduate training. I lived in a predominantly white university town of approximately ninety thousand people, attending a large research university with twenty-five thousand students. Soon after arriving, I sought out the local hardcore scene and began attending shows. The setting's richness and my interests led me to take advantage of this opportunistic research situation (Riemer 1977). My four-year absence from the scene allowed me to approach the setting with a relatively fresh perspective, while my personal involvement and knowledge of the sXe ideology enabled me to gain entrée into the local scene very quickly. Since fall 1996, I have participated in the sXe scene as a complete member (Adler and Adler 1987).

I gathered data primarily through longitudinal participant observation (Agar 1996) with sXers from 1996 to 2001. The sXers I studied were mostly area high school or university students from middle-class backgrounds. My contacts grew to include approximately sixty sXers in the local area and another thirty sXe and non-sXe acquaintances associated with the larger metropolitan hardcore scene. My interaction with the group occurred primarily at hardcore shows and simply socializing at sXers' houses.

To supplement my participant observation, I conducted unstructured, in-depth interviews with seventeen sXe men and eleven women between the ages of seventeen and thirty. To learn from a variety of individuals, I selected sXers with differing levels of involvement in the scene, including new and old adherents, and individuals who had made the movement central or peripheral to their lives. I conducted in-depth interviews at sXers' homes or at public places free from disturbances, recording and later transcribing each session. Though I organized the sessions around particular themes, I left the interviews unstructured enough that individuals could share exactly what sXe meant to them. I sometimes asked for referrals in a snowball fashion (Biernacki and Waldorf 1981), though I knew most participants well enough to approach them on my own. The variety of participants allowed me continually to cross-check reports and seek out evidence disconfirming my findings (Campbell 1975; Stewart 1998; see also Douglas 1976). Through participant observation, I was able to examine how participants' behaviors differed from their stated intentions. I consciously distanced myself from the setting to maintain a critical outlook by continually questioning my observations and consulting with colleagues to gain an outsider perspective. I was especially attentive to variations on the patterns I discovered.

In an effort to expand my knowledge of sXe beyond my primary circle of contacts, I sought interviews with adherents from outside of the local scene, including individuals from other cities and members of touring out-of-state bands who played in Clearweather. I sometimes contacted other individuals around the country via e-mail with specific questions. I also spent several days in New York City, Los Angeles, and Connecticut to experience the scenes there, taking field notes and conducting informal interviews. In addition to participant observation, casual conversation, and interviews, I examined a variety of other sources including newspaper stories, music lyrics, World Wide Web pages, and sXe 'zines,[7] coding relevant snippets of information into my field notes.

To record and organize my data, I took brief notes at shows and other events that I immediately afterward expanded into more full field notes on computer. Using headings and subheadings, I coded data according to particular topics of interest, beginning the process of organizing data into useful and interesting categories (Charmaz 1983). Throughout my research, I sought patterns and emerging typologies of data (Lofland and Lofland 1995). Reexamining the coded field notes and transcribed interviews led me to analyze several themes, including the subculture's core values. I continually refined these themes as I gathered more data through emergent, inductive analysis (Becker and Geer 1960).

## Straight Edge Core Values

A core set of sXe values and ideals guided and gave meaning to members' behavior: positivity/clean living, reserving sex for caring relationships, self-realization, spreading the message, and involvement in progressive causes. Adherents maintained that sXe meant something different to each person assuming the identity, and as with any group, individual members' dedication to these ideals varied. However, while individuals were free to follow the philosophy in various ways, often adding their own interpretations, these fundamental values underlay the entire movement.

T-shirt slogans, song lyrics, tattoos, and other symbols constantly reminded sXers of their mission and dedication: "It's OK Not to Drink," "True till Death," and "One Life Drug Free" were among the more popular messages. The "X," sXe's universal symbol, emerged in the early 1980s, when music club owners marked the hands of underage concertgoers with an X to ensure that bartenders would not serve them alcohol (see Lahickey 1997, 99). Soon, the kids intentionally marked their own hands both to signal club workers of their intention not to drink and, more importantly, to make a statement of pride and defiance to other kids at the shows. The movement appropriated the X, a symbol meant to be negative, transforming its meaning into discipline and commitment to a drug-free lifestyle.[8] Youth wore Xs on their backpacks, shirts, and necklaces; they tattooed them on their bodies and drew them on their school folders, skateboards, cars, and other possessions. The X united youth around the world, communicating a common set of values and experiences. Straight edgers found strength, camaraderie, loyalty, and encouragement in their sXe friends, valuing them above all else.[9] For many, sXe became a "family," a "brotherhood," a supportive space to be different together. A powerful sense of community, based in large part on the hardcore music scene, was the glue that held sXe and its values together for twenty years.

Like the other youth movements, sXe was a product of the times and culture that it resisted; oppositional subcultures do not emerge in a vacuum (Kaplan and Lööw 2002). The lifestyle reflects the group's emergence during a time of increasing conservatism and religious fundamentalism, an escalating drug war, and Nancy Reagan's "Just Say No" campaign. The rise of the New Christian Right in the late 1970s and early 1980s contributed to a more conservative national climate that influenced youth values (Liebman and Wuthnow 1983). Fundamentalism gained appeal among populations who felt they were losing control of their way of life (Hunter 1987). The unyielding, black-and-white strictures on behavior of sXe were similar to fundamentalist religion's rigid, clear-cut beliefs (Marty and Appleby 1993).

In particular, sXe's emphasis on clean living, sexual purity, lifetime commitment, and meaningful community was reminiscent of youth evangelical movements, while the focus on self-control suggested Puritanical roots. In addition to these conservative influences, sXe was, in many ways, a continuation of New Left middle-class radicalism oriented toward "issues of a moral or humanitarian nature," a radicalism whose payoff is "in the emotional satisfaction derived from expressing personal values in action" (Parkin 1968, 41). The movement's core values reflect this curious blend of conservative and progressive influences.

## Positive, Clean Living

The foundation underlying the sXe identity was positive, clean living. It was, as Darrell Irwin (1999) suggested, fundamentally about subverting the drug scene and creating an alternative, drug-free environment. Clean living was the key precursor to a positive life. Many sXers shunned caffeine and medicinal drugs, and most members were committed vegetarians or vegans.[10] Positive living had broad meaning, including questioning and resisting society's norms, having a positive attitude, being an individual, treating people with respect and dignity, and taking action to make the world a better place. Straight edgers claimed that one could not fully question dominant society while under the influence of drugs, and once one questioned social convention, substance use, eating meat, and promiscuous sex were no longer appealing. Therefore, clean living and positivity were inseparable; they reinforced one another and constituted the foundation for all other sXe values. "Joe,"[11] an eighteen-year-old high school senior, explained how the "positivity" he gained from sXe shaped his life:

> To me, I guess what I've gotten from [sXe] is living a more positive lifestyle. Striving to be more positive in the way you live. Because where I was at when I found it was really (laughs) I was really negative myself. I was negative around people and influenced them to be negative. I was surrounded by negativity. Then I found this and it was like something really positive to be a part of. Also, like the ethics, drug free, alcohol free, no promiscuous sex. It's just saying no to things that are such a challenge for people my age, growing up at that time. It's a big thing for some people to say "No."

Refusing drugs and alcohol had a variety of meanings for individual sXers, including purification, control, and breaking abusive family patterns. Purification literally meant being free from toxins that threatened one's health and potentially ruined lives. Popular T-shirt slogans proclaimed

"Purification—vegan straight edge" and "Straight edge—my commitment against society's poisons." Straight edgers believed that drugs and alcohol influenced people to do things they would normally not do, such as have casual sex, fight, and harm themselves. By labeling themselves as more "authentic" than their peers who used alcohol and drugs, sXers created an easy way to distinguish themselves. They experienced a feeling of uniqueness, self-confidence, and sometimes superiority by rejecting the typical teenage life. Refusing alcohol and drugs symbolized refusing the "popular" clique altogether as well as the perceived nihilism of punks, hippies, and skinheads.

The movement provided young people a way to feel more in control of their lives. Many youth felt peer pressure to drink alcohol, smoke cigarettes, or try illegal drugs. For some, this pressure created feelings of helplessness and lack of control; acceptance often hinged on substance use. Straight edgers reported that the group gave them a way to feel accepted without using and helped them maintain control over their personal situations. Many sXers celebrated the fact that they would never wake up after a night of binge drinking wondering what had happened the previous evening. Adherents reported that sXe allowed them to have a "clear" mind and be free to make choices without artificial influence. Walter, a reserved twenty-one-year-old university student, explained,

> I don't make any stupid decisions. . . . I like to have complete control of my mind, my body, my soul. I like to be the driver of my body, not some foreign substance that has a tendency to control other people. I get a sense of pride from telling other people, "I don't need that stuff. It might be for you but I don't need that stuff." And people are like, "Whoa! I respect that. That's cool."

In addition to the personalized meanings the identity held for adherents, sXers viewed their abstinence as a collective challenge. The group offered a visible means of separating oneself from most youth and taking a collective stand against youth culture and previous youth subcultures, including punks, skinheads, and hippies. Furthermore, for many positivity and refusing drugs and alcohol were symbolic of a larger resistance to other societal problems including racism, sexism, and greed.

Straight edgers made a lifetime commitment to positive, clean living. They treated their abstinence and adoption of the sXe identity as a sacred vow, calling it an "oath," "pledge," or "promise." Members made no exceptions to this rule. Patrick, an easy-going twenty-year-old musician and ex-football player, said, "If you just sip a beer, or take a drag off of a cigarette, you can never call yourself straight edge again. There's no slipping up in straight edge." Ray, raised in an alcoholic family and already heavily tattooed at age nineteen, compared the sXe vow to vows of matrimony: "It's true till death. Once you put the X on your hand, it's not like a wedding ring. You can always take a wedding ring off, but you can't wash the ink from your hands." Ray proceeded to show me a tattoo on his chest depicting a heart with "True till Death" written across it. Many sXe youth had similar tattoos, signifying the permanence of their commitment.

Some sXers took their commitment so seriously they labeled people who broke their vows of abstinence as traitors or "sellouts." Despite their vehement insistence they would "stay true" forever, relatively few sXers maintained the identity beyond their early to midtwenties. Many maintained the values and rarely used alcohol or drugs, but "adult" responsibilities and relationships infringed on their involvement in the scene. When formerly sXe individuals began drinking, smoking, or using drugs, adherents claimed they had "sold out" or "lost the edge." While at times losing the edge caused great conflict, I observed that more often the youth's bonds of friendship superseded resentment and disappointment, and they remained friends. However, a former sXer's sXe friends often expressed deep regret and refused to allow the transgressor to claim the identity ever again. Brent, a serious and outspoken twenty-two-year-old vegan, said, "It's frustrating to see people who you think are your friends make such heavy decisions without consulting you. . . . It's not a betrayal like turning around. It's just that you feel abandoned. . . . It's demoralizing." Kate, a twenty-two-year-old activist, explained her frustration with sellouts:

> It was hard for me at first because I think when people do that it takes away the power of sXe. When people are like, "I'm sXe" and then the next day they're not. It—not delegitimizes completely—in away it takes away some of the legitimacy of the movement. . . . It definitely upset me a little bit. How can you go from claiming sXe one day and the next day just forget about it completely? That was the main thing, I just didn't understand it.

When particularly outspoken or well-known members of the scene sold out, sXers spoke as if another hero had fallen. A very small minority of individuals did base their friendships on adherence to the movement and almost practiced "shunning," the religious equivalent of casting someone out. It was this type of action, despite its rarity, that contributed to outsiders' conceptions of sXe as a judgmental, dogmatic group. Straight edge youth were less likely to socialize regularly with people who used simply because of the incompatibility of the lifestyles. Straight edgers rarely openly criticized friends who had sold out, but during interviews participants expressed to me a deeper frustration and sense of betrayal than they would ever publicly show.

# Reserving Sex for Caring Relationships

Reserving sex for caring relationships was an extension of the positive, clean lifestyle. Straight edgers viewed casual sex as yet another downfall of dominant society, their counterparts in other youth subcultures, and their more mainstream peers. It carried the possibility of sexually transmitted diseases and feelings of degradation and shame. Whereas hippies viewed liberated sex as revolutionary, punks saw it as just another pleasure, and skinheads valued sex as a supreme expression of masculinity, sXers saw abstinence from "promiscuous" sex as a powerful form of resistance. Rejecting the casualness of many youth sexual encounters, they believed that sexual relationships entailed much more than physical pleasure. They were particularly critical of their image of the "predatory," insatiable male, searching for sex wherever he could get it. Kent, a twenty-one-year-old university student with several colorful tattoos, said, "My personal views have to do with self-respect, with knowing that I'm going to make love with someone I'm really into, not a piece of meat." Kyle, a twenty-three-year-old senior architecture major at Clearweather University, said, "For me personally, I won't sleep around with a bunch of people just for health's sake. A good positive influence. [Sex] doesn't mean anything if you don't care about a person." Walter, the university student, said,

> For me it's just choosing how I want to treat my body. It's not something I'm just going to throw around. I'm not going to smoke or use drugs. My body is something that I honor. It's something we should respect. I think sex, if you're gonna do it you should do it, but you shouldn't throw your body around and do it with as many people as you want. If you love your body so much as to not do those things to your body you should have enough respect to treat women and sex how they deserve to be treated.

Though sXe values regarding sexuality appeared conservative when compared to many other youth subcultures, sXers were neither antisex nor homophobic as a group. Premarital sex was not wrong or "dirty" in the sense of some traditional religious views, and numerous sXers and sXe bands took a strong stance against homophobia.[12] Sex could be a positive element of a caring relationship. Believing that sex entailed power and emotional vulnerability, sXers strove to minimize potentially negative experiences by rejecting casual sex. Kevin, a twenty-seven-year-old martial artist who had dropped out of high school, said,

> To this day I'm by no means celibate; however . . . in the last eight years I've had sex with three girls. I'm not celibate by any means but I also don't believe

in fuckin' bullshit meaningless sex. So those tenets kind of took place in my life even though I didn't take it to the actual celibacy extreme. . . . It should be on an emotional level. It's an addiction like everything else. My first understanding of sXe was to not be addicted.

There was no direct religious basis for sXe views on sex. In fact, many of the sXers I associated with grew up with no formal religious involvement, and almost none of them were presently involved in formal religion. While a fewsXers connected their sXe and Christian identities, the group advocated no form of religion, and most adherents were deeply suspicious or critical of organized faiths.

Most sXers also believed that objectifying women was pervasive and wrong, rejecting the stereotypical image of high school males. A local sXe band (five male members) decried sexual abuse and rape: "This song is the most important song we play. It's about the millions of women who have suffered rape. One out of four women will be the victim of a sexual assault in her lifetime. We've got to make it stop." The movement's "rule" against promiscuous sex was more difficult for members to enforce, and thus there was greater variation in belief regarding sex than substance use. Several of my participants, both males and females aged twenty-one to twenty-three, had consciously decided to postpone sex because they had not found someone with whom they felt an intimate emotional attachment. Most of the young women believed not drinking reduced their risk of being sexually assaulted or otherwise put in a compromising situation. Jenny, an eighteen-year-old college freshman and activist, said,

> Like I said, it's all about control over your own body, over your own life. It's about reclaiming, claiming your dignity and self-respect. Saying I'm not going to put this stuff into my body. I'm not going to have you inside of my body if I don't want you in there. It all just very much ties together. I like sXe because it allows me to make very rational, intelligent decisions. That's one of the decisions I think it's really important to think through very carefully. I'm not against premarital sex at all. But personally, I've got to be in love.

Some adherents insisted that sex should be reserved for married couples, while a few believed sXe placed no strictures on sexual activity. Only one young man with relatively little connection to the Clearweather scene had a reputation as a "player." A minority of sXe men were little different than the hypermasculine stereotype they sought to reject. Most insisted that sex between strangers or near strangers was potentially destructive, emotionally and possibly physically, and that positivity demanded that sex should be part of an emotional relationship based on trust.

## Self-Realization

Like members of other subcultures, sXers sought to create and express a "true" or "authentic" identity amid a world that they felt encouraged conformity and mediocrity. Straight edgers claimed that resisting social standards and expectations allowed them to follow their own, more meaningful path in life toward greater self-realization. Like punks, they abhorred conformity and insisted on being "true to themselves." Similar to hippies, sXers believed that as children we have incredible potential that is "slowly crushed and destroyed by a standardized society and mechanical teaching" (Berger 1967, 19). Subcultures, like social movements, engage in conflict over cultural reproduction, social integration, and socialization; they are often especially concerned with quality of life, self-realization, and identity formation (Habermas 1984-87; Buechler 1995). Straight edgers believed toxins such as drugs and alcohol inhibited people from reaching their full potential. This view sharply contrasted with the hip version of self-realization through dope (Davis 1968). For sXers, drugs of any kind inhibited rather than enabled self-discovery; they believed people were less genuine and true to themselves while high. A clear, focused mind helped sXers achieve their highest goals. Kate, the activist, said, "If you have a clear mind you're more likely to be aware of who you are and what things around you really are rather than what somebody might want you to think they are. A little bit more of an honest life, being true to yourself." Elizabeth, a twenty-six-year-old with an advanced degree who had been sXe and vegetarian for many years, said,

> You're not screwed up on drugs and alcohol and you can make conscientious decisions about things. You're not letting some drug or alcohol subdue your emotions and thoughts. You're not desensitizing yourself to your life. And if you're not desensitizing your life, then yeah, you're gonna feel more things. The more you feel, the more you move, the more that you grow. . . . I truly believe [sXers] are living and feeling and growing, and it's all natural growth. It's not put off. That's a unique characteristic.

Like adherents of previous subcultures, sXers constructed a view of the world as mediocre and unfulfilling, believing society encouraged people to medicate themselves with crutches such as drugs, alcohol, and sex to forget their unhappiness. Straight edgers felt the punks', skinheads', and hippies' associations with these things blunted their opportunities to offer meaningful resistance. Substances and social pressures clouded clear thought and individual expression. Claiming that many people used substances as a means to escape their problems, the movement encouraged members to avoid escapism, confront problems with a clear mind, and create their own positive, fulfilling lives. Brent emphatically insisted that self-realization did not require drugs:

> There are ways to open your mind without drinking and smoking. . . . You definitely don't have to take mushrooms and sit out in the desert to have a spiritual awakening or a catharsis of any sort. People don't accept that. People think you're uptight. . . . There is a spiritual absence in the world I know right now, in America. To be money driven is the goal. It's one of the emptiest, least fulfilling ways to live your life. . . . The way people relieve themselves of the burdens of their spiritual emptiness is through drugs and alcohol. The way people see escape is sometimes even through a shorter lifespan, through smoking. To be sXe and to understand and believe that means you have opened the door for yourself to find out why we're really on this earth, or what I want to get out of a relationship with a person, or what I want my kids to think of me down the line.

Straight edgers rarely spoke openly about self-realization, and they would likely scoff at anything that suggested mysticism or enlightenment (which they would connect to hippies and therefore drugs). Nevertheless, for many, underlying the ideology was an almost spiritual quest for a genuine self, a "truth." Some connected sXe to other identities: "queer edge," feminism, and activism, for example. For others, sXe offered a means of overcoming abusive family experiences. Mark, a quiet sixteen-year-old new to the scene, claimed sXe as a protest: "Straight edge to me, yeah, it's a commitment to myself, but to me it's also a protest. I don't want to give my kids the same life I had from my father."

## Spreading the Message

Straight edge efforts at resistance transcended members' simple abstention. Straight edgers often actively encouraged other young people to become drug and alcohol free. Some hippies believed their "ultimate social mission is to 'turn the world on'—i.e. Make everyone aware of the potential virtues of LSD for ushering in an era of universal peace, freedom, brotherhood and love" (Davis 1968, 157). Likewise, many sXers undertook a mission to convince their peers that resisting drugs, rather than using them, would help create a better world. A minority of sXers, labeled "militant" or "hardline" by other sXers, were very outspoken, donning Xs and sXe messages at nearly all times and confronting their peers who used. While sXe promoted individuality and clear, free thought, for some adherents the rigid lifestyle requirements created conformity, close-mindedness, and intolerance, a far cry from the "positivity" the movement promulgated. There was an ongoing tension within the movement over how much members should promote their lifestyle. At one extreme was the "live and let live" faction— individuals should make their own choices, and sXers have no right to infringe on that choice. At the other end was the more militant branch, often composed of new adherents, who believed sXers' duty lay in showing users the possibilities of a drug-free lifestyle. Most sXers maintained that their example was enough. Jenny, the student-activist, said,

> I wanna show people there's a community out there that it doesn't make you a fucking dork to be sXe. There are other people out there who are really, really into it. There's a whole group of people you can belong to. You don't have to belong to just them obviously. I just think it can be a really positive thing for people. I go to a dorm where you walk down every fucking hall and the smell of pot knocks you upside the head. I just think that in that case it's really important to get your message out there. . . . I think the best political, social, personal statement you can make is to live by example. That's definitely what I try to do.

Cory, an artist and veteran of the scene at age twenty-one, explained why sXers should set an example for others:

> It's all about calling yourself straight edge. You could be drug free and you can not drink and not smoke and go to parties and do whatever, but you're not helping out. There's a pendulum in society and it's tilted one way so far, and sitting in the middle of the pendulum isn't going to help it swing back. There needs to be more straight edgers on the other side to help even it out, at the least.

Thus, while adherents maintained that sXe was a personal lifestyle choice rather than a movement directed toward others, many members "wore their politics on their sleeves" in a not-so-subtle attempt to encourage others to follow their path. Wearing a shirt with an sXe message may be a personal stylistic decision, but when an entire group of people wears such shirts that so clearly defy the norm, style has the potential to become collective challenge.

Straight edge resistance also targeted the corporate interests of alcohol and tobacco, which adherents claimed profit from people's addictions and suffering. Kate, who clearly connected sXe with her activism, said, "By rejecting Miller Lite and Coors, they have less control over me and my life because I'm not giving them my money; I'm not supporting them." Brent, the outspoken vegan, said,

Each individual in society is connected to one another. When you hurt your-self, you're hurting your society. You're leading by example; your kids will see what you're doing and they'll pick it up. . . . Resisting temptation, resisting what's thrown at you day after day, by your peers, by your parents, by their generation, by businesspeople, by what's hip and cool on MTV. Resistance is huge. That's why sXe is a movement. . . . It's all connected: resisting drugs, resisting rampant consumerism, resisting voting Democrat when you can vote third party.

By focusing their message at their families, subcultural peers, mainstream youth, and the larger society, sXe created a multilayered resistance that indi-viduals could customize to their own interests.

## Involvement in Social Change

Like members of the other subcultures, sXers often became involved in a variety of social causes. The sXe youth with whom I associated insisted that working for social change was not a prerequisite of sXe. Indeed, only a few belonged to the substantial activist community in our city. However, many viewed involvement in social change as a logical progression from clean liv-ing that led them to embrace progressive concerns and become directly involved at some level. Clean living and positivity led to clear thinking, which in turn created a desire to resist and self-realize. This entire process opened them up to the world's problems, and their concerns grew.[13] Tim, twenty-seven, the singer of a very popular sXe band, explained,

The reasoning behind [sXe] is to have a clear mind and to use that clear mind to reach out to other people and do what you can to start thinking about fair-ness, thinking about how to make things more just in society and the world as a whole. . . . It's about freedom. It's about using that freedom that clarity of mind that we have as a vehicle for progression, to make ourselves more peace-ful people. And by making ourselves more peaceful people we make the world a more just place. (Sersen 1999)

Jenny considered sXe central to her activism:

I think every element of my life philosophy is very much interconnected. They all sort of fit together like a puzzle piece. The connection I make between sXe and political activism is sort of that whole attitude like you see something wrong, fix it. I don't like the things that drugs and drinking bring about in society so I fix it by fixing myself. When I see other problems in society as well, I have the same drive to fix it by doing everything that I can do. It's all about claiming power, saying, "All right, I'm in charge of my life. I can do as much good as I want to do."

Kevin, the martial artist, believed that sXe was fundamentally about becoming a strong person in every aspect of life. Strength included rejecting stereotypes and prejudices:

Technically, according to the "rules," you can be homophobic and racist and fuckin' sexist and shit like that and still technically be sXe. You're not drinking; you're not smoking; you're not doing drugs. But I don't personally, on a personal level, I wouldn't consider that person sXe. Because they're weak. I don't think you can be sXe and weak.

Again contrasting against the hippies, punks, and skinheads, for sXers, a clear, drug-free mind was pivotal to developing a consciousness of resistance. The movement provided a general opening up or expansion of social awareness. Kent, the rather quiet young man with many tattoos, said, "I would never have even considered being vegetarian or vegan if it wasn't for sXe. Once you go sXe, I don't really think you're supposed to stop there. It's supposed to open you up to more possibilities. . . . It just makes me think differently. It makes you not so complacent."

In the mid-1980s to late 1980s, sXe became increasingly concerned with animal rights and environmental causes. Influential leaders in bands called for an end to cruelty against animals and a general awareness of eco-destruction. At least three out of four sXers were vegetarian, and many adopted completely cruelty-free, or vegan, lifestyles. Among the approximately sixty sXers I associated with regularly, only fifteen ate meat. Several individuals had "vegan" tattooed on their bodies. Others led or actively participated in a campus animal defense organization. Essentially, the movement framed (see Snow, Rochford, Worden, and Benford 1986) animal rights as a logical extension of the positivity frame underpinning the entire lifestyle, much like reserving sex for caring relationships and self-realization. Brian, an extremely positive and fun-loving twenty-one-year-old, explained vegetarianism's connection to sXe: "sXe kids open their minds a lot more. They're more conscious of what's around them. . . . Some people think it's healthier and other people like me are more on the animal liberation thing." Elizabeth, the older veteran, said,

If you are conscientious and care about the environment or the world, which perhaps more sXe people are than your average population, then [animal rights is] just going to be a factor. You're going to consider "How can I make the world a better place?" Well, being vegetarian is another place you can start. . . . I'm glad it's usually a part of the sXe scene because it just goes along with awareness and choices. What kind of things are you doing to yourself and how is that impacting the world and the environment? The big corporate-owned beef lots and cutting down the rainforests . . . the most impactful thing you can do for the environment is to stop eating meat.

Some sXe youth involved themselves in social justice causes such as home-lessness, human rights, and women's rights. They organized benefit concerts to raise money for local homeless shelters, and often the price of admission to shows included a canned good for the local food pantry or a donation to a women's shelter. I observed several sXers participating in local protests against the World Bank and International Monetary Fund in conjunction with the large 1999-2000 protests in Seattle and Washington, D.C., and others took part in a campus antisweatshop campaign. Similar to progres-sive punks, some sXe youth printed 'zines on prisoners' rights, fighting neo-Nazism, challenging police brutality, and various human rights and environmental issues.

Many sXe women disdained more traditional female roles and appreci-ated the scene as a space in which they felt less pressure to live up to gender expectations, and the movement encouraged men to reject certain hyper-masculine traits and challenge sexism on a personal level. A majority of bands wrote songs against sexism, and many young sXe men demonstrated an exceptional understanding of gender oppression given their ages and experiences. However, despite the movement's claims of community and inclusivity, some sXe women felt isolated and unwelcome in the scene. Men significantly outnumbered women, often creating a "boys club" mentality exemplified by the masculine call for "brotherhood." The almost complete lack of female musicians in bands, the hypermasculine dancing at shows, and the male cliques reinforced the movement's own unspoken gender assumptions that women were not as important to the scene as men and ensured that many women would never feel completely at home.

While some sXers joined animal rights, women's rights, environmental, and other groups, most strove to live out their values in everyday life rather than engage in more conventional "political" protest (e.g., picketing, civil disobedience, petitioning). Instead of challenging tobacco, beer, or beef com-panies directly, for example, a sXer refuses their products and might boycott Kraft (parent company of cigarette manufacturer Phillip Morris), adopt a vegetarian lifestyle, or wear a shirt to school reading "It's OK not to drink. Straight Edge" or "Go Vegan!" In sXe and other youth movements, the per-sonal was political. Subcultures are themselves politically meaningful, and they often serve as a bridge to further political involvement.

Appendix E: An Ethnography

# Conclusion

Straight edgers' understandings of the group's core values show that resistance is much more complex than a stylistic reaction to mainstream culture. I conclude by discussing an analytical framework for understanding the individual and collective meanings, multiple sites, and personal and political methods of resistance of any subculture.

Members of youth subcultures construct both individualized and collective meanings for their participation. Participants may hold individualized meanings that are not central to the group's ideology while simultaneously maintaining collective understandings of the subculture's significance. Widdicombe and Wooffitt (1995), for example, found that "punk may be constituted both through shared goals, values and so on, and through individual members" (p. 204). Subcultures help define "who I am" during the uncertainty of coming of age (p. 25). They offer a space for experimentation and a place to wrestle with questions about the world, creating a "home" for identity in a modern era when personal identity suffers a homelessness brought about by the forces of modernity (Melucci 1989; Giddens 1991). Thus, at the individual level, resistance entails staking out an individual identity and asserting subjectivity in an adversarial context. In addition, for most participants, individualized resistance is symbolic of a larger collective oppositional consciousness. The collective meanings central to the sXe identity included defying the stereotypical "jock" image, setting a collective example for other youth, supporting a drug-free social setting, and avoiding society's "poisons" that dull the mind. Youth claimed the sXe label rather than simply remaining "drug free" specifically because they believed their individual choices would add up to a collective challenge. Here, resistance involves collectively showing disapproval for some aspect of culture, questioning dominant goals, making an invisible ideology visible, and creating an alternative.

Members of youth subcultures understand their resistance at the macro, meso, and micro levels.[14] Past theorizing on resistance has privileged mainstream hegemonic adult culture, the class structure, or the state as the macrolevel target of subcultural resistance (Hall 1972). Indeed, sXers rejected aspects of a culture they believed marketed alcohol and tobacco products to youth, established alcohol use as the norm, promoted conformity, and glorified casual sexual encounters. In addition to challenging culture at the macro level, youth movements offer resistance at the meso level. Straight edgers focused much, if not most, of their message toward their fellow youth, reacting against mainstream youth and perceived contradictions in other subcultures. Overall, sXe illustrates that subcultures form in reaction to other subcultures as well as the larger social structure. Members resisted what they saw as youth culture's fixation on substance use and sex; punks' "no future" and nihilistic tendencies; skinheads' patriotism, sexism, and working-class ideology, as well as some members' racism; and hippies' drug use, passivity, and escapism—believing that these undermine the resistance potential each of these groups share. However, despite its insistence on countering counterculture, sXe co-opted many values of the previous youth movements, clearly owing its "question everything" mentality and aggressive music to punk, its intimation of self-realization and cultural challenge to hippies, and its clean-cut image, personal accountability, and sense of pride to skinheads. Analyzing youth movements at the meso level in terms of their relationship to other youth cultures is vital to an accurate understanding of these groups, as is recognizing the identity battles within the group. Youth reflexively examine their own groups and often attempt to resolve intragroup contradictions. Leblanc (1999, 160) noted, for example, that female punks "subvert the punks' subversion" just as some sXers resisted militant "tough guys" within their scene. All youth movements share disdain for the mainstream; how they express their contempt and challenge existing structures depends in large part on current and previous youth subcultures that often become meso-level targets for change. No doubt the contradictions in sXe will provoke new innovations both within sXe and from other subcultures seeking to transcend sXe's limitations.

Finally, sXers also reported resistance at the micro level as they rejected the substance abuse within their families and made changes in their individual lives. Many sXers claimed that they abstained from drugs and alcohol at least in part in defiance of family members' substance abuse or their own addictive tendencies. Clearly, meanings of subcultural involvement extend beyond contradictions in adult culture and the class structure.

Furthermore, sXe demonstrated that subcultures use many methods of resistance, both personal and political. Distrustful of political challenges and organized social activism, subcultures often embody a more individualistic opposition. Many sXers did seek to change youth culture, but their primary methods were very personal: leading by example, personally living the changes they sought, expressing a personal style, and creating a space to be "free" from their perceived constraints of peer pressure and conformity to mainstream culture.[15] As Widdicombe and Wooffitt (1995) noted in their study of punk identity, "We observed in particular that these oppositional narratives do not invoke radical activities or public displays of resistance; rather, they are fashioned around the routine, the personal and the everyday" (p. 204). Everyday resistance has political consequences (Scott 1985), and (collective) resistance and (individual) authenticity/realization are not mutually exclusive (Muggleton 2000). Buechler (1999, 151) wrote, "In the case of life politics, the politicized self and the self-actualizing self become one and the same. The microphysics of power also points to identity as the battleground in contemporary forms of resistance" (see also Giddens 1991).

Though focused on personal methods of resistance, sXers understood their involvement in political terms as well.[16] Their abstinence from drugs, alcohol, and casual sex was an essential component of a broader resistance to dominant society and mainstream youth culture. As Buechler (1999) pointed out, "Although this form of politics originates on the microlevel of personal identity, its effects are not likely to remain confined to this level" (p. 150). The movement engages in what Giddens (1991, 214-15) called "life politics"—a "politics of choice," a "politics of lifestyle," a "politics of self-actualization," and a "politics of life decisions." Through their individual actions, sXers seek a "remoralizing of social life" (Buechler 1999, 150). For example, becoming a vegetarian or vegan may be an individualistic dietary choice, but when a subculture does so and advocates their choice, it opens up possibilities for other youth. As Leblanc (1999) noted, the intent to influence others is an important component of resistance: "Accounts of resistance must detail not only resistant acts, but the subjective intent motivating these as well. . . . Such resistance includes not only behaviors, but discursive and symbolic acts" (p. 18).

Looking at resistance through the lens of meanings, sites, and methods forces us to reexamine the "success" of subcultural resistance. Analyzing sXe's core values shows that members' understandings of resistance are many layered and contextual. The issue of resistance goes beyond whether a subculture resists dominant culture to how members construct resistance in particular situations and contexts. Certainly, sXe, like other subcultures, has illusory tendencies; the movement's contradictions include its antisexist yet male-centered ideology. However, examining sXe with the framework I suggest shows that involvement has real consequences for the lives of its members, other peer groups, and possibly mainstream society. Personal realization and social transformation are not mutually exclusive (Calhoun 1994). Although sXe has not created a revolution in either youth or mainstream culture, it has for more than twenty years, however, provided a haven for youth to contest these cultures and create alternatives.

# Notes

1. Straight edgers abbreviate straight edge as sXe. The s and the e stand for straight edge, and the X is the straight edge symbol.

2. Hardcore is a more aggressive, faster style of punk. Though punk and hardcore overlap, in the 1990s the two scenes increasingly became distinct. While present in both scenes, sXe is considerably more prevalent in the hardcore scene. The hardcore style is more clean-cut than punk.

3. Punks and sXers draw a sharp distinction between "shows" and "concerts." Shows attract a much smaller crowd, are less expensive, feature underground bands, often showcase local bands, and are set up by local kids in the scene at little or no profit. Concerts are large, commercialized, for-profit ventures typically featuring more mainstream bands.

4. Straight edge individuals never refer to themselves as *straight edgers* and find the term quite funny. It likely comes from media portrayals of the group. Adherents call themselves sXe "kids," no matter their ages. I use *straight edger* in this article simply for ease of communication.

5. See Muggleton (2000) for a discussion on the importance of grounding any subcultural analysis in members' subjective experiences.

6. I encountered one antiracist skinhead who also claimed to be sXe. He eventually dropped out of both groups, however. An older Latino sXer I knew, a veteran of the scene, claimed he was a skinhead many years ago.

7. Individuals or small groups produce 'zines filled with artwork, stories, record and concert reviews, band interviews, and columns on everything from police brutality and animal rights to homelessness and freeing journalist and former Black Panther Mumia Abu-Jamal from prison. 'Zines, like concerts, are generally DIY; that is, kids create them at home, distribute them, and rarely make any money off of them (in fact, 'zines often cost the producers a great deal of money).

8. Movements often appropriate and modify their oppressors' symbols. The gay and lesbian liberation movement changed the pink triangle from a Nazi death camp label for homosexuals into a symbol for unity and pride. The American Indian movement turned the American flag upside down to demonstrate its disgust with the U.S. government.

9. The community in Clearweather was very tight knit. In addition to shows, frequent potlucks, movie nights, parties, hanging out at popular campus locations, involvement in local animal rights activism, and even the occasional sleepover kept members in regular contact. Many sXe youth lived together. With the advent of e-mail and the Internet, sXe kids communicated via a virtual community around the country and sometimes the globe.

10. Veganism had become such a significant part of sXe by the late 1990s that many sXers gave it equal importance to living drug and alcohol free. Thus, many sXe vegans would self-identify as "vegan straight edge," and some bands identify as "vegan straight edge" rather than simply "straight edge." Veganism, while still widely practiced, had a declining presence after 2000.

11. All names are pseudonyms.

12. The popular bands Earth Crisis, Outspoken, and Good Clean Fun encouraged listeners to challenge homophobia. At one time, there was even a Web site dedicated to "Queer Edge."

13. Earth Crisis, one of the most popular sXe bands, sings, "An effective revolutionary, with the clarity of mind that I've attained."

14. Leblanc's (1999) work with punk girls illustrates multiple sites of resistance to hegemonic gender constructions. At the macro level, these young women resist society's dominant constructions of femininity; at the meso level, they resist gender roles in punk; and at the micro level, they challenge gender constructions in their families and focus on personal empowerment and self-esteem.

15. Leblanc (1999, 17) wrote, "Whereas subculture theorists conceptualize resistance as stylistic, and feminist theorists consider discursive accounts, recent critics of resistance theorizing have begun to examine the behavioral forms of resistance constructed by oppressed individuals in their everyday lives."

16. "To an increasing degree, problems of individual identity and collective action become meshed together: the solidarity of the group is inseparable from the personal quest" (Melucci 1996, 115).

# References

Adler, P. A., and P. Adler. 1987. *Membership roles in field research* .Newbury Park, CA: Sage.

Agar, M. H. 1996. *The professional stranger: An informal introduction to ethnography.* 2nd ed. San Diego, CA: Academic Press.

Becker, H. S., and B. Geer. 1960. Participant observation: The analysis of qualitative field data. In *Human organization research: Field relations and techniques,* edited by Richard N. Adams and Jack J. Preiss, 267-89. Homewood, IL: Dorsey.

Berger, B. M. 1967. Hippie morality—More old than new. *Trans-Action* 5 (2): 19–26.

Biernacki, P., and D. Waldorf. 1981. Snowball sampling. *Sociological Research and Methods* 10:141–63.

Bjorgo,T., and R.Wilte, eds. 1993. *Racist violence in Europe.* NewYork: St. Martin's.

Blackman, S. J. 1995. *Youth: Positions and oppositions—Style, sexuality and schooling.* Aldershot, UK: Avebury.

Brake, M. 1985. *Comparative youth culture: The sociology of youth culture and youth subcultures in America, Britain, and Canada.* London: Routledge Kegan Paul.

Buechler, S. M. 1995. New social movement theories. *Sociological Quarterly* 36:441–64.

———. 1999. *Social movements in advanced capitalism: The political economy and cultural construction of social activism.* New York: Oxford University Press.

Calhoun, C. 1994. *Social theory and the politics of identity.* Oxford, UK: Blackwell.

Campbell, D. T. 1975. Degrees of freedom and the case study. *Comparative Political Studies* 8:178–93.

Charmaz, K. 1983. The grounded theory method: An explication and interpretation. In *Contemporary field research: A collection of readings,* edited by R. M. Emerson, 109–26. Boston: Little, Brown.

Clarke, G. [1981] 1997. Defending ski-jumpers: A critique of theories of youth subcultures. In *The subcultures reader,* edited by K. Gelder and S. Thornton. London: Routledge.

Clarke, J., S. Hall, T. Jefferson, and B. Roberts. 1975. Subcultures, cultures, and class: A theoretical overview. In *Resistance through rituals: Youth subcultures in post-war Britain*, edited by S. Hall and T. Jefferson. London: Hutchinson.

Cohen, S. 1972. *Folk devils and moral panics: The creation of the mods and the rockers*. Oxford, UK: Martin Robertson.

———.1980. Symbols of trouble. In *Folk devils and moral panics: The creation of the mods and the rockers*. Oxford, UK: Martin Robertson.

Davis, F. 1967. Focus on the flower children. Why all of us may be hippies someday. *Trans-Action* 5 (2): 10–18.

———. 1968. Heads and freaks: Patterns and meanings of drug use among hippies. *Journal of Health and Social Behavior* 9 (2): 156–64.

Douglas, J. D. 1976. *Investigative social research*. Beverly Hills, CA: Sage.

Earisman, D. L. 1968. *Hippies in our midst*. Philadelphia: Fortress.

Earth Crisis. 1995. The discipline. On *Destroy the machines*. Compact disc. Chicago: Victory Records.

Fox, K. J. 1987. Real punks and pretenders: The social organization of a counterculture. *Journal of Contemporary Ethnography* 16 (3): 344–70.

Giddens, A. 1991. *Modernity and self-identity: Self and society in the late modern age*. Stanford, CA: Stanford University Press.

Good Clean Fun. 2001. Today the scene, tomorrow the world. On *Straight outta hardcore*. Compact disc. Washington, DC: Phyte Records.

Gottschalk, S. 1993. Uncomfortably numb: Countercultural impulses in the postmodern era. *Symbolic Interaction* 16 (4): 351–78.

Grossberg, L. 1992. *We gotta get out of this place: Popular conservatism and postmodern culture*. New York: Routledge.

Habermas, J. 1984-87. *The theory of communicative action*. Translated by T. McCarthy. Boston: Beacon.

Hall, S. 1972. Culture and the state. In *The state and popular culture*, edited by Milton Keynes. Berkshire, UK: Open University Press.

Hall, S., and T. Jefferson, eds. 1976. *Resistance through rituals: Youth subcultures in post-war Britain*. London: Hutchinson.

Hebdige, D. 1979. *Subcultures: The meaning of style*. London: Methuen.

Henry, T. 1989. *Break all rules! Punk rock and the making of a style*. Ann Arbor: University of Michigan Research Press.

Heylin, C. 1998. *Never mind the bollocks, here's the Sex Pistols: The Sex Pistols*. New York: Schirmer Books.

Hunter, J. D. 1987. *Evangelism: The coming generation*. Chicago: University of Chicago Press.

Irwin, D. 1999. The straight edge subculture: Examining the youths' drug-free way. *Journal of Drug Issues* 29 (2): 365–80.

Irwin, J. 1977. *Scenes*. Beverly Hills, CA: Sage.

Kaplan, J., and H. Lööw, eds. 2002. *The cultic milieu: Oppositional subcultures in an age of globalization*. Walnut Creek, CA: AltaMira Press.

Lahickey, B. 1997. *All ages: Reflections on straight edge*. Huntington Beach, CA: Revelation Books.

Leblanc, L. 1999. *Pretty in punk: Girls' gender resistance in a boys' subculture*. New Brunswick, NJ: Rutgers University Press.

Liebman, R. C., and R. Wuthnow, eds. 1983. *The new Christian Right: Mobilization and legitimation*. Hawthorne, NY: Aldine.

Lofland, J., and L. Lofland. 1995. *Analyzing social settings: A guide to qualitative observation and analysis*. 3rd ed. Belmont, CA: Wadsworth.

Marty, M. E., and R. S. Appleby, eds. 1993. *Fundamentalism and the state*. Chicago: University of Chicago Press.

Melucci, A. 1989. *Nomads of the present: Social movements and individual needs in contemporary society*. Philadelphia: Temple University Press.

———. 1996. *Challenging codes: Collective action in the information age*. New York: Cambridge University Press.

Miller, T. 1991. *The hippies and American values*. Knoxville: University of Tennessee Press.

———. 1999. *The 60s communes: Hippies and beyond*. Syracuse, NY: Syracuse University Press.

Moore, D. 1994. *The lads in action: Social process in an urban youth subculture*. Aldershot, UK: Arena.

Muggleton, D. 2000. *Inside subculture: The postmodern meaning of style*. Oxford, UK: Berg.

O'Hara, C. 1999. *The philosophy of punk: More than noise*. London: AK Press.

Parkin, F. 1968. *Middle class radicalism: The social bases of the British campaign for nuclear disarmament*. New York: Praeger.

Riemer, J. 1977. Varieties of opportunistic research. *Urban Life* 5 (4): 467–77.

Rose, T. 1994. *Black noise: Rap music and black culture in contemporary America*. New York: Routledge.

Scott, J. 1985. *Weapons of the weak: Everyday forms of peasant resistance*. New Haven, CT: Yale University Press.

Sersen, B., producer and director. 1999. *Release*. Film. Chicago: Victory Records.

Snow, D. A., E. B. Rochford Jr., S. K. Worden, and R. D. Benford. 1986. Frame alignment processes, micromobilization, and movement participation. *American Sociological Review* 51:464–81.

Stewart, A. 1998. *The ethnographer's method*. Thousand Oaks, CA: Sage.

Strife. 1997. *To an end*. On *In this defiance*. Compact disc. Chicago: Victory Records.

Taylor, V., and N. E. Whittier. 1992. Collective identity in social movement communities: Lesbian feminist mobilization. In *Frontiers in social movement theory*, edited by A. D. Morris and C. M. Mueller, 104–29. New Haven, CT: Yale University Press.

Widdicombe, S., and R. Wooffitt. 1995. *The language of youth subcultures: Social identity in action*. London: Harvester Wheatsheaf.

Willis, P. 1977. *Learning to labor: How working class kids get working class jobs*. New York: Columbia University Press.

Wooden, W. S., and R. Blazak. 2001. *Renegade kids, suburban outlaws: From youth culture to delinquency*. 2nd ed. Belmont, CA: Wadsworth.

Young, K., and L. Craig. 1997. Beyond white pride: Identity, meaning and contradiction in the Canadian skinhead subculture. *Canadian Review of Sociology and Anthropology* 34 (2): 175–206.

Youth of Today. 1986. Youth crew. On *Can't close my eyes*. LP record. Huntington Beach, CA: Revelation Records.

# Appendix F

## A Case Study

### Campus Response to a Student Gunman

*Kelly J. Asmussen*

*John W. Creswell*

With increasingly frequent incidents of campus violence, a small, growing scholarly literature about the subject is emerging. For instance, authors have reported on racial [12], courtship and sexually coercive [3, 7, 8], and hazing violence [24]. For the American College Personnel Association, Roark [24] and Roark and Roark [25] reviewed the forms of physical, sexual, and psychological violence on college campuses and suggested guidelines for prevention strategies. Roark [23] has also suggested criteria that high-school students might use to assess the level of violence on college campuses they seek to attend. At the national level, President Bush, in November 1989, signed into law the "Student Right-to-Know and Campus Security Act" (P.L. 101-542), which requires colleges and universities to make available to students, employees, and applicants an annual report on security policies and campus crime statistics [13].

One form of escalating campus violence that has received little attention is student gun violence. Recent campus reports indicate that violent crimes from thefts and burglaries to assaults and homicides are on the rise at colleges and universities [13]. College campuses have been shocked by killings such as those at The University of Iowa [16], The University of Florida [13], Concordia University in Montreal, and the University of Montreal-Ecole Polytechnique [22]. Incidents such as these raise critical concerns, such as psychological trauma, campus safety, and disruption of campus life. Aside from an occasional newspaper report, the postsecondary literature is silent on campus reactions to these tragedies; to understand them one must turn to studies about gun violence in the public school literature. This literature addresses strategies for school intervention [21, 23], provides case studies of incidents in individual schools [6, 14, 15], and discusses the problem of students who carry weapons to school [1] and the psychological trauma that results from homicides [32].

A need exists to study campus reactions to violence in order to build conceptual models for future study as well as to identify campus strategies and protocols for reaction. We need to understand better the psychological dimensions and organizational issues of constituents involved in and affected by these incidents. An in-depth qualitative case study exploring the context of an incident can illuminate such conceptual and pragmatic understandings. The study presented in this article is a qualitative case analysis [31] that describes and interprets a campus response to a gun incident. We asked the following exploratory research questions: What happened? Who was involved in response to the incident? What themes of response emerged during the eight-month period that followed this incident? What theoretical constructs helped us understand the campus response, and what constructs were unique to this case?

## The Incident and Response

The incident occurred on the campus of a large public university in a Midwestern city. A decade ago, this city had been designated an "all-American city," but more recently, its normally tranquil environment has been disturbed by an increasing number of assaults and homicides. Some of these violent incidents have involved students at the university.

The incident that provoked this study occurred on a Monday in October. A forty-three-year-old graduate student, enrolled in a senior-level actuarial science class, arrived a few minutes before class, armed with a vintage Korean War military semiautomatic rifle loaded with a thirty-round clip of thirty caliber ammunition. He carried another thirty-round clip in his pocket. Twenty of the thirty-four students in the class had already gathered for class, and most of them were quietly reading the student newspaper. The instructor was en route to class.

The gunman pointed the rifle at the students, swept it across the room, and pulled the trigger. The gun jammed. Trying to unlock the rifle, he hit the butt of it on the instructor's desk and quickly tried firing it again. Again it did not fire. By this time, most students realized what was happening and dropped to the floor, overturned their desks, and tried to hide behind them. After about twenty seconds, one of the students shoved a desk into the gunman, and students ran past him out into the hall and out of the building. The gunman hastily departed the room and went out of the building to his parked car, which he had left running. He was captured by police within the hour in a nearby small town, where he lived. Although he remains incarcerated at this time, awaiting trial, the motivations for his actions are unknown.

Campus police and campus administrators were the first to react to the incident. Campus police arrived within three minutes after they had received a telephone call for help. They spent several anxious minutes outside the building interviewing students to obtain an accurate description of the gunman. Campus administrators responded by calling a news conference for 4:00 P.M. the same day, approximately four hours after the incident. The police chief as well as the vice-chancellor of Student Affairs and two students described the incident at the news conference. That same afternoon, the Student Affairs office contacted Student Health and Employee Assistance Program (EAP) counselors and instructed them to be available for any students or staff requesting assistance. The Student Affairs office also arranged for a new location, where this class could meet for the rest of the semester. The Office of Judicial Affairs suspended the gunman from the university. The next day, the incident was discussed by campus administrators at a regularly scheduled campuswide cabinet meeting. Throughout the week, Student Affairs received several calls from students and from a faculty member about "disturbed" students or unsettling student relations. A counselor of the Employee Assistance Program consulted a psychologist with a specialty in dealing with trauma and responding to educational crises. Only one student immediately set up an appointment with the student health counselors. The campus and local newspapers continued to carry stories about the incident.

When the actuarial science class met for regularly scheduled classes two and four days later, the students and the instructor were visited by two county attorneys, the police chief, and two student mental health counselors who conducted "debriefing" sessions. These sessions focused on keeping students fully informed about the judicial process and having the students and the instructor, one by one, talk about their experiences and explore their feelings about the incident. By one week after the incident, the students in the class had returned to their standard class format. During this time, a few students, women who were concerned about violence in general, saw Student Health Center counselors. These counselors also fielded questions from several dozen parents who inquired about the counseling services and the level of safety on campus. Some parents also called the campus administration to ask about safety procedures.

In the weeks following the incident, the faculty and staff campus newsletter carried articles about post-trauma fears and psychological trauma. The campus administration wrote a letter that provided facts about the incident to the board of the university. The administration also mailed campus staff and students information about crime prevention. At least one college dean sent out a memo to staff about "aberrant student behavior," and one academic department chair requested and held an educational group session with counselors and staff on identifying and dealing with "aberrant behavior" of students.

Three distinctly different staff groups sought counseling services at the Employee Assistance Program, a program for faculty and staff, during the next several weeks. The first group had had some direct involvement with the assailant, either by seeing him the day of the gun incident or because they had known him personally. This group was concerned about securing professional help, either for the students or for those in the group who were personally experiencing effects of the trauma. The second group consisted of the "silent connection," individuals who were indirectly involved and yet emotionally traumatized. This group recognized that their fears were a result of the gunman incident, and they wanted to deal with these fears before they escalated. The third group consisted of staff who had previously experienced a trauma, and this incident had retriggered their fears. Several employees were seen by the EAP throughout the next month, but no new groups or delayed stress cases were reported. The EAP counselors stated that each group's reactions were normal responses. Within a month, although public discussion of the incident had subsided, the EAP and Student Health counselors began expressing the need for a coordinated campus plan to deal with the current as well as any future violent incidents.

## The Research Study

We began our study two days after the incident. Our first step was to draft a research protocol for approval by the university administration and the Institutional Review Board. We made explicit that we would not become involved in the investigation of the gunman or in the therapy to students or staff who had sought assistance from counselors. We also limited our study to the reactions of groups on campus rather than expand it to include off-campus groups (for example, television and newspaper coverage). This bounding of the study was consistent with an exploratory qualitative case study design [31], which was chosen because models and variables were not available for assessing a campus reaction to a gun incident in higher education. In the constructionist tradition, this study incorporated the paradigm assumptions of an emerging design, a context-dependent inquiry, and an inductive data analysis [10]. We also bounded the study by time (eight months) and by a single case (the campus community). Consistent with case study design [17, 31], we identified campus administrators and student newspaper reporters as multiple sources of information for initial interviews. Later we expanded interviews to include a wide array of campus informants, using a semi-structured interview protocol that consisted of five questions: What has been your role in the incident? What has happened since the event that you have been involved in? What has been the impact of this incident on the university community? What larger ramifications, if any, exist from the incident? To whom should we talk to find out more about the campus reaction to the incident? We also gathered observational data, documents, and visual materials (see table 1 for types of information and sources).

The narrative structure was a "realist" tale [28], describing details, incorporating edited quotes from informants, and stating our interpretations of events, especially an interpretation within the framework of organizational and psychological issues. We verified the description and interpretation by taking a preliminary draft of the case to select informants for feedback and later incorporating their comments into the final study [17, 18]. We gathered this feedback in a group interview where we asked: Is our description of the incident and the reaction accurate? Are the themes and constructs we have identified consistent with your experiences? Are there some themes and constructs we have missed? Is a campus plan needed? If so, what form should it take?

## Themes

### Denial

Several weeks later we returned to the classroom where the incident occurred. Instead of finding the desks overturned, we found them to be neatly in order; the room was ready for a lecture or discussion class. The hallway outside the room was narrow, and we visualized how students, on that Monday in October, had quickly left the building, unaware that the gunman, too, was exiting through this same passageway. Many of the

**Table 1**    Data Collection Matrix: Type of Information by Source

| Information/ Information Source | Interviews | Observations | Documents | Audio-Visual Materials |
|---|---|---|---|---|
| Students involved | Yes | | Yes | |
| Students at large | Yes | | | |
| Central administration | Yes | | Yes | |
| Campus police | Yes | Yes | | |
| Faculty | Yes | Yes | Yes | |
| Staff | Yes | | | |
| Physical plant | | Yes | Yes | |
| News reporters/ papers/television | Yes | | Yes | Yes |
| Student health counselors | Yes | | | |
| Employee Assistance Program counselors | Yes | | | |
| Trauma expert | Yes | | Yes | Yes |
| Campus businesses | | | Yes | |
| Board members | | | Yes | |

students in the hallway during the incident had seemed unaware of what was going on until they saw or heard that there was a gunman in the building. Ironically though, the students had seemed to ignore or deny their dangerous situation. After exiting the building, instead of seeking a hiding place that would be safe, they had huddled together just outside the building. None of the students had barricaded themselves in classrooms or offices or had exited at a safe distance from the scene in anticipation that the gunman might return. "People wanted to stand their ground and stick around," claimed a campus police officer. Failing to respond to the potential danger, the class members had huddled together outside the building, talking nervously. A few had been openly emotional and crying. When asked about their mood, one of the students had said, "Most of us were kidding about it." Their conversations had led one to believe that they were dismissing the incident as though it were trivial and as though no one had actually been in danger. An investigating campus police officer was not surprised by the students' behavior:

It is not unusual to see people standing around after one of these types of incidents. The American people want to see excitement and have a morbid curiosity. That is why you see spectators hanging around bad accidents. They do not seem to understand the potential danger they are in and do not want to leave until they are injured.

This description corroborates the response reported by mental health counselors: an initial surrealistic first reaction. In the debriefing by counselors, one female student had commented, "I thought the gunman would shoot out a little flag that would say 'bang'." For her, the event had been like a dream. In this atmosphere no one from the targeted class had called the campus mental health center in the first twenty-four hours following the incident, although they knew that services were available. Instead, students described how they had visited with friends or had gone to bars; the severity of the situation had dawned on them later. One student commented that he had felt fearful and angry only after he had seen the television newscast with pictures of the classroom the evening of the incident.

Though some parents had expressed concern by phoning counselors, the students' denial may have been reinforced by parent comments. One student reported that his parents had made comments like, "I am not surprised you were involved in this. You are always getting yourself into things like this!" or "You did not get hurt. What is the big deal? Just let it drop!" One student expressed how much more traumatized he had been as a result of his mother's dismissal of the event. He had wanted to have someone whom he trusted willing to sit down and listen to him.

## Fear

Our visit to the classroom suggested a second theme: the response of fear. Still posted on the door several weeks after the incident, we saw the sign announcing that the class was being moved to another undisclosed building and that students were to check with a secretary in an adjoining room about the new location. It was in this undisclosed classroom, two days after the incident, that two student mental health counselors, the campus police chief, and two county attorneys had met with students in the class to discuss fears, reactions, and thoughts. Reactions of fear had begun to surface in this first "debriefing" session and continued to emerge in a second session.

The immediate fear for most students centered around the thought that the alleged assailant would be able to make bail. Students felt that the assailant might have harbored resentment toward certain students and that he would seek retribution if he made bail. "I think I am going to be afraid when I go back to class. They can change the rooms, but there is nothing stopping him from finding out where we are!" said one student. At the first debriefing session the campus police chief was able to dispel some of this fear by announcing that during the initial hearing the judge had denied bail. This announcement helped to reassure some students about their safety. The campus police chief thought it necessary to keep the students informed of the gunman's status, because several students had called his office to say that they feared for their safety if the gunman were released.

During the second debriefing session, another fear surfaced: the possibility that a different assailant could attack the class. One student reacted so severely to this potential threat that, according to one counselor, since the October incident, "he had caught himself walking into class and sitting at a desk with a clear shot to the door. He was beginning to see each classroom as a 'battlefield'." In this second session students had sounded angry, they expressed feeling violated, and finally [they] began to admit that they felt unsafe. Yet only one female student immediately accessed the available mental health services, even though an announcement had been made that any student could obtain free counseling.

The fear students expressed during the "debriefing" sessions mirrored a more general concern on campus about increasingly frequent violent acts in the metropolitan area. Prior to this gun incident, three young females and a male had been kidnapped and had later been found dead in a nearby city. A university football player who experienced a psychotic episode had severely beaten a woman. He had later suffered a relapse and was shot by police in a scuffle. Just three weeks prior to the October gun incident, a female university student had been abducted and brutally murdered, and several other homicides had occurred in the city. As a student news reporter commented, "This whole semester has been a violent one."

## Safety

The violence in the city that involved university students and the subsequent gun incident that occurred in a campus classroom shocked the typically tranquil campus. A counselor aptly summed up the feelings of many: "When the students walked out of that classroom, their world had become very chaotic; it had become very random, something had happened that robbed them of their sense of safety." Concern for safety became a central reaction for many informants.

When the chief student affairs officer described the administration's reaction to the incident, he listed the safety of students in the classroom as his primary goal, followed by the needs of the news media for details about the case, helping all students with psychological stress, and providing public information on safety. As he talked about the safety issue and the presence of guns on campus, he mentioned that a policy was under consideration for the storage of guns used by students for hunting. Within four hours after the incident, a press conference was called during which the press was briefed not only on the details of the incident, but also on the need to ensure the safety of the campus. Soon thereafter the university administration initiated an informational campaign on campus safety. A letter, describing the incident, was sent to the university board members. (One board member asked, "How could such an incident happen at this university?") The Student Affairs Office sent a letter to all students in which it advised them of the various dimensions of the campus security office and of the types of services it provided. The Counseling and Psychological Services of the Student Health Center promoted their services in a colorful brochure, which was mailed to students in the following week. It emphasized that services were "confidential, accessible, and professional." The Student Judiciary Office advised academic departments on various methods of dealing with students who exhibited abnormal behavior in class. The weekly faculty newsletter stressed that staff needed to respond quickly to any post-trauma fears associated with this incident. The campus newspaper quoted a professor as saying, "I'm totally shocked that in this environment, something like this would happen." Responding to the concerns about disruptive students or employees, the campus police department sent plainclothes officers to sit outside offices whenever faculty and staff indicated concerns.

An emergency phone system, Code Blue, was installed on campus only ten days after the incident. These thirty-six ten-foot-tall emergency phones, with bright blue flashing lights, had previously been approved, and specific spots had already been identified from an earlier study. "The phones will be quite an attention getter," the director of the Telecommunications Center commented. "We hope they will also be a big detractor [to crime]." Soon afterwards, in response to calls from concerned students, trees and shrubbery in poorly lit areas of campus were trimmed.

Students and parents also responded to these safety concerns. At least twenty-five parents called the Student Health Center, the university police, and the Student Affairs Office during the first week after the incident to inquire what kind of services were available for their students. Many parents had been traumatized by the news of the event and immediately demanded answers from the university. They wanted assurances that this type of incident would not happen again and that their child[ren were] safe on the campus. Undoubtedly, many parents also called their children during the weeks immediately following the incident. The students on campus responded to these safety concerns by forming groups of volunteers who would escort anyone on campus, male or female, during the evening hours.

Local businesses profited by exploiting the commercial aspects of the safety needs created by this incident. Various advertisements for self-defense classes and protection devices inundated the newspapers for several weeks. Campus and local clubs [that] offered self-defense classes filled quickly, and new classes were formed in response to numerous additional requests. The campus bookstore's supply of pocket mace and whistles was quickly depleted. The campus police received several inquiries by students who wanted to purchase handguns to carry for protection. None [was] approved, but one wonders whether some guns were not purchased by students anyway. The purchase of cellular telephones from local vendors increased sharply. Most of these purchases were made by females; however, some males also sought out these items for their safety and protection. Not unexpectedly, the price of some products was raised as much as 40 percent to capitalize on the newly created demand. Student conversations centered around the purchase of these safety products: how much they cost, how to use them correctly, how accessible they would be if students should need to use them, and whether they were really necessary.

## Retriggering

In our original protocol, which we designed to seek approval from the campus administration and the Institutional Review Board, we had outlined a study that would last only three months—a reasonable time, we thought, for this incident to run its course. But during early interviews with counselors, we were referred to a psychologist who specialized in dealing with "trauma" in educational settings. It was this psychologist who mentioned the theme of "retriggering." Now, eight months later, we begin to understand how, through "retriggering," that October incident could have a long-term effect on this campus.

This psychologist explained retriggering as a process by which new incidents of violence would cause individuals to relive the feelings of fear, denial, and threats to personal safety that they had experienced in connection with the original event. The counseling staffs and violence expert also stated that one should expect to see such feelings retriggered at a later point in time, for example, on the anniversary date of the attack or whenever newspapers or television broadcasts mentioned the incident again. They added that a drawn-out judicial process, during which a case were "kept alive" through legal maneuvering, could cause a long period of retriggering and thereby greatly thwart the healing process. The fairness of the judgment of the court as seen by each victim, we were told, would also influence the amount of healing and resolution of feelings that could occur.

As of this writing, it is difficult to detect specific evidence of retriggering from the October incident, but we discovered the potential consequences of this process firsthand by observing the effects of a nearly identical violent gun incident that had happened some eighteen years earlier. A graduate student carrying a rifle had entered a campus building with the intention of shooting the department chairman. The student was seeking revenge, because several years earlier he had flunked a course taught by this professor. This attempted attack followed several years of legal maneuvers to arrest, prosecute, and incarcerate this student, who, on more than one occasion, had tried to carry out his plan but each time had been thwarted by quick-thinking staff members who would not reveal the professor's whereabouts. Fortunately, no shots were ever fired, and the student was finally apprehended and arrested.

The professor who was the target of these threats on his life was seriously traumatized not only during the period of these repeated incidents, but his trauma continued even after the attacker's arrest. The complex processes of the criminal justice system, which, he believed, did not work as it should have, resulted in his feeling further victimized. To this day, the feelings aroused by the original trauma are retriggered each time a gun incident is reported in the news. He was not offered professional help from the university at any time; the counseling services he did receive were secured through his own initiative. Eighteen years later his entire department is still affected in that unwritten rules for dealing with disgruntled students and for protecting this particular professor's schedule have been established.

## Campus Planning

The question of campus preparedness surfaced during discussions with the psychologist about the process of "debriefing" individuals who had been involved in the October incident [19]. Considering how many diverse groups and individuals had been affected by this incident, a final theme that emerged from our data was the need for a campuswide plan. A counselor remarked, "We would have been inundated had there been twenty-five to thirty deaths. We need a mobilized plan of communication. It would be a wonderful addition to the campus considering the nature of today's violent world." It became apparent during our interviews that better communication could have occurred among the constituents who responded to this incident. Of course, one campus police officer noted, "We can't have an officer in every building all day long!" But the theme of being prepared across the whole campus was mentioned by several individuals.

The lack of a formal plan to deal with such gun incidents was surprising, given the existence of formal written plans on campus that addressed various other emergencies: bomb threats, chemical spills, fires, earthquakes, explosions, electrical storms, radiation accidents, tornadoes, hazardous material spills, snowstorms, and numerous medical emergencies. Moreover, we found that specific campus units had their own protocols that had actually been used during the October gun incident. For example, the police had a procedure and used that procedure for dealing with the gunman and the students at the scene; the EAP counselors debriefed staff and faculty; the Student Health counselors used a "debriefing" process when they visited the students twice in the classroom following the incident. The question that concerned us was, what would a campuswide plan consist of, and how would it be developed and evaluated?

As shown in table 2, using evidence gathered in our case, we assembled the basic questions to be addressed in a plan and cross-referenced these questions to the literature about post-trauma stress, campus violence, and the disaster literature (for a similar list drawn from the public school literature, see Poland and Pitcher [21]). Basic elements of a campus plan to enhance communication across units should include determining what the rationale for the plan is; who should be involved in its development; how it should be coordinated; how it should be staffed; and what specific procedures should be followed. These procedures might include responding to an immediate crisis, making the campus safe, dealing with external groups, and providing for the psychological welfare of victims.

## Discussion

The themes of denial, fear, safety, retriggering, and developing a campuswide plan might further be grouped into two categories, an organizational and a psychological or social-psychological response of the campus community to the gunman incident. Organizationally, the campus units responding to the crisis exhibited both a loose coupling [30] and an interdependent communication. Issues such as leadership, communication, and authority emerged during the case analysis. Also, an environmental response developed, because the campus was transformed into a safer place for students and staff. The need for centralized planning, while allowing for autonomous operation of units in response to a crisis, called for organizational change that would require cooperation and coordination among units.

Sherrill [27] provides models of response to campus violence that reinforce as well as depart from the evidence in our case. As mentioned by Sherrill, the disciplinary action taken against a perpetrator, the group

**Table 2**    Evidence From the Case, Questions for a Campus Plan, and References

| Evidence From the Case | Question for the Plan | References Useful |
|---|---|---|
| Need expressed by counselors | Why should a plan be developed? | Walker (1990); Bird et al. (1991) |
| Multiple constitutes reacting to incident | Who should be involved in developing the plan? | Roark & Roark (1987); Walker (1990) |
| Leadership found in units with their own protocols | Should the leadership for coordinating be identified within one office? | Roark & Roark (1987) |
| Several unit protocols being used in incident | Should campus units be allowed their own protocols? | Roark & Roark (1987) |
| Questions raised by students reacting to case | What types of violence should be covered in the plan? | Roark (1987); Jones (1990) |
| Groups/individuals surfaced during our interviews | How are those likely to be affected by the incident to be identified? | Walker (1990); Bromet (1990) |
| Comments from campus police, central administration | What provisions are made for the immediate safety of those in the incident? | |
| Campus environment changed after incident | How should the physical environment be made safer? | Roark & Roark (1987) |
| Comments from central administration | How will the external publics (e.g., press, businesses) be apprised of the incident? | Poland & Pitcher (1990) |
| Issue raised by counselors and trauma specialist | What are the likely sequelae of psychological events for victims? | Bromet (1990); Mitchell (1983) |
| Issue raised by trauma specialist | What long-term impact will the incident have on victims? | Zelikoff (1987) |
| Procedure used by Student Health Center counselors | How will the victims be debriefed? | Mitchell (1983); Walker (1990) |

counseling of victims, and the use of safety education for the campus community were all factors apparent in our case. However, Sherrill raises issues about responses that were not discussed by our informants, such as developing procedures for individuals who are first to arrive on the scene, dealing with non-students who might be perpetrators or victims, keeping records and documents about incidents, varying responses based on the size and nature of the institution, and relating incidents to substance abuse such as drugs and alcohol.

Also, some of the issues that we had expected after reading the literature about organizational response did not emerge. Aside from occasional newspaper reports (focused mainly on the gunman), there was little campus administrative response to the incident, which was contrary to what we had expected from Roark and Roark [25], for example. No mention was made of establishing a campus unit to manage future incidents—for example, a campus violence resource center—reporting of violent incidents [25], or conducting annual safety audits [20]. Aside from the campus police mentioning that the State Health Department would have been prepared to send a team of trained trauma experts to help emergency personnel cope with the tragedy, no discussion was reported about formal linkages with community agencies that might assist in the event of a tragedy [3]. We also did not hear directly about establishing a "command center" [14] or a crisis coordinator [21], two actions recommended by specialists on crisis situations.

On a psychological and social-psychological level, the campus response was to react to the psychological needs of the students who had been directly involved in the incident as well as to students and staff who had been indirectly affected by the incident. Not only did signs of psychological issues, such as denial, fear, and retriggering, emerge, as expected [15], gender and cultural group issues were also mentioned, though they were not discussed enough to be considered basic themes in our analysis. Contrary to assertions in the literature that violent behavior is often accepted in our culture, we found informants in our study to voice concern and fear about escalating violence on campus and in the community.

Faculty on campus were conspicuously silent on the incident, including the faculty senate, though we had expected this governing body to take up the issue of aberrant student or faculty behavior in their classrooms [25]. Some informants speculated that the faculty might have been passive about this issue because they were unconcerned, but another explanation might be that they were passive because they were unsure of what to do or whom to ask for assistance. From the students we failed to hear that they responded to their post-traumatic stress with "coping" strategies, such as relaxation, physical activity, and the establishment of normal routines [29]. Although the issues of gender and race surfaced in early conversations with informants, we did not find a direct discussion of these issues. As Bromet [5] comments, the sociocultural needs of populations with different mores must be considered when individuals assess reactions to trauma. In regard to the issue of gender, we did hear that females were the first students to seek out counseling at the Student Health Center. Perhaps our "near-miss" case was unique. We do not know what the reaction of the campus might have been had a death (or multiple deaths) occurred, although, according to the trauma psychologist, "the trauma of no deaths is as great as if deaths had occurred." Moreover, as with any exploratory case analysis, this case has limited generalizability [17], although thematic generalizability is certainly a possibility. The fact that our information was self-reported and that we were unable to interview all students who had been directly affected by the incident so as to not intervene in student therapy or the investigation also poses a problem.

Despite these limitations, our research provides a detailed account of a campus reaction to a violent incident with the potential for making a contribution to the literature. Events emerged during the process of reaction that could be "critical incidents" in future studies, such as the victim response, media reporting, the debriefing process, campus changes, and the evolution of a campus plan. With the scarcity of literature on campus violence related to gun incidents, this study breaks new ground by identifying themes and conceptual frameworks that could be examined in future cases. On a practical level, it can benefit campus administrators who are looking for a plan to respond to campus violence, and it focuses attention on questions that need to be addressed in such a plan. The large number of different groups of people who were affected by this particular gunman incident shows the complexity of responding to a campus crisis and should alert college personnel to the need for preparedness.

## Epilogue

As we conducted this study, we asked ourselves whether we would have had access to informants if someone had been killed. This "near-miss" incident provided a unique research opportunity, which could, however, only approximate an event in which a fatality had actually occurred. Our involvement in this study was serendipitous, for one of us had been employed by a correctional facility and therefore had direct experience with gunmen such as the individual in our case; the other was a University of Iowa graduate and thus familiar with the setting and circumstances surrounding another violent incident there in 1992. These experiences obviously affected our assessment of this

Appendix F: A Case Study

case by drawing our attention to the campus response in the first plan and to psychological reactions like fear and denial. At the time of this writing, campus discussions have been held about adapting the in-place campus emergency preparedness plan to a critical incident management team concept. Counselors have met to discuss coordinating the activities of different units in the event of another incident, and the police are working with faculty members and department staff to help identify potentially violence-prone students. We have the impression that, as a result of this case study, campus personnel see the interrelatedness and the large number of units that may be involved in a single incident. The anniversary date passed without incident or acknowledgment in the campus newspaper. As for the gunman, he is still incarcerated awaiting trial, and we wonder, as do some of the students he threatened, if he will seek retribution against us for writing up this case if he is released. The campus response to the October incident continues.

# References

Asmussen, K. J. "Weapon Possession in Public High Schools." *School Safety* (Fall 1992), 28–30.

Bird, G. W., S. M. Stith, and J. Schladale. "Psychological Resources, Coping Strategies, and Negotiation Styles as Discriminators of Violence in Dating Relationships." *Family Relations*, 40 (1991), 45–50.

Bogal-Allbritten, R., and W. Allbritten. "Courtship Violence on Campus: A Nationwide Survey of Student Affairs Professionals." *NASPA Journal*, 28 (1991), 312–18.

Boothe, J. W., T. M. Flick, S. P. Kirk, L. H. Bradley, and K. E. Keough. "The Violence at Your Door." *Executive Educator* (February 1993), 16–22.

Bromet, E. J. "Methodological Issues in the Assessment of Traumatic Events." *Journal of Applied Psychology*, 20 (1990), 1719–24.

Bushweller, K. "Guards with Guns." *American School Board Journal* (January 1993), 34–36.

Copenhaver, S., and E. Grauerholz. "Sexual Victimization among Sorority Women." *Sex Roles: A Journal of Research*, 24 (1991), 31–41.

Follingstad, D., S. Wright, S. Lloyd, and J. Sebastian. "Sex Differences in Motivations and Effects in Dating Violence." *Family Relations*, 40 (1991), 51–57.

Gordon, M. T., and S. Riger. *The Female Fear*. Urbana: University of Illinois Press, 1991.

Guba, E., and Y. Lincoln. "Do Inquiry Paradigms Imply Inquiry Methodologies?" In *Qualitative Approaches to Evaluation in Education*, edited by D. M. Fetterman. New York: Praeger, 1988.

Johnson, K. "The Tip of the Iceberg." *School Safety* (Fall 1992), 24–26.

Jones, D. J. "The College Campus as a Microcosm of U.S. Society: The Issue of Racially Motivated Violence." *Urban League Review*, 13 (1990), 129–39.

Legislative Update. "Campuses Must Tell Crime Rates." *School Safety* (Winter 1991), 31.

Long, N. J. "Managing a Shooting Incident." *Journal of Emotional and Behavioral Problems*, 1 (1992), 23–26.

Lowe, J. A. "What We Learned: Some Generalizations in Dealing with a Traumatic Event at Cokeville." Paper presented at the Annual Meeting of the National School Boards Association, San Francisco, 4–7 April 1987.

Mann, J. *Los Angeles Times Magazine*, 2 June 1992, pp. 26–27, 32, 46–47.

Merriam, S. B. *Case Study Research in Education: A Qualitative Approach*. San Francisco: Jossey-Bass, 1988.

Miles, M. B., and A. M. Huberman. *Qualitative Data Analysis: A Sourcebook of New Methods*. Beverly Hills, Calif.: Sage, 1984.

Mitchell, J. "When Disaster Strikes." *Journal of Emergency Medical Services* (January 1983), 36–39.

NSSC Report on School Safety. "Preparing Schools for Terroristic Attacks." *School Safety* (Winter 1991), 18–19.

Poland, S., and G. Pitcher. *Crisis Intervention in the Schools*. New York: Guilford, 1992.

Quimet, M. "The Polytechnique Incident and Imitative Violence against Women." *SSR*, 76 (1992), 45–47.

Roark, M. L. "Helping High School Students Assess Campus Safety." *The School Counselor*, 39 (1992), 251–56.

——. "Preventing Violence on College Campuses." *Journal of Counseling and Development*, 65 (1987), 367–70.

Roark, M. L., and E. W. Roark. "Administrative Responses to Campus Violence." Paper presented at the annual meeting of the American College Personnel Association/National Association of Student Personnel Administrators, Chicago, 15–18 March 1987.

"School Crisis: Under Control," 1991 [video]. National School Safety Center, a partnership of Pepperdine University and the United States Departments of Justice and Education.

Sherill, J. M., and D. G. Seigel (eds.). *Responding to Violence on Campus*. New Directions for Student Services, No. 47. San Francisco: Jossey-Bass, 1989.

Van Maanen, J. *Tales of the Field*. Chicago: University of Chicago Press, 1988.

Walker, G. "Crisis-Care in Critical Incident Debriefing." *Death Studies*, 14 (1990), 121–33.

Weick, K. E. "Educational Organizations as Loosely Coupled Systems." *Administrative Science Quarterly*, 21 (1976), 1–19.

Yin, R. K. *Case Study Research, Design and Methods*. Newbury Park, Calif.: Sage, 1989.

Zelikoff, W. I., and I. A. Hyman. "Psychological Trauma in the Schools: A Retrospective Study." Paper presented at the annual meeting of the National Association of School Psychologists, New Orleans, La., 4–8 March 1987.

# References

Aanstoos, C. M. (1985). The structure of thinking in chess. In A. Giorgi (Ed.), *Phenomenology and psychological research* (pp. 86–117). Pittsburgh, PA: Duquesne University Press.

Agar, M. H. (1980). *The professional stranger: An informal introduction to ethnography.* San Diego, CA: Academic Press.

Agar, M. H. (1986). *Speaking of ethnography.* Beverly Hills, CA: Sage.

Agger, B. (1991). Critical theory, poststructuralism, postmodernism: Their sociological relevance. In W. R. Scott & J. Blake (Eds.), *Annual Review of Sociology* (Vol. 17, pp. 105–131). Palo Alto, CA: Annual Reviews.

American Anthropological Association. (1967). *Statement on problems of anthropological research and ethics.* Adopted by the Council of the American Anthropological Association. Arlington, VA: American Anthropological Association.

American Psychological Association. (2001). *Publication manual of the American Psychological Association* (5th ed.). Washington, DC: American Psychological Association.

Anderson, E. H., & Spencer, M. H. (2002). Cognitive representations of AIDS: A phenomenological study. *Qualitative Health Research, 12,* 1338–1352.

Angen, M. J. (2000, May). Evaluating interpretive inquiry: Reviewing the validity debate and opening the dialogue. *Qualiative Health Research, 10,* 378–395.

Angrosino, M. V. (1989a). *Documents of interaction: Biography, autobiography, and life history in social science perspective.* Gainesville: University of Florida Press.

Angrosino, M. V. (1989b). Freddie: The personal narrative of a recovering alcoholic—Autobiography as case history. In M. V. Angrosino, *Documents of interaction: Biography, autobiography, and life history in social science perspective* (pp. 29–41). Gainesville: University of Florida Press.

Angrosino, M. V. (1994). On the bus with Vonnie Lee. *Journal of Contemporary Ethnography, 23,* 14–28.

Armstrong, D., Gosling, A., Weinman, J., & Marteau, T. (1997). The place of inter-rater reliability in qualitative research: An empirical study. *Sociology, 31,* 597–606.

Asmussen, K. J., & Creswell, J. W. (1995). Campus response to a student gunman. *Journal of Higher Education, 66,* 575–591.

Atkinson, P., Coffey, A., & Delamont, S. (2003). *Key themes in qualitative research: Continuities and changes.* Walnut Creek, CA: AltaMira.

Atkinson, P., & Hammersley, M. (1994). Ethnography and participant observation. In N. K. Denzin & Y. S. Lincoln (Eds.), *Handbook of qualitative research* (pp. 248–261). Thousand Oaks, CA: Sage.

Barbour, R. S. (2000). The role of qualitative research in broadening the "evidence base" for clinical practice. *Journal of Evaluation in Clinical Practice, 6*(2), 155–163.

Barritt, L. (1986). Human sciences and the human image. *Phenomenology and Pedagogy, 4*(3), 14–22.

Bazeley, P. (2002). The evolution of a project involving an integrated analysis of structured qualitative and quantitative data: From N3 to NVivo. *International Journal of Social Research Methodology, 5,* 229–243.

Bernard, H. R. (1994). *Research methods in anthropology: Qualitative and quantitative approaches* (2nd ed.). Thousand Oaks, CA: Sage.

Beverly, J. (2005). *Testimonio,* subalternity, and narrative authority. In N. K. Denzin & Y. S. Lincoln (Eds.), *The Sage handbook of qualitative research* (3rd ed., pp. 547–558). Thousand Oaks, CA: Sage.

Bloland, H. G. (1995). Postmodernism and higher education. *Journal of Higher Education, 66,* 521–559.

Bogdan, R. C., & Biklen, S. K. (1992). Qualitative research for education: An introduction to theory and methods. Boston: Allyn & Bacon.

Bogdan, R., & Taylor, S. (1975). *Introduction to qualitative research methods.* New York: John Wiley.

Bogdewic, S. P. (1992). Participant observation. In B. F. Crabtree & W. L. Miller (Eds.), *Doing qualitative research* (pp. 45–69). Newbury Park, CA: Sage.

Borgatta, E. F., & Borgatta, M. L. (Eds.). (1992). *Encyclopedia of sociology* (Vol. 4). New York: Macmillan.

Boyle, J., & McKay, J. (1995). "You leave your troubles at the gate": A case study of the exploitation of older women's labor and "leisure" in sport. *Gender & Society, 9,* 556–575.

Brickhous, N., & Bodner, G. M. (1992). The beginning science teacher: Classroom narratives of convictions and constraints. *Journal of Research in Science Teaching, 29,* 471–485.

Brown, J., Sorrell, J. H., McClaren, J. & Creswell, J. W. (2006). Waiting for a liver transplant. *Qualitative Health Research, 16*(1), 119–136.

Burrell, G., & Morgan, G. (1979). *Sociological paradigms and organizational analysis.* London: Heinemann.

Carspecken, P. F., & Apple, M. (1992). Critical qualitative research: Theory, methodology, and practice. In M. L. LeCompte, W. L. Millroy, & J. Preissle (Eds.), *The handbook of qualitative research in education* (pp. 507–553). San Diego, CA: Academic Press.

Carter, K. (1993). The place of a story in the study of teaching and teacher education. *Educational Researcher, 22,* 5–12, 18.

Casey, K. (1995/1996). The new narrative research in education. *Review of Research in Education, 21,* 211–253.

Charmaz, K. (1983). The grounded theory method: An explication and interpretation. In R. Emerson (Ed.), *Contemporary field research* (pp. 109–126). Boston: Little, Brown.

Charmaz, K. (2005). Grounded theory in the 21st century: Applications for advancing social justice studies. In N. K. Denzin & Y. S. Lincoln, *The Sage handbook of qualitative research* (3rd ed., pp. 507–536). Thousand Oaks, CA: Sage.

Charmaz, K. (2006). *Constructing grounded theory*. London: Sage.

Chase, S. (2005). Narrative inquiry: Multiple lenses, approaches, voices. In N. K. Denzin and Y. S. Lincoln (Eds.), *The Sage handbook of qualitative research* (3rd ed., pp. 651–680). Thousand Oaks, CA: Sage.

Cheek, J. (2004). At the margins? Discourse analysis and qualitative research. *Qualitative Health Research, 14*, 1140–1150.

Chenitz, W. C., & Swanson, J. M. (1986). *From practice to grounded theory: Qualitative research in nursing*. Menlo Park, CA: Addison-Wesley.

Cherryholmes, C. H. (1992). Notes on pragmatism and scientific realism. *Educational Researcher, 14*, 13–17.

Clandinin, D. J. (Ed.). (2006). *Handbook of narrative inquiry: Mapping a methodology*. Thousand Oaks, CA: Sage.

Clandinin, D. J., & Connelly, F. M. (2000). *Narrative inquiry: Experience and story in qualitative research*. San Francisco: Jossey-Bass.

Clarke, A. E. (2005). *Situational analysis: Grounded theory after the postmodern turn*. Thousand Oaks, CA: Sage.

Clifford, J. (1970). *From puzzles to portraits: Problems of a literary biographer*. Chapel Hill: University of North Carolina Press.

Clifford, J., & Marcus, G. E. (Eds.). (1986). *Writing culture: The poetics and politics of ethnography*. Berkeley: University of California Press.

Colaizzi, P. F. (1978). Psychological research as the phenomenologist views it. In R. Vaile & M. King (Eds.), *Existential phenomenological alternatives for psychology* (pp. 48–71). New York: Oxford University Press.

Cole, A. (1994, April). *Doing life history research in theory and in practice*. Paper prepared for the annual meeting of the American Educational Research Association, New Orleans, LA.

Connelly, F. M., & Clandinin, D. J. (1990). Stories of experience and narrative inquiry. *Educational Researcher, 19*(5), 2–14.

Conrad, C. F. (1978). A grounded theory of academic change. *Sociology of Education, 51*, 101–112.

Corbin, J., & Morse, J. M. (2003). The unstructured interactive interview: Issues of reciprocity and risks when dealing with sensitive topics. *Qualitative Inquiry, 9*, 335–354.

Corbin, J., & Strauss, A. (1990). Grounded theory research: Procedures, canons, and evaluative criteria. *Qualitative Sociology, 13*(1), 3–21.

Cortazzi, M. (1993). *Narrative analysis*. London: Falmer Press.

Crabtree, B. F., & Miller, W. L. (1992). *Doing qualitative research*. Newbury Park, CA: Sage.

Creswell, J. W. (1994). *Research design: Qualitative and quantitative approaches.* Thousand Oaks, CA: Sage.

Creswell, J. W. (2003). *Research design: Qualitative, quantitative, and mixed methods approaches.* (2nd ed.). Thousand Oaks, CA: Sage.

Creswell, J. W. (2005). *Educational research: Planning, conducting, and evaluating quantitative and qualitative research.* (2nd ed.). Upper Saddle River, NJ: Pearson Education.

Creswell, J. W., & Brown, M. L. (1992). How chairpersons enhance faculty research: A grounded theory study. *Review of Higher Education, 16*(1), 41–62.

Creswell, J. W., & Maietta, R. C. (2002). Qualitative research. In D. C. Miller & N. J. Salkind (Eds.), *Handbook of social research* (pp. 143–184). Thousand Oaks, CA: Sage.

Creswell, J. W., & Miller, D. L. (2000). Determining validity in qualitative inquiry. *Theory Into Practice 39,* 124–130.

Creswell, J. W., & Plano Clark, V. L. (2007). *Designing and conducting mixed methods research.* Thousand Oaks, CA: Sage.

Crotty, M. (1998). *The foundations of social research: Meaning and perspective in the research process.* London: Sage.

Cunningham, J. W., & Fitzgerald, J. (1996). Epistemology and reading. *Reading Research Quarterly, 31*(1), 36–60.

Czarniawska, B. (2004). *Narratives in social science research.* Thousand Oaks, CA: Sage.

Daiute, C., & Lightfoot, C. (Eds.). (2004). *Narrative analysis: Studying the development of individuals in society.* Thousand Oaks, CA: Sage.

Damschroder, L. J. (2006, March). Personal communication, VA Ann Arbor Health Care System, Center for Practice Management and Outcomes Research, Ann Arbor, Michigan.

Davidson, F. (1996). *Principles of statistical data handling.* Thousand Oaks, CA: Sage.

Deem, R. (2002). Talking to manager-academics: Methodological dilemmas and feminist research strategies. *Sociology, 36*(4), 835–855.

Denzin, N. K. (1989a). *Interpretive biography.* Newbury Park, CA: Sage.

Denzin, N. K. (1989b). *Interpretive interactionism.* Newbury Park, CA: Sage.

Denzin, N. K., & Lincoln, Y. S. (1994). *The handbook of qualitative research.* Thousand Oaks, CA: Sage.

Denzin, N. K., & Lincoln, Y. S. (2000). *Handbook of qualitative research* (2nd ed.). Thousand Oaks, CA: Sage.

Denzin, N. K., & Lincoln, Y. S. (2005). *The Sage handbook of qualitative research* (3rd ed.). Thousand Oaks, CA: Sage.

Dey, I. (1993). *Qualitative data analysis: A user-friendly guide for social scientists.* London: Routledge.

Dey, I. (1995). Reducing fragmentation in qualitative research. In U. Keele (Ed.), *Computer-aided qualitative data analysis* (pp. 69–79). Thousand Oaks, CA: Sage.

Dukes, S. (1984). Phenomenological methodology in the human sciences. *Journal of Religion and Health, 23*(3), 197–203.

Edel, L. (1984). *Writing lives: Principia biographica.* New York: Norton.

Edwards, L. V. (2006). Perceived social support and HIV/AIDS medication adherence among African American women. *Qualitative Health Research, 16,* 679–691.

Eisner, E. W. (1991). *The enlightened eye: Qualitative inquiry and the enhancement of educational practice.* New York: Macmillan.

Elliot, J. (2005). *Using narrative in social research: Qualitative and quantitative approaches.* London: Sage.

Ellis, C. (1993). "There are survivors": Telling a story of sudden death. *The Sociological Quarterly, 34,* 711–738.

Ellis, C. (2004). *The ethnographic it: A methodological novel about autoethnography.* Walnut Creek, CA: AltaMira.

Ely, M. (2006). Re-forming re-presentations. In D. J. Clandinin (Ed.), *Handbook of narrative research.* Thousand Oaks, CA: Sage.

Ely, M., Anzul, M., Friedman, T., Garner, D., & Steinmetz, A. C. (1991). *Doing qualitative research: Circles within circles.* New York: Falmer Press.

Emerson, R. M., Fretz, R. I., & Shaw, L. L. (1995). *Writing ethnographic fieldnotes.* Chicago: University of Chicago Press.

Erlandson, D. A., Harris, E. L., Skipper, B. L., & Allen, S. D. (1993). *Doing naturalistic inquiry: A guide to methods.* Newbury Park, CA: Sage.

Ezeh, P. J. (2003). Integration and its challenges in participant observation. *Qualitative Research, 3,* 191–205.

Fay, B. (1987). *Critical social science.* Ithaca, NY: Cornell University Press.

Ferguson, M., & Wicke, J. (1994). *Feminism and postmodernism.* Durham, NC: Duke University Press.

Fetterman, D. M. (1998). *Ethnography: Step by step* (2nd ed.). Thousand Oaks, CA: Sage.

Fischer, C. T., & Wertz, F. J. (1979). An empirical phenomenology study of being criminally victimized. In A. Giorgi, R. Knowles, & D. Smith (Eds.), *Duquesne studies in phenomenological psychology* (Vol. 3, pp. 135–158). Pittsburgh, PA: Duquesne University Press.

Flinders, D. J., & Mills, G. E. (1993). *Theory and concepts in qualitative research.* New York: Teachers College Press.

Foucault, M. (1972) (A. M. Sheridan Smith, Trans.). *The archeology of knowledge and the discourse on language.* New York: Harper.

Fox-Keller, E. (1985). *Reflections on gender and science.* New Haven, CT: Yale University Press.

Gamson, J. (2000). Sexualities, queer theory and qualitative research. In N. K. Denzin & Y. S. Lincoln (Eds.), *Handbook of qualitative research* (2nd ed.). London: Sage.

Geertz, C. (1973). Deep play: Notes on the Balinese cockfight. In C. Geertz (Ed.), *The interpretation of cultures: Selected essays* (pp. 412–435). New York: Basic Books.

Geiger, S. N. G. (1986). Women's life histories: Method and content. *Signs: Journal of Women in Culture and Society, 11,* 334–351.

Gergen, K. (1994). *Realities and relationships: Soundings in social construction.* Cambridge, MA: Harvard University Press.

Gilchrist, V. J. (1992). Key informant interfviews. In B. F. Crabtree & W. L. Miller (Eds.), *Doing qualitative research* (pp. 70–89). Newbury Park, CA: Sage.

Gilgun, J. F. (2005). "Grab" and good science: Writing up the results of qualitative research. *Qualitative Health Research, 15,* 256–262.

Gioia, D. A., & Pitre, E. (1990). Multiparadigm perspectives on theory building. *Management Review, 15,* 584–602.

Giorgi, A. (Ed.). (1985). *Phenomenology and psychological research.* Pittsburgh, PA: Duquesne University Press.

Giorgi, A. (1994). A phenomenological perspective on certain qualitative research methods. *Journal of Phenomenological Psychology, 25,* 190–220.

Glaser, B. G. (1978). *Theoretical sensitivity.* Mill Valley, CA: Sociology Press.

Glaser, B. G. (1992). *Basics of grounded theory analysis.* Mill Valley, CA: Sociology Press.

Glaser, B., & Strauss, A. (1965). *Awareness of dying.* Chicago: Aldine.

Glaser, B., & Strauss, A. (1967). *The discovery of grounded theory.* Chicago: Aldine.

Glaser, B., & Strauss, A. (1968). *Time for dying.* Chicago: Aldine.

Glesne, C., & Peshkin, A. (1992). *Becoming qualitative researchers: An introduction.* White Plains, NY: Longman.

Goffman. E. (1989). On fieldwork. *Journal of Contemporary Ethnography, 18,* 123–132.

Grigsby, K. A., & Megel, M. E. (1995). Caring experiences of nurse educators. *Journal of Nursing Research, 34,* 411–418.

Gritz, J. I. (1995). *Voices from the classroom: Understanding teacher professionalism.* Unpublished manuscript, Administration, Curriculum, and Instruction, University of Nebraska-Lincoln.

Guba, E. G. (1990). The alternative paradigm dialog. In E. G. Guba (Ed.), *The paradigm dialog* (pp. 17–30). Newbury Park, CA: Sage.

Guba, E. G., & Lincoln, Y. S. (1988). Do inquiry paradigms imply inquiry methodologies? In D. M. Fetterman (Ed.), *Qualitative approaches to evaluation in education* (pp. 89–115). New York: Praeger.

Guba, E. G., & Lincoln, Y. S. (1989). *Fourth generation evaluation.* Newbury Park, CA: Sage.

Guba, E. G., & Lincoln, Y. S. (2005). Paradigmatic controversies, contradictions, and emerging confluences. In N. K. Denzin & Y. S. Lincoln, *The Sage handbook of qualitative research* (3rd ed., pp. 191–215). Thousand Oaks, CA: Sage.

Gubrium, J. F., & Holstein, J. A. (2003). *Postmodern interviewing.* Thousand Oaks, CA: Sage.

Haenfler, R. (2004). Rethinking subcultural resistance: Core values of the straight edge movement. *Journal of Contemporary Ethnography, 33,* 406–436.

Hamel, J., Dufour, S., & Fortin, D. (1993). *Case study methods.* Newbury Park, CA: Sage.

Hammersley, M., & Atkinson, P. (1995). *Ethnography: Principles in practice* (2nd ed.). New York: Routledge.

Harding, S. (1987). *Feminism and methodology.* Bloomington: Indiana University Press.

Harper, W. (1981). The experience of leisure. *Leisure Sciences, 4*, 113–126.

Harris, C. (1993). Whiteness as property. *Harvard Law Review, 106,* 1701–1791.

Harris, M. (1968). *The rise of anthropological theory: A history of theories of culture.* New York: T. Y. Crowell.

Hatch, J. A. (2002). *Doing qualitative research in education settings.* Albany: State University of New York Press.

Heilbrun, C. G. (1988). *Writing a woman's life.* New York: Ballantine.

Heinrich, K. T. (1995). Doctoral advisement relationships between women. *Journal of Higher Education, 66,* 447–469.

Heron, J., & Reason, P. (1997). A participatory inquiry paradigm. *Qualitative Inquiry, 3,* 274–294.

Hill, B., Vaughn, C., & Harrison, S. B. (1995, September/October). Living and working in two worlds: Case studies of five American Indian women teachers. *The Clearinghouse, 69*(1), 42–48.

Hoshmand, L. L. S. T. (1989). Alternative research paradigms: A review and teaching proposal. *The Counseling Psychologist, 17*(1), 3–79.

Howe, K., & Eisenhardt, M. (1990). Standards for qualitative (and quantitative) research: A prolegomenon. *Educational Researcher, 19*(4), 2–9.

Huber, J., & Whelan, K. (1999). A marginal story as a place of possibility: Negotiating self on the professional knowledge landscape. *Teaching and Teacher Education, 15,* 381–396.

Huberman, A. M., & Miles, M. B. (1994). Data management and analysis methods. In N. K. Denzin & Y. S. Lincoln (Eds.), *Handbook of qualitative research* (pp. 428–444). Thousand Oaks, CA: Sage.

Husserl, E. (1931). *Ideas: General introduction to pure phenomenology* (D. Carr, Trans.). Evanston, IL: Northwestern University Press.

Husserl, E. (1970). *The crisis of European sciences and transcendental phenomenology* (D. Carr, Trans.). Evanston, IL: Northwestern University Press.

Jacob, E. (1987). Qualitative research traditions: A review. *Review of Educational Research, 57,* 1–50.

Jorgensen, D. L. (1989). *Participant observation: A methodology for human studies.* Newbury Park, CA: Sage.

Josselson, R., & Lieblich, A. (Eds.). (1993). *The narrative study of lives* (Vol. 1). Newbury Park, CA: Sage.

Karen, C. S. (1990, April). *Personal development and the pursuit of higher education: An exploration of interrelationships in the growth of self-identity in returning women students—summary of research in progress.* Paper presented at the annual meeting of the American Educational Research Association, Boston.

Kearney, M. H., Murphy, S., & Rosenbaum, M. (1994). Mothering on crack cocaine: A grounded theory analysis. *Social Science Medicine, 38*(2), 351–361.

Kelle, E. (Ed.). (1995). *Computer-aided qualitative data analysis.* Thousand Oaks, CA: Sage.

Kemmis, S., & Wilkinson, M. (1998). Participatory action research and the study of practice. In B. Atweh, S. Kemmis, & P. Weeks (Eds.), *Action research in practice: Partnerships for social justice in education* (pp. 21–36). New York: Routledge.

Kerlinger, F. N. (1964). *Foundations of behavioral research: Educational and psychological inquiry.* New York: Holt, Rinehart & Winston.

Kerlinger, F. N. (1979). *Behavioral research: A conceptual approach.* New York: Holt, Rinehart & Winston.

Kidder, L. (1982). Face validity from multiple perspectives. In D. Brinberg & L. Kidder (Eds.), *New directions for methodology of social and behavioral science: Forms of validity in research* (pp. 41–57). San Francisco: Jossey-Bass.

Kincheloe, J. L. (1991). *Teachers as researchers: Qualitative inquiry as a path of empowerment.* London: Falmer Press.

Koro-Ljungberg, M., & Greckhamer, T. (2005). Strategic turns labeled "ethnography": From description to openly ideological production of cultures. *Qualitative Research, 5*(3), 285–306.

Krueger, R. A. (1994). *Focus groups: A practical guide for applied research* (2nd ed.). Thousand Oaks, CA: Sage.

Kus, R. J. (1986). From grounded theory to clinical practice: Cases from gay studies research. In W. C. Chenitz & J. M. Swanson (Eds.), *From practice to grounded theory* (pp. 227–240). Menlo Park, CA: Addison-Wesley.

Kvale, S. (1996). *InterViews: An introduction to qualitative research interviewing.* Thousand Oaks, CA: Sage.

Kvale, S. (2006). Dominance through interviews and dialogues. *Qualitative Inquiry, 12,* 480–500.

Labaree, R. V. (2002). The risk of "going observationalist": Negotiating the hidden dilemmas of being an insider participant observer. *Qualitative Research, 2,* 97–122.

Ladson-Billings, G., & Donnor, J. (2005). The moral activist role in critical race theory scholarship. In N. K. Denzin & Y. S. Lincoln (Eds.), *The Sage handbook of qualitative research* (3rd ed., pp. 279–201). Thousand Oaks, CA: Sage.

Lancy, D. F. (1993). *Qualitative research in education: An introduction to the major traditions.* New York: Longman.

Landis, M. M. (1993). *A theory of interaction in the satellite learning classroom.* Unpublished doctoral dissertation, University of Nebraska-Lincoln.

Lather, P. (1991). *Getting smart: Feminist research and pedagogy with/in the postmodern.* New York: Routledge.

Lather, P. (1993). Fertile obsession: Validity after poststructuralism. *Sociological Quarterly, 34,* 673–693.

Lauterbach, S. S. (1993). In another world: A phenomenological perspective and discovery of meaning in mothers' experience with death of a wished-for baby: Doing phenomenology. In P. L. Munhall & C. O. Boyd (Eds.), *Nursing research: A qualitative perspective* (pp. 133–179). New York: National League for Nursing Press.

LeCompte, M. D., & Goetz, J. P. (1982). Problems of reliability and validity in ethnographic research. *Review of Educational Research, 51,* 31–60.

LeCompte, M. D., Millroy, W. L., & Preissle, J. (1992). *The handbook of qualitative research in education.* San Diego, CA: Academic Press.

LeCompte, M. D., & Schensul, J. J. (1999). *Designing and conducting ethnographic research* (Ethnographer's toolkit, Vol. 1). Walnut Creek, CA: AltaMira.

Leipert, B. D., & Reutter, L. (2005). Developing resilience: How women maintain their health in northern geographically isolated settings. *Qualitative Health Research, 15,* 49–65.

LeVasseur, J. J. (2003). The problem of bracketing in phenomenology. *Qualitative Health Research, 13*(3), 408–420.

Lieblich, A., Tuval-Mashiach, R., & Zilber, T. (1998). *Narrative research: Reading, analysis, and interpretation.* Thousand Oaks, CA: Sage.

Lincoln, Y. S. (1995). Emerging criteria for quality in qualitative and interpretive research. *Qualitative Inquiry, 1,* 275–289.

Lincoln, Y. S., & Guba, E. G. (1985). *Naturalistic inquiry.* Beverly Hills, CA: Sage.

Lincoln, Y. S., & Guba, E. G. (2000). Paradigmatic controversies, contradictions, and emerging confluences. In N. K. Denzin & Y. S. Lincoln (Eds.), *Handbook of qualitative research* (2nd ed., pp. 163–188). Thousand Oaks, CA: Sage.

Lipson, J. G. (1994). Ethical issues in ethnography. In J. M. Morse (Eds.), *Critical issues in qualitative research methods* (pp. 333–355). Thousand Oaks, CA: Sage.

Lofland, J. (1974). Styles of reporting qualitative field research. *American Sociologist, 9,* 101–111.

Lofland, J., & Lofland, L. H. (1995). *Analyzing social settings: A guide to qualitative observation and analysis* (3rd ed.). Belmont, CA: Wadsworth.

Lomask, M. (1986). *The biographer's craft.* New York: Harper & Row.

Lopez, K. A., & Willis, D. G. (2004). Descriptive versus interpretive phenomenology: Their contributions to nursing knowledge. *Qualitative Health Reseaerch, 14*(5), 726–735.

Luck, L., Jackson, D., & Usher, K. (2006). Case study: A bridge across the paradigms. *Nursing Inquiry, 13,* 103–109.

Madison, D. S. (2005). *Critical ethnography: Methods, ethics, and performance.* Thousand Oaks, CA: Sage.

Marshall, C., & Rossman, G. B. (2006). *Designing qualitative research* (4th ed.). Thousand Oaks, CA: Sage.

Martin, J. (1990). Deconstructing organizational taboos: The suppression of gender conflict in organizations. *Organization Science, 1,* 339–359.

Mastera, G. (1995). *The process of revising general education curricula in three private baccalaureate colleges.* Unpublished manuscript, Administration, Curriculum, and Instruction, University of Nebraska-Lincoln.

Maxwell, J. (2005). *Qualitative research design: An interactive approach* (2nd ed.). Thousand Oaks, CA: Sage.

May, K. A. (1986). Writing and evaluating the grounded theory research report. In W. C. Chenitz & J. M. Swanson (Eds.), *From practice to grounded theory* (pp. 146–154). Menlo Park, CA: Addison-Wesley.

McCracken, G. (1988). *The long interview.* Newbury Park, CA: Sage.

McVea, K., Harter, L., McEntarffer, R., & Creswell, J. W. (1999). Phenomenological study of student experiences with tobacco use at City High School. *High School Journal, 82*(4), 209–222.

Merleau-Ponty, M. (1962). *Phenomenology of perception* (C. Smith, Trans.). London: Routledge & Kegan Paul.

Merriam, S. (1988). *Case study research in education: A qualitative approach.* San Francisco: Jossey-Bass.

Merriam, S. B. (1998). *Qualitative research and case study applications in education.* San Francisco: Jossey-Bass.

Mertens, D. M. (1998). *Research methods in education and psychology: Integrating diversity with quantitative and qualitative approaches.* Thousand Oaks, CA: Sage.

Mertens, D. M. (2003). Mixed methods and the politics of human research: The transformative-emancipatory perspective. In A. Tashakkori & C. Teddlie (Eds.), *Handbook of mixed methods in social & behavioral research* (pp. 135–164). Thousand Oaks, CA: Sage.

Miles, M. B., & Huberman, A. M. (1994). *Qualitative data analysis: A sourcebook of new methods* (2nd ed.). Thousand Oaks, CA: Sage.

Miller, D. W., Creswell, J. W., & Olander, L. S. (1998). Writing and retelling multiple ethnographic tales of a soup kitchen for the homeless. *Qualitative Inquiry, 4*(4), 469–491.

Miller, T. (2000). Losing the plot: Narrative construction and longitudinal childbirth research. *Qualiative Health Research, 10,* 309–323.

Miller, W. L., & Crabtree, B. F. (1992). Primary care research: A multimethod typology and qualitative road map. In B. F. Crabtree & W. L. Miller (Eds.), *Doing qualitative research* (pp. 3–28). Newbury Park, CA: Sage.

Morgan, D. L. (1988). *Focus groups as qualitative research.* Newbury Park, CA: Sage.

Morrow, R. A., & Brown, D. D. (1994). *Critical theory and methodology.* Thousand Oaks, CA: Sage.

Morrow, S. L., & Smith, M. L. (1995). Constructions of survival and coping by women who have survived childhood sexual abuse. *Journal of Counseling Psychology, 42,* 24–33.

Morse, J. M. (1994). Designing funded qualitative research. In N. K. Denzin & Y. S. Lincoln (Eds.), *Handbook of qualitative research* (pp. 220–235). Thousand Oaks, CA: Sage.

Morse, J. M., & Field, P. A. (1995). *Qualitative research methods for health professionals* (2nd ed.). Thousand Oaks, CA: Sage.

Morse, J. M., & Richards, L. (2002). *README FIRST for a user's guide to qualitative methods.* Thousand Oaks, CA: Sage.

Moss, P. (2006). Emergent methods in feminist research. In S. N. Hesse-Biber (Ed.), *Handbook of feminist research methods.* Thousand Oaks, CA: Sage.

Moustakas, C. (1994). *Phenomenological research methods.* Thousand Oaks, CA: Sage.

Munhall, P. L., & Oiler, C. J. (Eds.). (1986). *Nursing research: A qualitative perspective.* Norwalk, CT: Appleton-Century-Crofts.

Murphy, J. P. (with Rorty, R.). (1990). *Pragmatism: From Peirce to Davidson.* Boulder, CO: Westview Press.

Natanson, M. (Ed.). (1973). *Phenomenology and the social sciences.* Evanston, IL: Northwestern University Press.

National Academy of Sciences. (2000). *Scientific research in education.* Washington, DC: National Research Council.

Nelson, L. W. (1990). Code-switching in the oral life narratives of African-American women: Challenges to linguistic hegemony. *Journal of Education, 172,* 142–155.

Neuman, W. L. (2000). *Social research methods: Qualitative and quantitative approaches* (4th ed.). Boston: Allyn & Bacon.

Nielsen, J. M. (Ed.). (1990). *Feminist research methods: Exemplary readings in the social sciences.* Boulder, CO: Westview Press.

Nieswiadomy, R. M. (1993). *Foundations of nursing research* (2nd ed.). Norwalk, CT: Appleton & Lange.

Nunkoosing, K. (2005). The problems with interviews. *Qualitative Health Research, 15,* 698–706.

Oiler, C. J. (1986). Phenomenology: The method. In P. L. Munhall & C. J. Oiler (Eds.), *Nursing research: A qualitative perspective* (pp. 69–82). Norwalk, CT: Appleton-Century-Crofts.

Olesen, V. (1994). Feminisms and models of qualitative research. In N. K. Denzin & Y. S. Lincoln (Eds.), *Handbook of qualitative research* (pp. 158–174). Thousand Oaks, CA: Sage.

Olesen, V. (2005). Early millennial feminist qualitative research: Challenges and contours. In N. K. Denzin & Y. S. Lincoln (Eds.), *The Sage handbook of qualitative research* (3rd ed., pp. 235–278). Thousand Oaks, CA: Sage.

Ollerenshaw, J. A., & Creswell, J. W. (2002). Narrative research: A comparison of two restorying data analysis approaches. *Qualitative Inquiry, 8,* 329–347.

Olson, L. N. (2004). The role of voice in the (re)construction of a battered woman's identity: An autoethnography of one woman's experiences of abuse. *Women's Studies in Communication, 27,* 1–33.

Parker, L., & Lynn, M. (2002). What race got to do with it? Critical race theory's conflicts with and connections to qualitative research methodology and epistemology. *Qualitative Inquiry, 8*(1), 7–22.

Patton, M. Q. (1980). *Qualitative evaluation methods.* Beverly Hills, CA: Sage.

Patton, M. Q. (1990). *Qualitative evaluation and research methods.* Newbury Park, CA: Sage.

Personal Narratives Group. (1989). *Interpreting women's lives.* Bloomington: Indiana University Press.

Phillips, D. C., & Burbules, N. C. (2000). *Postpositivism and educational research.* Lanham, MD: Rowman & Littlefield.

Pink, S. (2001). *Doing visual ethnography.* London: Sage.

Pinnegar, S., & Daynes, J. G. (2006). Locating narrative inquiry historically: Thematics in the turn to narrative. In D. J. Clandinin (Ed.), *Handbook of narrative inquiry.* Thousand Oaks, CA: Sage.

Plummer, K. (1983). *Documents of life: An introduction to the problems and literature of a humanistic method.* London: George Allen & Unwin.

Plummer, K. (2005). Critical humanism and queer theory: Living with the tensions. In N. K. Denzin & Y. S. Lincoln (Eds.), *The Sage handbook of qualitative research* (3rd ed., pp. 357–373). Thousand Oaks, CA: Sage.

Polkinghorne, D. E. (1989). Phenomenological research methods. In R. S. Valle & S. Halling (Eds.), *Existential-phenomenological perspectives in psychology* (pp. 41–60). New York: Plenum Press.

Polkinghorne, D. E. (1995). Narrative configuration in qualitative analysis. *Qualitative Studies in Education, 8,* 5–23.

Prior, L. (2003). *Using documents in social research.* London: Sage.

Redfield, R. (1963). *The little community: Viewpoints for the study of a human whole.* Chicago: University of Chicago Press.

Reinharz, S. (1992). *Feminist methods in social research.* New York: Oxford University Press.

Rhoads, R. A. (1995). Whales tales, dog piles, and beer goggles: An ethnographic case study of fraternity life. *Anthropology and Education Quarterly, 26,* 306–323.

Richards, L., & Morse., J. M. (2007). *README FIRST for a users guide to qualitative methods* (Second edition). Thousand Oaks, CA: Sage.

Richardson, L. (1990). *Writing strategies: Reaching diverse audiences.* Newbury Park, CA: Sage.

Richardson, L. (1994). Writing: A method of inquiry. In N. K. Denzin & Y. S. Lincoln (Eds.), *Handbook of qualitative research* (pp. 516–529). Thousand Oaks, CA: Sage.

Richardson, L., & St. Pierre, E. A. (2005). Writing: A method of inquiry. In N. K. Denzin & Y. S. Lincoln (Eds.), *The Sage handbook of qualitative research* (3rd ed., pp. 959–978). Thousand Oaks, CA: Sage.

Riemen, D. J. (1986). The essential structure of a caring interaction: Doing phenomenology. In P. M. Munhall & C. J. Oiler (Eds.), *Nursing research: A qualitative perspective* (pp. 85–105). Norwalk, CT: Appleton-Century-Crofts.

Riessman, C. K. (1993). *Narrative analysis.* Newbury Park, CA: Sage.

Roman, L. G. (1992). The political significance of other ways of narrating ethnography: A feminist materialist approach. In M. L. LeCompte, W. L. Millroy, & J. Preissle (Eds.), *The handbook of qualitative research in education* (pp. 555–594). San Diego, CA: Academic Press.

Rorty, R. (1983). *Consequences of pragmatism.* Minneapolis: University of Minnesota Press.

Rorty, R. (1990). Pragmatism as anti-representationalism. In J. P. Murphy (Ed.), *Pragmatism: From Pierce to Davidson* (pp. 1–6). Boulder, CO: Westview Press.

Rosenau, P. M. (1992). *Post-modernism and the social sciences: Insights, inroads, and intrusions.* Princeton, NJ: Princeton University Press.

Rossman, G. B, & Wilson, B. L. (1985). Numbers and words: Combining quantitative and qualitative methods in a single large-scale evaluation study. *Evaluation Review, 9*(5), 627–643.

Roulston, K., deMarrais, K., & Lewis, J. B. (2003). Learning to interview in the social sciences. *Qualitative Inquiry, 9,* 643–668.

Rubin, H. J., & Rubin, I. S. (1995). *Qualitative interviewing.* Thousand Oaks, CA: Sage.

Sampson, H. (2004). Navigating the waves: The usefulness of a pilot in qualitative research. *Qualitative Research, 4,* 383–402.

Sanjek, R. (1990). *Fieldnotes: The makings of anthropology.* Ithaca, NY: Cornell University Press.

Schwandt, T. A. (2001). *Dictionary of qualitative inquiry* (2nd ed.). Thousand Oaks, CA: Sage.

Silverman, D. (2005). *Doing qualitative research: A practical handbook* (2nd ed.). London: Sage.

Slife, B. D., & Williams, R. N. (1995). *What's behind the research? Discovering hidden assumptions in the behavioral sciences.* Thousand Oaks, CA: Sage.

Smith, L. M. (1987). The voyage of the Beagle: Field work lessons from Charles Darwin. *Educational Administration Quarterly, 23*(3), 5–30.

Smith, L. M. (1994). Biographical method. In N. K. Denzin & Y. S. Lincoln (Eds.), *Handbook of qualitative research* (pp. 286–305). Thousand Oaks, CA: Sage.

Solorzano, D. G., & Yosso, T. J. (2002). Critical race methodology: Counter-storytelling as an analytical framework for education research. *Qualitative Inquiry, 8*(1), 23–44.

Sparkes, A. C. (1992). The paradigms debate: An extended review and celebration of differences. In A. C. Sparkes (Ed.), *Research in physical education and sport: Exploring alternative visions* (pp. 9–60). London: Falmer Press.

Spiegelberg, H. (1982). *The phenomenological movement* (3rd ed.). The Hague, Netherlands: Martinus Nijhoff.

Spindler, G., & Spindler, L. (1987). Teaching and learning how to do the ethnography of education. In G. Spindler & L. Spindler (Eds.), *Interpretive ethnography of education: At home and abroad* (pp. 17–33). Hillsdale, NJ: Lawrence Erlbaum.

Spradley, J. P. (1979). *The ethnographic interview.* New York: Holt, Rinehart & Winston.

Spradley, J. P. (1980). *Participant observation.* New York: Holt, Rinehart & Winston.

Stake, R. (1995). *The art of case study research.* Thousand Oaks, CA: Sage.

Stake, R. E. (2005). Qualitative case studies. In N. K. Denzin & Y. S. Lincoln (Eds.), *The Sage handbook of qualitative research* (3rd ed., pp. 443–466). Thousand Oaks, CA: Sage.

Stake, R. E. (2006). *Multiple case study analysis.* New York: Guilford Press.

Stewart, A. J. (1994). Toward a feminist strategy for studying women's lives. In C. E. Franz & A. J. Stewart (Eds.), *Women creating lives: Identities, resilience and resistance* (pp. 11–35). Boulder, CO: Westview Press.

Stewart, D., & Mickunas, A. (1990). *Exploring phenomenology: A guide to the field and its literature* (2nd ed.). Athens: Ohio University Press.

Stewart, D. W., & Shamdasani, P. N. (1990). *Focus groups: Theory and practice.* Newbury Park, CA: Sage.

Stewart, K., & Williams, M. (2005). Researching online populations: The use of online focus groups for social research. *Qualitative Research, 5,* 395–416.

Strauss, A. (1987). *Qualitative analysis for social scientists.* New York: Cambridge University Press.

Strauss, A., & Corbin, J. (1990). *Basics of qualitative research: Grounded theory procedures and techniques.* Newbury Park, CA: Sage.

Stringer, E. T. (1993). Socially responsive educational research: Linking theory and practice. In D. J. Flinders & G. E. Mills (Eds.), *Theory and concept in qualitative research: Perspectives from the field* (pp. 141–162). New York: Teachers College Press.

Sudnow, D. (1978). *Ways of the hand.* New York: Knopf.

Suoninen, E., & Jokinen, A. (2005). Persuasion in social work interviewing. *Qualitative Social Work, 4,* 469–487.

Swingewood, A. (1991). *A short history of sociological thought.* New York: St. Martin's Press.

Tashakkori, A., & Teddlie, C. (Eds.). (2003). *Handbook of mixed methods in the social and behavioral sciences.* Thousand Oaks, CA: Sage.

Taylor, S. J., & Bogdan, R. (1998). *Introduction to qualitative research methods: A guidebook and resource* (3rd ed.). New York: John Wiley.

Tesch, R. (1988). *The contribution of a qualitative method: Phenomenological research.* Unpublished manuscript, Qualitative Research Management, Santa Barbara, CA.

Tesch, R. (1990). *Qualitative research: Analysis types and software tools.* Bristol, PA: Falmer Press.

Thomas, J. (1993). *Doing critical ethnography.* Newbury Park, CA: Sage.

Thomas, W. I., & Znaniecki, F. (1958). *The Polish peasant in Europe and America.* New York: Dover. (Originally published 1918–1920)

Tierney, W. G. (1995). (Re)presentation and voice. *Qualitative Inquiry, 1,* 379–390.

Tierney, W. G. (1997). Academic outlaws: Queer theory and cultural studies in the academy. London: Sage.

Trujillo, N. (1992). Interpreting (the work and the talk of) baseball. *Western Journal of Communication, 56,* 350–371.

Turner, W. (2000). *A genealogy of queer theory.* Philadelphia: Temple University Press.

Valerio, M. (1995). *The lived experience of teenagers who are pregnant: A grounded theory study.* Unpublished manuscript, Administration, Curriculum, and Instruction, University of Nebraska-Lincoln.

Van Kaam, A. (1966). *Existential foundations of psychology.* Pittsburgh, PA: Duquesne University Press.

Van Maanen, J. (1988). *Tales of the field: On writing ethnography.* Chicago: University of Chicago Press.

van Manen, M. (1990). *Researching lived experience: Human science for an action sensitive pedagogy.* London, Ontario, Canada: The University of Western Ontario.

van Manen, M. (2006). Writing qualitatively, or the demands of writing. *Qualitative Health Research, 16,* 713–722.

Wallace, A. F. C. (1970). *Culture and personality* (2nd ed.). New York: Random House.

Watson, K. (2005). Queer theory. *Group Analysis, 38*(1), 67–81

Weis, L., & Fine, M. (2000). *Speed bumps: A student-friendly guide to qualitative research.* New York: Teachers College Press.

Weitzman, E. A., & Miles, M. B. (1995). *Computer programs for qualitative data analysis.* Thousand Oaks, CA: Sage.

Whittemore, R. Chase, S. K., & Mandle, C. L. (2001) Validity in qualitative research. *Qualitative Health Research, 11,* 522–537.

Willis, P. (1977). *Learning to labour: How working class kids get working class jobs.* Westmead, UK: Saxon House.

Winthrop, R. H. (1991). *Dictionary of concepts in cultural anthropology.* Westport, CT: Greenwood Press.

Wolcott, H. F. (1983). Adequate schools and inadequate education: The life history of a sneaky kid. *Anthropology and Education Quarterly, 14*(1), 2–32.

Wolcott, H. F. (1987). On ethnographic intent. In G. Spindler & L. Spindler (Eds.), *Interpretive ethnography of education: At home and abroad* (pp. 37–57). Hillsdale, NJ: Lawrence Erlbaum.

Wolcott, H. F. (1990a). On seeking—and rejecting—validity in qualitative research. In E. W. Eisner & A. Peshkin (Eds.), *Qualitative inquiry in education: The continuing debate* (pp. 121–152). New York: Teachers College Press.

Wolcott, H. F. (1990b). *Writing up qualitative research.* Newbury Park, CA: Sage.

Wolcott, H. F. (1992). Posturing in qualitative research. In M. D. LeCompte, W. L. Millroy, & J. Preissle (Eds.), *The handbook of qualitative research in education* (pp. 3–52). San Diego, CA: Academic Press.

Wolcott, H. F. (1994a). The elementary school principal: Notes from a field study. In H. F. Wolcott, *Transforming qualitative data: Description, analysis, and interpretation* (pp. 115–148). Thousand Oaks, CA: Sage.

Wolcott, H. F. (1994b). *Transforming qualitative data: Description, analysis, and interpretation.* Thousand Oaks, CA: Sage.

Wolcott, H. F. (1996, November 15). Personal communication.

Wolcott, H. F. (1999). *Ethnography: A way of seeing.* Walnut Creek, CA: AltaMira.

Wolcott, H. F. (2001). *Writing up qualitative research.* (2nd ed.). Thousand Oaks, CA: Sage.

Yin, R. K. (2003). *Case study research: Design and method* (3rd ed.). Thousand Oaks, CA: Sage.

Yussen, S. R., & Ozcan, N. M. (1997). The development of knowledge about narratives. *Issues in Educational Psychology: Contributions From Educational Psychology, 2,* 1–68.

Ziller, R. C. (1990). *Photographing the self: Methods for observing personal orientation.* Newbury Park, CA: Sage.

# Author Index

# Subject Index

# About the Author

 John W. Creswell, PhD, is the Clifton Institute Professor and has been Professor of Educational Psychology at the University of Nebraska–Lincoln since 1978. He specializes in research methods and writes, teaches, and conducts research on mixed methods research, qualitative research, and research designs. At the University of Nebraska, he co-directs the Office of Qualitative and Mixed Methods Research, a service and research unit that provides methodological support for proposal development and funded projects. In addition, he has been Adjunct Professor of Family Medicine at the University of Michigan Health System (2001–2005) and serves as a consultant on many family medicine and Department of Veterans Administration large-scale funded projects. He was recently appointed co-editor of the new Sage Publications journal, the *Journal of Mixed Methods Research*. He has authored 10 books, including his bestselling book, *Research Design: Qualitative, Quantitative, and Mixed Methods* (2003), and his most recent book, *Designing and Conducting Mixed Methods Research* (2007), coauthored with Vicki Plano Clark. Many of his books have been translated into different languages, and they are widely used around the world.